Knowledge & Information

Knowledge & Information

Studies in Information Science

Knowledge and Information (K&I) is a peer-reviewed information science book series. The scope of information science comprehends representing, providing, searching and finding of relevant knowledge including all activities of information professionals (e.g., indexing and abstracting) and users (e.g., their information behavior). An important research area is information retrieval, the science of search engines and their users. Topics of knowledge representation include metadata as well as methods and tools of knowledge organization systems (folksonomies, nomenclatures, classification systems, thesauri, and ontologies). Informetrics is empirical information science and consists, among others, of the domain-specific metrics (e.g., scientometrics, webometrics), user and usage research, and evaluation of information systems. The sharing and the distribution of internal and external information in organizations are research topics of knowledge management. The information market can be defined as the exchange of digital information on networks, especially the World Wide Web. Further important research areas of information science are information ethics, information law, and information sociology.

De Gruyter Saur

Nguyễn Thanh Tuyên

Knowledge Economy and Sustainable Economic Development

A critical review

De Gruyter Saur

ISBN 978-3-598-25181-8
e-ISBN 978-3-598-44157-8
ISSN 1868-842X

Library of Congress Cataloging-in-Publication Data

Nguyễn, Thanh Tuyên.
Knowledge economy and sustainable economic development : a critical review / Thanh Tuyên Nguyễn.
p. cm. -- (Knowledge & information : studies in information science)
ISBN 978-3-598-25181-8
1. Knowledge management--Economic aspects--Vietnam. 2. Information technology--Economic aspects--Vietnam. 3. Sustainable development--Technological innovations--Vietnam. I. Title.
HD30.2.N498 2010
303.48'3309597--dc22
2010019661

Bibliographic information published by the Deutsche Nationalbibliothek
The Deutsche Nationalbibliothek lists this publication in the Deutsche Nationalbibliografie; detailed bibliographic data are available in the Internet at http://dnb.d-nb.de.

Printing: Hubert & Co. GmbH & Co. KG, Göttingen
∞ Printed on acid-free paper

Printed in Germany

www.degruyter.com

For my lovely children Thanh Quỳnh and Xuân Cường
May future never let them down

Contents

Figures and Tables

Figures

List of Tables

Acknowledgements

The odyssey in conceiving this book originated with my experience in the Vietnam-Canada Information Technology Project where I co-worked with Garth Graham and other Canadian and Vietnamese colleagues in formulating national ICT strategies and policies for Vietnam. Through the project which ended in 2000 I became deeply interested in the intersecting issues of knowledge, ICTs and sustainable development for Vietnam. Garth's thoughtful points made in our discussions helped to crystallize my ideas for this study.

First and foremost I wish to thank Emeritus Professor Don Schauder, my main supervisor at the Faculty of Information Technology, Monash University, for his special guidance, encouragement, creativity and support through all the stages of the work. Without Don, the goal of completing this manuscript would not have been attainable.

I am also deeply grateful to Associate Professor Graeme Johanson and Dr Steve Wright, my co-supervisors at the Centre for Community Networking Research (CCNR) for their thoughtful academic advice and weekly personal support and encouragement. I was fortunate to have in Dr Wright someone who could point me to literature on Marxism and other issues which played a large role in developing my thinking on some of the economic and social aspects of the study.

Many thanks to AusAID (particularly Kinda Say at Monash International) and to the Ministry of Training and Education of Vietnam for granting me a full scholarship. In addition, the three rounds of fieldwork in Vietnam would not have been possible without financial support from both AusAID and CCNR. My special thanks go to Dr Nguyen Anh Tuan and the Ministry of Posts and Telematics of Vietnam (now Ministry of Information and Communications) for permission to undertake my long absence from work.

I would like to express my graditude to Professor Michel J. Menou for encouraging me to send the manuscript to K. G. Saur Verlag for publication.

As English is not my first language, much editorial work was needed on the final draft of the manuscript. I am grateful to both Cherryl Schauder and Don Schauder for their tireless efforts in this regard. I would like to thank also Dr des. Sonja Gust von Loh, Heinrich-Heine University and Ms Claudia Heyer of de Gruyter Publisher for the final editorial work so that book could be published.

Thanks go also to the friendly and supportive staff of the Faculty of Information Technology and Monash Library, Monash University where I was made to feel at home over the four years of study.

And of course, none of this research could have been done without the participation of the large number of people who were willing to contribute their thoughts and ideas to the study both in Vietnam and Australia. Though they are not named in the manuscript, I will always owe them my deep gratitude.

On a sad note, three outstanding scholars who participated with great generosity in the interview and discussion phases of the study have subsequently passed away through illness: Prof Tran Luu Chuong of Hoalac Hi-Tech Park, Prof Nguyen Dinh Ngoc of Thanglong University, and Dr Pham Ngoc Len of UNICEF Hanoi. The manuscript conclusions owe much to their contributions.

I acknowledge here Cu Thi Ngoc Ha and Tran Kim Anh who assisted me in transcribing recorded conversations into words; and also Nguyen Ho Nam, Mai Hoang

Khanh Minh and Pham Xuan Phuong who helped me as local guides to arrange interviews outside Hanoi.

There are no words by which to express my gratitude to my parents, Vu Thi Lan and Nguyen Xuan Loc who have always loved and encouraged me in my life and career. My thanks for their spiritual support also go to my brothers Thanh Binh, Thanh Thai and Quoc Trinh, my sisters in law Thu Ha and Xuan Hoa, my nieces Thuy Linh, Thanh Ngan, Xuan Thanh and Phuong Thao, and nephew Minh Hoang.

Finally, to my dearest daughter Thanh Quynh and little son Xuan Cuong, who are always in my mind and heart, thank you for bearing my absences during these long years. This manuscript is dedicated to you with all my gratitude and love!

Chapter 1
The Thachsanh connection

1.1 The utopia of inexhaustible Thachsanh's rice bowl

The title of this beginning chapter and the spirit in which the book is inspired are based on the author's personal interpretation of the ancient Vietnamese legend of Thachsanh's rice bowl. Thachsanh is a humble, everyman kind of hero who, by his courage frees from captivity the son of the Emperor of the Ocean. In gratitude the Emperor offers Thachsanh any gifts he wishes. The Prince of the Ocean counsels Thachsanh to choose an old rice bowl in the King's store room rather than gold or precious stones. The Emperor congratulates him on his choice. The miracle is that as the rice in the bowl is eaten, it is constantly replenished. In one adventure Thachsanh even brings peace between two hungry armies, by feeding all the soldiers from the rice bowl, whereupon "The former enemies feasted and celebrated together long into the night" (N. C. Nguyen & Sachs, 2003, p. 113).

Asked what makes Vietnamese folklore particularly Vietnamese, the folklorist Nguyen Dong Chi named four features including that:

> The protagonists in Vietnamese folktales often display dissatisfaction with the state of the world. They aim toward a new society that is more just and more reasonable (N. C. Nguyen & Sachs, 2003, p. 2).

Consistent with this tradition, the present author's interpretation of the Thachsanh rice bowl legend is that the sea, air and land are the perpetual source of nourishment for humankind as long as they are used with moderation, respect and humility. The book originates from the metaphor of "Thachsanh's rice bowl" because over the last two centuries in particular, the magical powers of the earth to sustain humankind have been jeopardized almost to the point of no return. If humankind is to change its ways and learn to treat the earth with the wisdom of Thachsanh, information and knowledge are crucial to that transformation.

1.2 Aim of the book

The aim of this book is to explore how the deployment and use of information and communication technologies (referred to as ICTs) can support economic sustainability in Vietnam, focusing on its current condition as a developing country.

This aim derives from the author's own experience as a public servant in the ICT field in Vietnam. The book constitutes a personal odyssey that emerged from his experience in the Vietnam-Canada Information Technology Project where he co-worked with Garth Graham and other Canadians to formulate national ICT strategies and policies for Vietnam.

The book draws on four sources: the author's experience, Vietnamese and international statistical data, research literature, and 135 interviews with 138 persons with a wide cross-section of characteristics in Vietnam to derive insights which form the basis of suggested guidelines for consideration by the government of Vietnam.

Unlike in many studies, this book does not commence with a set of specific and detailed questions. Rather it starts with a broad aim and seeks to comprehend a wide terrain of enquiry in order to yield grounded theoretical insights and strategic guidelines for policy.

1.3 Assumptions

The study proceeds from the following key assumptions:

- That economic development i.e. continuous wealth creation, is a vital factor in the advancement of human well being;
- That "sustainable development" is a meaningful and ultimately feasible objective. Sustainable development is a term so rich in mutual contradiction that it has been called an oxymoron (Redclift, 2006); and
- That technology, which is in great part responsible for the problems of ecosystem degradation being faced by the world can play a major role in their reversal.

Without these assumptions the book would need to address far more basic existential questions than it does, but this important task has been left to others.

1.4 Rationale

The rationale for this book project was well conveyed by one of the key statements of the World Summit on the Information Society (WSIS). In December 2003, heads of states and representatives from 175 countries gathered at WSIS Part 1 in Geneva to declare a desire and commitment to build:

> a people-centred, inclusive and development-oriented Information Society, where everyone can create, access, utilize and share information and knowledge, enabling individuals, communities and peoples to achieve their full potential in promoting their sustainable development and improving their quality of life (World Summit on Information Society, 2003).

This statement specifically brings together and underlines the importance of the nexus between information and knowledge on the one hand, and sustainable development on the other, in the hope and trust that appropriate knowledge and information strategies can help resolve the tensions between sustainability and economic development.

A reason why the book topic is not only justifiable but of urgent current relevance is that the stage has been reached where time is running out and there is little room for mistakes and miscalculations. Identifying sustainable economic development as a matter of existential risk for humankind, Bostrom offered this powerful conclusion:

> Our approach to existential risks cannot be one of trial-and-error. There is no opportunity to learn from errors. The reactive approach – see what happens, limit damages, and learn from experience – is unworkable. Rather, we must take a proactive approach. This requires foresight to anticipate new types of threats and a willingness to take decisive preventive action and to bear the costs (moral and economic) of such actions. (Bostrom, 2002)

1.5 Limitations

In addition to the usual constraints of time and funds, other limitations of the book relate to factors influencing the quality of interviews as a source of evidence. These include the inevitable sensitivities and subtleties of the interaction between the interviewer and interviewees, the level of grasp of questions and answers by both groups, and possible misunderstandings or gaps in reporting during the summarization and, in this case, language translation processes.

1.6 Contribution to knowledge

By selecting such a broad-ranging book topic, the author was well aware that he could not do justice to any one part of its coverage. However he felt that this trade-off was justified. The intersecting areas of information and knowledge, environmental sustainability and economic development are very rich in narrowly focused research reports and articles. This book attempts to add to such contributions by integrating the findings and interpretations of other scholars, connecting these with new grounded evidence, theorizing the key factors and relationships identified, and using insights gained to propose strategic policy guidelines. Certainly such an attempt had not been made in respect of issues facing Vietnam when the book project commenced, and the author was keenly aware of this lack from his perspective as a government official working in the field of ICT policy.

1.7 Setting the scene: major issues, themes and concepts

Many concepts are introduced in this section, as well as in the chapters of the book. In addition to the explanations provided in the text, a glossary of terms and acronyms is provided as Appendix 1 to assist the reader.

1.7.1 Vision of the knowledge economy as a potentially inexhaustible resource

Debates about the implications of an emerging revolution empowered by the new ICTs have appeared widely in the literature since the 1990s.

Castells (2000, p. 75) stated:

> A new economy emerged in the last quarter of the twentieth century on a worldwide scale. I call it informational, global, and networked to identify its fundamental distinctive features and to emphasize its intertwining.

This fundamental change, Castell argued, brought human history into a new age, the age of “network society” (also called “digital society“, “knowledge society” or “information society”), where society and its economic relations are no longer primarily based on physical materials. The new economy and society were seen to rely fundamentally on information and knowledge. For Castells the network society is the result of “informationalism” (Castells, 2000, pp. 99-100).

The claim that a new knowledge-based economy largely displaces traditional material-based production was supported by many others, such as, Kranzberg (1985), Neef (1998) and Tapscott and Caston (1993). To such writers, knowledge has become the most crucial factor of the social and economic world today. Stevens observed that in the OECD countries, the economy was becoming more dependent on the production and use of knowledge. “In the past decade, the high-technology share of OECD manufacturing production and exports has more than doubled, to reach 20-25%” (Stevens, 1996, p. 87).

Leadbeater, for instance, spoke of an increasingly “weightless economy”:

> These days most people in most advanced economies produce nothing that can be weighed; communications, software, advertising, financial services. They trade, write, design, talk, spin and create: rarely do they make anything (Leadbeater, 1999, p. 18).

In many western countries, knowledge seems to become a direct production force and the most crucial factor of the social and economic world today, as observed below.

The knowledge based economy was seen to be a key emergent phenomenon in developed economies, especially in North America where ICT penetration reached more than half of the population (53% in 2002) (Dutta & Jain, 2004).

Unlike finite physical resources such as minerals or water, knowledge has come to be viewed by many as an inexhaustible resource. Economies with well entrenched knowledge-based enterprise seemed to have the potential for open-ended economic growth. Drucker was a pioneer of this thinking:

> The basic economic resource – ‘the means of production’ to use the economist’s term – is no longer capital, nor natural resources (the economist’s ‘land’), nor ‘labour’ – *it is and will be knowledge* (Drucker, 1993, p. 7, Italics in original).

1.7.2 Can the knowledge economy be universal?

If countries in the First World could build knowledge economies they might achieve the elimination of poverty and an abundance of wealth.

Could knowledge-based prosperity extend throughout what Wallerstein (1984) called “the world system"? In Wallerstein’s classification the world system included not only the few OECD or “core” states, but also more than one hundred “semi-periphery” and “periphery” countries.

International organizations, and not least the movement that led to the two-part UN/ITU World Summit on the Information Society (Geneva 2003 and Tunis 2005) tend to assume that a knowledge economy can have beneficial effects in all countries.

Different organizations have proposed guidelines for the development of policies on ICTs in developing countries. For instance, the report by the United Nations Development Program, Accenture, and Markle Foundation (2001) proposed a framework to formulate digital opportunity initiatives which focus on ICTs as enablers for socio-economic activities, such as health, education, or economic activities. The framework comprises five factors to create an enabling environment for ICTs, namely infrastructure, human capacity, policy, enterprise, content and applications.

A report by the World Economic Forum (2004) recommended that national ICT policies should focus on three factors: (1) the development and use of ICTs; (2) the general macroeconomic and regulatory environment for ICTs in which the stakeholders play out their respective roles; and (3) the degrees of readiness to use and benefit from ICTs.

1.7.3 Questions and skepticism about the concept of a knowledge economy

Despite the general optimism about the new knowledge economy there was no common agreement among economists, sociologists and futurologists as to how the new economy actually operates in creating wealth and in supporting economic sustainability. For Hawken the key question was "to understand the changing ratio between mass [i.e. physical resources] and information in goods and services" (Hawken, 1983, p. 11), in other words, the achievement of "mind over mass" (Gilder, 1989). Economic theory since Keynes has not offered theories to explain the economic effects of knowledge (Bell & Kristol, 1981). Drucker made this point as follows:

> How knowledge behaves as an economic resource we do not yet fully understand. We have not had enough experience to formulate a theory and to test it. We can only say so far that we need such a theory. We need an economic theory that puts knowledge into the centre of the wealth-producing process. Such a theory alone can explain the present economy. It alone can explain economic growth. It alone can explain how the Japanese economy works and, above all, why it works (Drucker, 1993, p. 167).

Other authors such as May and Henwood have been skeptical about the notion of a knowledge economy. To May:

> The celebration of the information society, the celebration of the new age, is predicated on the novelty of today. However, this 'new age' is neither unprecedented nor necessarily as novel as often presumed (May, 2002, p. 47).

Henwood denied the very existence of a new economy because in his opinion the new economy is only another version of the old economy which operates in the same rules as before (Henwood, 2004).

Persistent growth of the physical economy, and the international division of labor

In the First World, the proportion of physical production has decreased relative to knowledge production, but the consumption of physical resources has risen.

As shown by data presented in Figure 1.1, the consumption of raw materials in the United States has increased dramatically throughout the 20th century. In the post WWII period this trend has increased sharply, even while computerized methods of work have been becoming ever more widely established. The consumption of non-renewable materials has grown rapidly from approximately 2 million tons (60% of the total) to 131 million tons (over 80% of the total) in one hundred years. This kind of data casts doubt on the notion that ICTs and the knowledge economy are friendly to the environment because knowledge-intensive economies still seem to consume a disproportionate share of natural resources.

An interpretation that Foster and Clark placed on such data was that raw materials for commodity production is organized around meeting the demands of the countries of the North, where approximately 25 per cent of the world's population lives, but which consumes 75 per cent of global resources (Foster & Clark, 2004, p. 194). It can be argued that, at least on a global level, the physical economy needs to develop more than ever to satisfy the world's growing hunger for food, water, energy and other resources.

The parallel increases in physical consumption and information intensity, combined with relative decreases in physical production in the economies of developed countries, gives rise to the following kinds of questions.

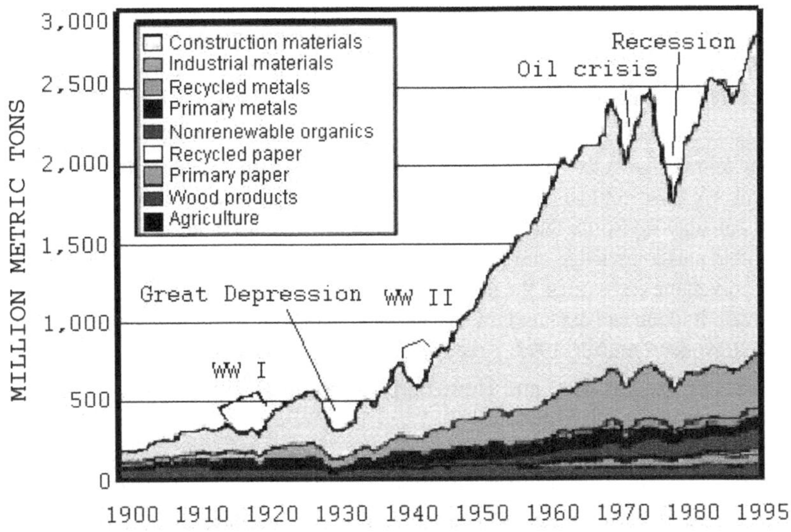

Figure 1.1 Measurement of the amount of raw materials consumed in the United States. Source: Adapted from Matos & Wagner (1998).

Has there emerged a new international division of labor in the last three decades that has shifted the physically-based economy from (in Wallerstein's terms) the core to the periphery? Do the knowledge economy and sustainability seem to be positively related in First World countries only because physical production has largely been relocated, together with its negative impacts on sustainability, to the developing countries of the periphery? In other words, is the knowledge economy really a label for the particular experience of First World countries only, where most citizens are now employed in non-physical production?

1.7.4 ICTs for sustainable development: potentials and doubts

A widely accepted definition of sustainability is:

> Meeting the needs of the present without compromising the ability of future generations to meet their needs" (World Commission on Environment and Development, 1987, p. 43).

Central to claims about the significance of the knowledge economy is the potential of ICTs to make more efficient almost any processes in sectors as diverse as public administration, business, education, health, and environment.

Such efficiencies can 'stretch' resources, allowing benefits to flow to many more people – mitigating 'digital divides' between haves and have-nots. Also the more that resources can be stretched, the greater is the apparent contribution of ICTs and the knowledge economy to sustainability: relatively higher economic growth might be achieved with relatively lower consumption of physical resources (Boisot, 1995).

ICTs came to be regarded by many nations as crucial in the development agenda. The United Nations Development Program (UNDP), for instance, launched an initiative promoting the use of ICTs for Development (ICT4D) by all nations to achieve the United Nations' Millennium Development Goals (MDGs) (Labelle, 2005).

Some commentators saw a knowledge based economy as a short-cut for developing countries to "leap-frog" into high value productive activity, and this has partially occurred in economies such as China (Dahlman, Aubert, & World Bank Institute, 2001) and India to be a "Knowledge Superpower" (Dahlman, Utz, & World Bank Institute, 2005).

Amidst claims that ICTs can support a global transformation from the traditional economy based on deteriorating material resources into an abundant knowledge based post-modernity, most governments around the world enthusiastically introduced new policies about using ICTs as enablers to achieve a social and economic leapfrog. Such hopes contributed to the convening of the WSIS by the International Telecommunication Union (ITU), the leading UN agency for ICTs (World Summit on Information Society, 2003).

By introducing supportive policies and legal regulations about ICTs, governments sought to create enabling foundations and environments for informationalism of their societies, including rapid improvement of so-called "e-readiness" among governments, businesses and citizens generally, catalysing the transformation of the status-quo toward a knowledge economy (World Summit on Information Society, 2003).

Not all writers have viewed the role of ICTS with optimism. For example, recent perceived failures of ICT4D projects have raised doubts about the relevance of applying ICTs for Millenium Development Goals (Heeks, 2005).

Taylor, writing from a broader perspective on the role of new technologies said that he did not believe that technologies could improve the sustainability of economic growth, because:

> The prime role of technology in our world is to create new products for new markets – for suburban shoppers, for generals, for doctors, for businesses, for farmers, for bankers – precisely the growth that is relentlessly leading to impasse (P. J. Taylor, 1996, p. 192).

1.7.5 The development of ICTs in Vietnam

Vietnam followed the global trend in making the development of ICTs one of its highest economic priorities in the first decade of the century. Many strategic initiatives and actions have been undertaken in the last 10 years to focus on national ICT goals and targets for the period 2001-2010.

Widely cited are the Vietnam Communist Party Political Bureau's Directive #58 (Vietnam Communist Party Central Committee, 2000) on the use and development of IT to support the country's industrialization and modernization goals; and Government's 5 year Plans on Information Technology (Government of Vietnam, 2002; T. T. Nguyen, 2003b; World Bank, 2002). Those policies aim at the following targets:

- IT shall be widely used in all sectors, firstly in government and party agencies, in major economic sectors and cities;
- The national information network shall be developed to reach nation-wide coverage. The Internet penetration rate shall reach that of the world average; and;
- The IT industry shall become a spearhead economic sector.

Table 1.1 Vietnam – Information and communication technology development in the period 2006-2009. Source: Ministry of Information and Communications (2009& 2010).

	2006	**2007**	**2008**	**2009**
Telephone main lines ('000)	8,567	11,165	14,767	17,427
Telephone main lines (per 100 population)	10.2	13.1	17.1	20.1
Mobile subscribers ('000)	18,892	45,024	74,872	98.2
Mobile subscribers (per 100 population)	22.4	52.9	86.9	113.4
Internet users ('000)	9,852	12,898	16,140	22,780
Internet users (per 100 population)	11.6	15.0	18.5	27.5
Broadband subscriber lines ('000)	407.2	618.2	2,885	3,214
Broadband subscriber lines (per 100 people)	0.5	0.7	3.33	3.71
Personal computers (stock per 100 households)	7.7	-	10.3	13.6
TV set (stock per 100 households)	78.2	-	82.3	86.9
IT hardware revenue (US$ m)	-	-	4,100.0	4,627
IT content revenue (US$ m)	-	-	440.0	690
Software revenue (US$ m)			680	850
Total IT revenue (US$ m)			5,220	690

The introduction of these policies is one of main drivers for the acceleration of ICT development in Vietnam in the last decade, especially in the last 5 years. The ICT industry has experienced rapid growth. The government facilitates the upgrading of infrastructure for Internet and telecommunications by opening the ICT industry to greater competition. The numbers of personal computers, Internet users and mobile phone subscribers have increased rapidly (see Table 1.1). However, the country received a low score (65th out of 69 countries) in the annual E-readiness Ranking in 2007, conducted by The Economist Intelligence Unit (2007).

1.7.6 Ownership versus free sharing of knowledge

In Roman mythology Janus, the god of portals, had two faces each looking a different way. To many authors a key to understanding the knowledge economy lies in the two contrasting ways that knowledge is treated as property in the age of the Internet.

ICTs facilitate the proprietary ownership of information and the development of international regimes to protect intellectual property rights (IPR). This aspect is fundamental to the role of the World Trade Organization (WTO).

However ICTs also create new opportunities for people to share knowledge and network together to create more knowledge, such as in the form of emergence of commons-based peer production (CBPP), exemplified by the open source software movement.

Proprietary knowledge

In the view of some authors, the key to generating wealth in a knowledge economy is the capacity to establish and enforce proprietary rights over information. For developed countries regimes to enforce such rights are crucial, both domestically and internationally.

To enter the global marketplace for buying and selling information, nations are required to conform to the rules of the WTO. Several writers, including the present author, have argued that international IPR regimes such as the WTO, TRIPS (Trade Related Aspects of International Property Rights) and the Bern Convention extend into the knowledge sphere the traditional economic advantages of developed over developing countries.

May (2000), Nelkin (1984) and N.T Nguyen (2001) viewed the IPR regimes as a new form of appropriation of wealth, "the key method to assert ownership over knowledge resources…, a tool of commodification or enclosure" (May, 2000, p. 13).

In nature, IPRs make information and knowledge artificially scarce through mechanisms of legal intervention such as the TRIPS or the Bern Convention.

Knowledge begins as something that has the potential to be infinitely shareable. IPRs confer the ability for the owner or vendor to place a price on knowledge, and such pricing is the means of creating artificial scarcity. It can be argued that whereas the availability of (and therefore price) of, say, platinum ore is ultimately influenced by its natural scarcity, knowledge can be as scarce and expensive, or as ubiquitous and cheap, as its owners decide to make it.

Deardorff (1990, p. 505) argued that IPR enforcement extending patent protection to poor countries involves a transfer of welfare from the poor to the rich.

Applying Deardorff's argument to all forms of IPR, TRIPS can be portrayed as a legal protector that allows advanced countries to engage in a form of asymmetrical appropriation. This appropriation process enables artificially scarce knowledge resources to be exchanged for naturally scarce physical resources from less advanced countries. Some authors see this hidden knowledge based appropriation as a new form of imperialism (Emmanuel, 1972; Foster & Clark, 2004).

All this raises the question of whether a knowledge economy based on IPRs can assist sustainable growth in developing countries in the medium term, or the world as a whole in the long term.

Free sharing of knowledge

Operating in parallel with IPR regimes is the fast-growing phenomenon of free software production. Benkler called it the commons-based peer production (CBPP) of knowledge in which the creative energy of large numbers of people is coordinated (usually with the aid of the internet) into large, meaningful projects, mostly without traditional hierarchical organization or financial compensation (Benkler, 2002).

An example of CBPP is the production of open source software. The Linux kernel, the GNU suite of software applications, and the combined GNU/Linux system are prominent products of CBPP. By extension, CBPP could also include "networking communities" where the new mode of computer-mediated cooperation interconnects communities and people with common interests whether are living in a particular area or not, such as the sharing of information in myspace.com or youtube.com. Benkler suggested that different from firms and markets, CBPP can:

> make large-scale collaborations in many information production fields sustainable and productive in the digitally networked environment without reliance either on markets or managerial hierarchy (Benkler, 2002).

The economic and social impacts of this rising phenomenon, often accessed in developing countries via low-cost cyber-cafes, are as yet unknown, Robinson argued:

> The ubiquity of these cyber-cafes suggests this network of access to the contemporary, always evolving digital tools and the wealth of information they access constitute a resource for personal and community empowerment and poverty reduction" (Robinson, 2004).

Such modes of collaboration might become very critical for the whole human race not only to counter the institutionalisation of IPRs, but also to bring people in finding novel solutions to the problems of sustainable growth in a world whose natural resources are finite, and increasingly scarce to the point of total depletion.

1.8 Methodology

The book report is divided into three main sections – foundational analysis, grounded accounts and theorizations. The foundational chapters (1 to 6) underlie the empirical and theoretical inquiries developed in the subsequent chapters.

1.8.1 The approach: exploratory grounded hermeneutic analysis

The priority in the research design of this book was deep insight ahead of broad generalization. The principal methodological approach chosen was interpretivist or hermeneutic, and the data collected was mainly qualitative.

It must be acknowledged that the methodological approach used in this book attracted skepticism from some of the people interviewed. For example during the author's fieldwork in Vietnam some of the knowledgeable government leaders questioned how a study that is not based on statistical probability sampling and standardized questions (preferably closed questions requiring, say, a tick on a Likert scale, rather than an open-ended comment) could be useful as the basis of input to policy development. Their fundamental doubt, expressed in different words, was whether a qualitative and ethnographic methodology and research design could be as demonstrably rigorous as a quantitative, positivist study. The overview offered below attempts to answer these criticisms, as well as to provide a systematic account of the research procedures followed.

The grounded interpretive or hermeneutic approach treats the phenomena being studied as a "text', within which the researcher strives to identify the patternings and coherences that constitute meaning. The book is exploratory and hypothesis generating, rather than hypothesis testing. It has sought to identify and explicate key factors and relationships relevant to the intersecting issues of knowledge, ICTs and sustainability.

Positivism, the fundamental epistemology of the natural sciences, employs mainly quantitative data, while interpretivism, the orienting epistemology of the humanities, rests mainly upon qualitative data. However both epistemologies in fact can and often do employ both kinds of data.

For a long period, quantitative studies have been the more familiar and preferred mode for empirical researchers, whether in the natural or social sciences. Traditionally, social sciences have adapted natural science methodology to social phenomena, wherever possible using standardized and quantifiable research designs.

Using large bodies of quantitative data, drawn from standardized surveys in random samples of population, social researchers expected that they could document and analyze the frequency and distribution of social phenomena in the population, and establish social-scientific "laws". The ultimate test of such laws within positivist social science, as in the natural science, is replication - that an identical study of further cases, undertaken in controlled conditions, produces identical results.

In recent decades it became increasingly realized that such an approach could be applied to only small subsets of social research problems. Society is an ever moving target, and the effect of what Giddens called "the double hermeneutic" is always at play: the research itself influences the phenomena under consideration. The mere fact that questions have been asked and answered changes social reality both for those who are questioned and those who ask the questions. Giddens explained the double hermeneutic as follows:

> The intersection of two frames of meaning as a logically necessary part of social science, the meaningful social world as constituted by lay actors and the metalanguages invented by social scientists; there is constant 'slippage' from one to the other involved in the social sciences (Giddens 1984, p.374).

Thus prediction in the social sciences is different from prediction in say, astronomy, where the system under study (for example, a distant galaxy) can in no way be affected by proceedings of the researchers, or actions taken as a result of the research (for example, the building of a more powerful telescope). In the social sciences the constantly changing conditions of the world and evolving insights of the social researcher are mutually influential, before, during and after any research project.

For such reasons there was some disenchantment with the application of mainstream scientific methods to social questions, especially to those involving many and complex variables in a state of high dynamism. Flick (2006, p. 13) observed:

> It has also become clear that social science results are rarely perceived and used in everyday life. In order to fulfil methodological standards, their investigations and findings often remain too far removed from every day questions and problems.

Qualitative-interpretive research is seen as increasingly relevant to the understanding of social phenomena in pluralized life-worlds. It has been perceived as a new way to explore the individualization of ways of living and biographical patterns (Beck, 1992) as well as the diversity of milieus, subcultures and lifestyles (Flick, 2006).

The main objective of this book was not primarily to expound what is already known (such as prevailing ICT policies), but to discover and develop new understandings. Evidence about the system state (as presented in statistical reports etc) was used as a backdrop to the task of exploring system processes. The book conclusions needed to be reconcilable with quantitative reporting to date, but even more importantly it needed to embody "grounded" insights about how people involved or affected are dealing with the impacts of the knowledge economy, ICTs and challenges of sustainability. A study of ICT policy-making at a national level needs to take into account also the country's socio-economic and global contexts. The main framework chosen to incorporate the complexity in this study was therefore a hermeneutic grounded theory approach (Bryman & Burgess, 1999; Charmaz, 2005; Glaser & Strauss, 1967).

1.8.2 Three rounds of fieldwork

The sample: inclusion, exclusion and saturation

In answer to critics of sampling in studies to build empirically grounded theories, Glaser and Strauss pointed out that qualitative research projects often involve fewer cases but elicits more in depth information than quantitative ones (Glaser & Strauss, 1967). With respect to the problem of generalization of the findings, the essential questions are not how many but which cases are appropriate for representing the issue under study and why are particular criteria for selection justified (Flick, 2006, p. 41).

The sampling technique used for selecting the cases was purposive sampling, a non-probability sampling approach "used when the investigator is interested only in obtaining a sufficient number of elements to satisfy research objectives. The resulting sample is ... 'balanced' on a range of variables, for example sex, age, education, income, location (or place of abode) ..." (Williamson, 2002). In other words, the sample is representative, but representativeness is not achieved through random selection as in a probability sample. Rather it is achieved through criteria-based selection of the individuals to be included in the sample. Representativeness means that the subset under study should manifest, in an unbiased way, selected characteristics of the superset, that are seen to be important to the research. The practical success of such an approach to sampling is daily demonstrated in the business world, for example in focus group work undertaken by marketing professionals for product and service development.

The sample sought representative balance according to the following criteria:

- Gender: Male, female
- Location: Rural, urban
- Economic status: Rich, middle, poor
- Region: North, Central, South
- Roles:
  + Providers, non-users, users (General users, Communal Post and Communal Point (CPCP)[1] users, game players)
  + Managers, financiers, consultants, researchers, trainers
- Age: Age groups
- Economic sector:
  + Government, private, NGO
  + ICT industry, other industries
- Mode of production: Traditional, modern
- Proprietorship: State-owned, national private, international

In Fieldwork Round 1 the emphasis was on obtaining commentary and critique from experts and senior officials about major economic sectors, social stratification (rural and urban, rich and poor), modes of production (traditional and modern) and proprietorship (state-owned, international or national private).

For Fieldwork Rounds 2 and 3, the emphasis was on representation of stakeholders in the ICT sector (providers; nonusers, users; government, private sector and NGO managers; financiers; researchers), the types of users (general user, CPCP user and game players), a variety of ICT providers (software developer, open source developer, game producer, hardware producer, seller, infrastructure provider, trainer, consultant, Internet cafe managers, CPCP managers, etc.), and a range of government managers (telecom, software, hardware, content, intellectual property, e-government, e-commerce).

A cross-cutting criterion for sampling was that respondents should be "trustworthy, observant, reflective, articulate and a good storyteller" (Johnson, 1990, p. 30).

For Fieldwork Round 1 - focusing on the macro level of the book topic - almost all respondents were leaders, senior managers or professionals: for example, ministers, vice ministers, senior managers or researchers, experienced reporters and web-moderators. Some interviewees were already known to the author. Some were introduced by the previous interviewees. Periods of fieldwork were planned to coincide with key conferences on sustainability and ICTs. Twelve leading professors and experts with PhD degrees from among the keynote speakers granted interviews for the project.

In contrast, Fieldwork Round 2 focused more on the practicality of ICTs for sustainability at a micro level for individuals and businesses. Interviewees, apart from seven government ICT managers or advisors, were encountered by the author opportunistically or located in the local telephone directory.

---

[1] Discussion about Communal Post and Communal Point will be given later.

In terms of exclusion criteria, tentative interviewees who were inaccessible or appeared hesitant to participate in conversation with the author were withdrawn, and wherever possible were replaced with comparable interviewees.

In total, the author selected from more than 300 possible participants, 152 respondents, including eight international citizens, who were actually interviewed. To decide how many cases were enough, the author used the criterion of "saturation", i.e. interviews were added until all new data fitted into the categories already devised from the previous cases (Charmaz, 2001, p. 6399; Morse, 1995). Also the guidance of Flick (1998) was followed about the "elasticity" of the criterion of saturation. He observed that when information from other sources showed that more categories might be possible than interviewing to date had revealed, additional cases should be added. Thus to cover a gap concerning CPCPs, six additional respondents were included in Fieldwork Round 3 for insight into new forms of communal post and cultural points in Vietnam.

Development of questionnaires

As already indicated, the questions that were asked in the interviews were developed from literature analysis and the author's experience in ICT policy work. During the pilot interviews, the questionnaires were refined and restructured so that they could: 1) provide sufficient information for the book 2) be coherent, relevant, clear and meaningful to the interviewees, and 3) focus on crucial themes so that the conversations would not be too long.

However the resulting questions were not carved in stone. As previously explained, the questionnaires continued to evolve reflexively – to "learn" – during use. During the interviews, the author discovered that the number of questions developed in the pilot phase were still too many to fit within a one hour interview. In several cases, only half of the set of questions were completed, and so the chance was missed to discuss critical issues from the second half of the questionnaire. (However the depth of discussion on the earlier questions contributed powerfully to category building – so there was a trade-off.) To solve the problem of time, in subsequent interviews afterwards, the author reframed the questionnaire so that four thematic questions covered the most important issues. The short form of questionnaire often allowed time for extending the conversation to include other questions, if the interviewee permitted

New questions inevitably arose during or after collection of data. For example, when the author began to write up his notes on ICT diffusion in rural areas, he realized that a new form of ICT provider, namely the communal post and cultural point (CPCP) had been recently set up, and had not been included in the initially-designed questions. This new phenomena could potentially play a crucial role in the model of ICTs for rural and agricultural development. Thus a new set of questions was built for the third round of fieldwork.

Interview procedure

Research participants were recruited purposively throughout Vietnam. For people in the groups of users and nonusers, the author, with an assistant and sometimes a local guide, met interviewees at site (their working place, home, or Internet café) to get their agreement for participation. For known persons (such as government ministers and officials), the author sent them a letter describing the study in detail and an appointment was made for an initial interview.

In other cases electronic communication, specifically Yahoo Messenger, was used. A pilot interview with a fellow Vietnamese PhD student, studying in Japan was conducted in this way. The author also undertook two other interviews using Yahoo Messenger – with super-moderators of Vn99.net, a music forum of Vietnamese students.

At the beginning of each interview, whether co-present or online – the author introduced the purpose and scope of the study. Informed consent was discussed in detail at the beginning of the interview, with an emphasis on confidentiality and any perceived risks relating to participation (e.g. inadvertent disclosure of sensitive facts or opinions).

Each of the 138 interviewees participated in a 30 to 90 minute in-depth interview. Each interview was audio-recorded. To facilitate the capture and transcription of the interviews, an assistant operated a digital recorder, and later helped to transcribe data into digital text files for narrative analysis. Sometimes a photograph was taken for additional documentation. Documentary evidence also consisted of the participants' personal writings or publications where available.

Participants signed a consent form before each interview began. Participants who asked to be anonymous were given a pseudonym or a code. Participants were promised the opportunity to review the notes taken and other information that might be used in the book, by emailing. This process in fact gave the author a great deal of further reflection and information from participants. In the event, as the book was being written up, a decision was made to anonymise all the interviewees (except in a very few instances). Appendices 2 and 4 however provide a listing of the interviewees by code number and affiliation where relevant. (The names of the interviewees are retained with the consent forms as required under the ethics and research regulations of Monash University).

Appendices 3, 4 and 5 are the questionnaires, statement of the study and informed consent letter that were actually used in the fieldwork.

Analysis and writing

The data consisted of over 200 hours of audio recordings which documented more than 170 hours of individual interviews, 5 hours of group sessions, 5 hours of online interview via the Yahoo Messenger and 20 hours of follow-up interactions with participants over a period of more than 26 months.

Almost all audio files were transcribed verbatim by the author's second assistants, Ms Cu Thi Ngoc Ha – for the first fieldwork, and Ms Tran Kim Anh – for the second one. The data corpus included over 3,000 pages of transcription, field notes and documents shared with the author by interviewees.

All data was in the Vietnamese language and therefore English-based computer assisted software such as Nvivo could not be used to assist data analysis. Instead the author summarized the data into Excel tables, with coding words, phrases and sentences to assist manipulation. The coding allowed the author to put data into a logical paradigm in order to identify key factors and relationships.

In keeping with Fine's suggestion that ethnographic researchers should seek further than the voices of researched informants when developing or refining models (Fine, 1992), the author invited 7 ICT advisors, policy makers and senior managers of Vietnam to challenge interim formulations and conclusions. During the second fieldwork round this exercise was undertaken mainly with leaders of the Ministry of Posts and Telematics. During the third fieldwork the same process was conducted with a senior international researcher attending the E-government Symposium 2006 in Hanoi.

All of the above processes culminate at the end of the book with the "Insights" derived from each chapter, and suggested guidelines for government policy development.

1.8.3 Detailed accounts of the three fieldwork rounds, and understandings sought from the interviews

Details of Fieldwork Round 1

The first round was to collect data on sustainable economic development and ICTs in Vietnam from October 2004 to May 2005 with 7 persons in the pilot phase in Australia, and 70 official interviewees in Vietnam.

Seventy interviewees were selected carefully to represent opinions of various groups from the government, business and "ordinary people", from North to South, from urban to rural, from Vietnamese to foreigner, from physically-based sectors such as ceramics, agriculture, aquaculture, forestry, to non-physically based sectors, such as tourism, education and IT. Many of them were champions of viewpoints, or members of think tanks, on development in Vietnam, or at the organization where they worked. They viewed development through a diverse "lenses", including economics and sociology, environmental policy, and ICTs.

The first round sample consisted of:

- Twenty nine interviewees categorized as "elite scholars and researchers". These included 10 with a primarily economic perspective, seven researchers in the natural sciences and IT, four sustainability researchers, four social scientists and four media journalists.
- Twenty three interviewees categorized as "policy-makers and managers". These included one member of the National Assembly, one senior official of the Communist Party, two ministers or vice ministers, eight senior officials of ministries, one official of local government, ten senior ICT managers, ten business managers, including four executives of large state owned enterprises, two executives of medium state companies, one executive of a large private enterprise, and three executives of small and medium private enterprises.
- Eight representatives of non-government organizations included four senior officer of international organizations, four representatives of business organizations, and two administrators or moderators of web blogs.

Initially, 23 questions (see Appendix 3) were grouped into three main parts: i) general information (date, occupation, gender, education, ii) economic development and sustainability (economic activities and challenges, and recommendations), and iii) knowledge and ICTs (knowledge, ICTs, challenges and recommendations). As described above, the content of the interviews was later shortened into four topics:

- What have been the main achievements of your business or the Vietnam economy in the last five years?
- What are the most critical factors for the sustainability of your business and/or the economy of Vietnam in the next 10 years?

- How could ICTs help to improve the sustainability of your business and/or the Vietnam economy?
- What are the main factors that prevent your enterprise/sector gaining benefit from ICTs and knowledge economy?

Details of Fieldwork Rounds 2 and 3

The second and third rounds, conducted between January and December 2006 focused more on ICT policies, ICT utilization, and ICT as an industry. These rounds involved 7 participants to pilot the questionnaire and 68 interviewees in the fieldwork proper. The 68 interviewees were selected from the four main stakeholders categories of the ICTs sector in Vietnam. The sample was comprised of 16 nonusers, 22 users, 23 ICT providers and 7 governmental ICT policy makers.

The sample sought to reflect a wide range of viewpoints across the social spectrum of Vietnam. Thus the author sought to balance the selection of respondents within the four major categories by keeping in mind the following diversities:

- North and South
- rural villagers and urban residents
- traditional artisans and mass production workers
- business, governmental and private users
- hardware and software providers
- general and game content providers
- Vietnamese and foreigners.

Six variants of the questionnaire were used for six different groups of interviewees: 1) nonusers, 2) general users, 3) users of communal posts and cultural points (CPCPs) 4) ICT providers 5) CPCP managers and 6) governmental ICT managers.

The interview for the group of nonusers aimed to gain insight on how well they were managing their knowledge needs without using ICTs, and what barriers they saw to the uptake of ICTs. The questions posed to them included:

- What are the problems of sustainable development perceived by ICT nonusers, for their own businesses and/or for Vietnam?
- What are the major channels of information through which they acquire new information and knowledge to tackle the problem?
- What are the practices they employ to maintain indigenous knowledge and promote indigenous knowledge?
- What are the barriers that prevent nonusers from enjoying the benefits of ICT?

Three types of ICT users were defined: general users, CPCP users and game players. General users were persons who were currently using computer or the Internet for any purpose. CPCP users were general users at communal posts and cultural points. Computer game players (from now on referred to as players, gamers or gamesters) were persons currently playing online or offline computer games.

Understandings sought in rounds 2 and 3: users and non-users

Interview conversations with non-users and users in rounds 2 and 3 sought to gain understanding of how people were living and working in relation to knowledge and ICTs. For what purposes were Vietnamese people connected to ICT networks? For what reasons were they not connected? What were the real changes, benefits and difficulties that people were experiencing with the advent of ICTs and the knowledge economy, and how did they handle these? What was the nature and extent of "digital divides" (Norris, 2001, p. 4) as perceived by Vietnamese? In addition, CPCP users were asked about the role and effectiveness of CPCPs in diffusion of information in rural areas. And finally above all, is the usage of ICTs really serving sustained growth?

Understandings sought in rounds 2 and 3: ICT providers and CPCP managers

Interviews conversations with ICT providers and CPCP managers aimed to explore major achievements, difficulties and challenges that the local and foreign ICT companies and other providers were facing, such as the legal environment, intellectual property, human resources, financial aspects, technology challenges, market, infrastructure, culture, security or international factors.

Understandings sought in rounds 2 and 3: government ICT policy makers and advisers

Finally, as an act of triangulation, the author selected 15 issues that had been of concern among the user/nonuser and provider participants to discuss with government ICT policy makers and advisers. The action was to elicit reflection on how these might be accommodated in the formulation of ICT policy to support development. These issues focused on the relationships between ICT policy and:

1. Facilitating sustainable and prosperous livelihoods for villagers
2. Preserving and promoting indigenous knowledge
3. Facilitating sustainable livelihoods for farmers
4. Ensuring affordable and universal access to the Internet and other ICT applications
5. Reversing the negative image of computer gaming
6. Turning the skills and interests of gamers to advantage for themselves and for Vietnam
7. Improving the telecom and Internet infrastructure
8. Optimizing policies and practices in relation to intellectual property
9. Building markets for ICT enterprises
10. Maximizing local ICT procurement by government
11. Improving low, medium and high-end ICT skills: both technical and managerial
12. Building an optimal environment for e-transactions
13. Optimizing Vietnam's positioning in relation to the open source software sector
14. Exploiting the potentials of virtual community building
15. Optimizing the legal environment for ICT enterprise and use.

Chapter 2
The global economy: Perspectives on the exchange of knowledge for material

2.1 Scope of the chapter

- Introduces key ideas about the global knowledge economy
- Maps and elaborates these ideas within Marx's twofold formulation of "social forces of capitalist production", comprising the personal activity of man, the subject of production and means of production; and "social relations of capitalist production", including division of labor, the way that capitalism rules the world economy and the way that wealth is made.
- Lays out the overarching issues to be explored in the book:
    - How material-based production interacts with knowledge-based production, and how this interaction
        - is affected by the advent of ICTs and
        - affects the exploitation of the natural world.
    - How the economic exchange between material and knowledge goods plays out in terms of the international division of labor between developing and developed countries, and is influenced by
        - rules and conventions about international property rights protection and
        - terms of trade.

2.2 Different views on the concept of a knowledge economy

The Organization for Economic Cooperation and Development (OECD) defined nations emerging as knowledge-based economies as "those where knowledge is the main source of wealth, growth and employment, with a strong reliance on information technologies" (cited in Debowski, 2006, p. 310).

The first use of the phrase "knowledge economy" was probably by Machlup in 1962 in his "The Production and Distribution of Knowledge in the United States" (Machlup, 1973; May, 2002, pp. 3-5). Until the mid 1970s, writings about the knowledge economy focused exclusively on the United States. From the late 1970s to the early 1990s many studies on the subject were published outside the United

States as ICTs came to be widely applied in other rich and developed countries (Poirier, 1990).

Castells believed that the new economy emerged in the 1990s from the United States, and started from information technology, finance, and biotechnology (Castells, 2000, p. 148), but Bell suggested it had happened even earlier, in the 1960s (1973). According to Henwood, the concept of a "knowledge economy" was mentioned most frequently before 2000 (Henwood, 2004).

A series of neologisms have appeared in the literature, for example "post-industrial society" (Bell, 1973), "digital economy" (Tapscott, 1996), "information society" (Organization for Economic Cooperation and Development, 1997), "new economy" (K. Kelly, 1999), "network society" (Castells, 2000), "era of man-made brainpower industries" (Thurow, 1998), "post-capitalist society" (Drucker, 1993) and "knowledge economy" (Drucker, 1998). For Castells, "[a] technological revolution, centred around information technologies, began to reshape, at accelerated pace, the material basis of society" (Castells, 2000, p. 1).

A common theme in these writings is great optimism about the implications of knowledge and ICTs for human development. For example Ronald Reagan believed that:

> In the new economy, human invention increasingly makes physical resources obsolete. We're breaking through the material conditions of existence to a world where man creates his destiny (…) In the beginning was the spirit, and it was from this spirit that the material abundance of creation issued forth (Henwood, 2004, p. 8).

Political leaders came to be convinced about the positive role of ICTs, for example, in the Okinawa Charter on Global Information Society (the forerunner of the World Summit on Information Society), the heads of G8 states proclaimed:

> Information and Communications Technology (IT) is one of the most potent forces in shaping the twenty-first century. Its revolutionary impact affects the way people live, learn and work and the way government interacts with civil society. IT is fast becoming a vital engine of growth for the world economy (Heads of G8 Nations, 2000).

When the Internet and the World Wide Web rose as major new channels for socio-economic development, and the possibility arose of affordable ICTs, not only in rich states or corporations, but also for poorer countries, claims about the information revolution found widespread interest in the developing countries. In December 2003, heads of states and representatives from 175 countries gathered at the WSIS in Geneva to declare a desire and commitment to build

> [a] people-centred, inclusive and development-oriented Information Society, where everyone can create, access, utilize and share information and knowledge, enabling individuals, communities and peoples to achieve their full potential in promoting their sustainable development and improving their quality of life...(World Summit on Information Society, 2003).

However not everyone believed in the new knowledge economy or the capacity of ICTs to facilitate human growth within the current social order. To Henwood, everything was almost the same as before and "the intoxicating miracles of the late 1990s" which were supposed to be brought about by the new economy did not last as long as expected (Henwood, 2004, p. 227).

May drew similar conclusions, for example:

> There is no compelling need to regard the information age as any different; no need to place it outside the continuing history of technology's interaction with society. After all, the history of technology is the history of technological innovation and its social impact (May, 2002, p. 155).

At a practical level, Heeks questioned whether the UN's Millennium Development Goals for all developing countries should give priority to the application of ICT projects when "[m]ost of these projects never properly work, and for those that might just get off the ground, go back two years later, and it's all crumbled to dust" (Heeks, 2005).

2.3 The social forces of production

In volume 1 of "Das Kapital", Marx and Engels developed the concept of "social forces of production" or "productive powers" from what Adam Smith had referred to as the "productive powers of labor" in "the Wealth of Nations" (A. Smith & Seligman, 1970). To Marx, the productive forces are a combination of both the means of production and labor, including: (1) the personal activity of man, (2) the subject of that work and (3) its instruments, including management and engineering functions technically indispensable for production (Marx, 1915, p. 174). Changes which have occurred in relation to the means of production and of labor are discussed in turn below.

2.3.1 Changes to the means of production

The invention of the steam engine and the invention of electricity

Over the last 40 years we have been witnessing pervasive changes on a global scale, brought about by new technologies, especially ICTs, including the converging set of technologies in micro-electronics, computing (software, hardware and recently, content), telecommunications/broadcasting, and opto-electronics (Castells, 2000, p. 29).

In reviewing human history from the agrarian age to the so-called post-modern age, historians have recorded at least three major changes in the world's means of production: the invention of the steam engine, the invention of electricity, and most recently, developments in ICTs.

The agrarian age was changed radically when, in 1776, James Watt finally improved the steam engine designed by Newcomen. Forbes, a historian of technology, remarked: "the invention of the steam engine is the central fact in the industrial revolution" (Forbes, 1958, p. 150). The steam engine quickly became a provider of power in manufacturing industries in England, such as textiles. Thirty-five years later, the first steam engine was put on a steamship on New York's Hudson River, and from about the 1850s, steam engines were present in almost the entire world (Drucker, 1993, p. 22). Watt's decisive invention is often argued to have been the foundation for the replacement of hand-work and simple tools by machines, the new means of production, and (together with the introduction of capitalism) for shifting the agricultural economy to the industrial one.

If the steam engine made possible the replacement of muscle labor by machinery, the widespread use of electricity at the end of the nineteenth century gave people invisibly lengthened hands to conquer nature. In a second industrial revolution, electricity was the central force because only through electricity generation and distribution could other fields and other applications be developed and connected together (Mokyr, 1990, p. 84). Electricity networks "made possible and induced large-scale organization of work in the industrial factory" (Castells, 2000, pp. 37-38). Humankind then had the enormous capacity to diffuse electrical power and messages invisibly and widely throughout the world using electricity and the telegraph network. This led to a radical reduction in the concept of geographical distance.

The application of the steam engine and electricity profoundly revolutionized the productive powers of the global economy and shifted it to the industrial epoch. Yet, it took more than two centuries for the two industrial revolutions to be diffused throughout the entire economic system and to permeate almost the entire world "with Tiber, Nepal and interior of tropical Africa the only exceptions" (Drucker, 1998, p. 17).

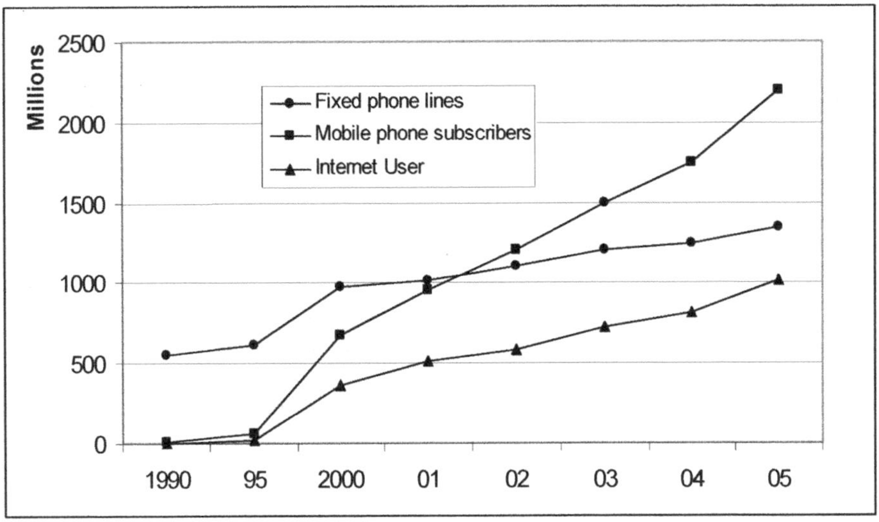

Figure 2.1 Growth in Telecommunications. Source: The chart is created by the author. Data of telephone lines and subscribers from ITU (2006a); and of Internet users from Internet World Stats (2007)

The development of information and communication technologies

Developments in ICTs did not take long to reach most parts of the world. The diffusion of three representatives of ICTs, namely the telephone, the mobile phone and the Internet, will be described here.

The telephone

The telephone was invented on 7 March 1876 by Alexander Graham Bell. For half a century, the telephone developed quite slowly. In 1905 the Bell System forecast that a telephone per five Americans would saturate the market. Until the 1960s, the capacity of the telephone network was very limited. The main routes which linked

America with Europe through four commercial circuits across the Atlantic could transmit only 35 conversations. Most telephone calls then were made locally (Cairncross, 2001, pp. 28-29). The explosion of telephone capacity really took place in the late 1980s with fibre-optic technology. Huge investment spending on a backbone network across continents and oceans resulted in increased capacity and much cheaper long-distance conversations. By 2001 the number of fixed phone lines across the world reached the first billion, 125 years after Bell's invention.

Mobile telephony

Mobile telephony has given the biggest boost to the diffusion of ICTs around the world, especially in developing countries where pre-paid services have been particularly popular. The concept of the mobile phone was initiated by the Bell Lab in 1947, but hand-held cellular radio devices were only available in the market since 1981. Due to their low establishment costs and rapid deployment, mobile phone networks have since spread rapidly throughout the world, outstripping the growth of fixed telephony. Figure 2.1 shows how the growth in the number of mobile phones has outpaced that of fixed lines. Mobile telephony has reached its first billion users in a period of 21 years (1981-2002), and the second billion users within just three years (2002-2005). The new challenge is to reach the next billion users (ITU, 2006b). The ubiquitous presence of the mobile phone can be seen by the appearance of a series of new terms beginning with m-, such as m-commerce, m-learning, m-government, m-literature, m-gaming and m-entertainment (Castells, Fernandez-Ardevol, Qui, & Sey, 2007, p. 77).

The Internet

In parallel with the development of network technologies and pervasive computing, the exploding development of the Internet in the last three decades is not less impressive. Originating in ARPANET, a computer network which was set up by the Advanced Research Projects Agency (ARPA) in 1969 as a military initiative to stimulate research in interactive computing and the sharing of computing time online among computer centers in the agency. The new technology grew up spectacularly in a "unique blending of military strategy, big science cooperation, technological entrepreneurship, and countercultural innovation" (Castells, 2000, p. 45). By 1990 several million computers were connected to the Internet, using the Internet protocol TCP/IP, but still, the use of it was limited because of difficulties in graphic transmission capacity and in the location of data. The invention of the World Wide Web by the Centre Europeen pour Recherche Nucleaire (CERN), and many other important contributions, such as the development of an Internet browser, made the Internet a new environment for working, living and entertainment in the post-modern society, the network society. Never has any new invention shot from obscurity to global fame in quite this way. In 1990, only a few academics had heard about this name. By 2000, it trapped 385 million people into this cobweb and five years later, one billion people around the world were using the Internet as a new way of communicating and a new global source of information (Internet World Stats, 2007). More than a mere communication media, this new technology is forming a series of "e-"phenomena, such as e-business, e-commerce, e-government, e-citizen, e-media, e-democracy, e-health, and e-mail. The list goes on, and the ubiquitous diffusion will continue to alter the way individuals, corporates, schools, organizations and governments work or operate.

The widening deployment of ICTs has had a great impact on the formulation of the knowledge society. As Drucker forecasted in 1968 in his work, "The Age of Discontinuity", the "impact of cheap, reliable, fast and universally available information will easily be as great as was the impact of electricity" (Drucker, 1969, p. 27).

2.3.2 Changes to the means of labor

Labor replaced by knowledge

Changes to the means of production could not take place without also changing the means of labor. In the developed countries in the last three decades the major and most valuable inputs for production have changed from natural materials to immaterial or "weightless" forms of production. Leadbeater claimed:

> These days most people in most advanced economies produce nothing that can be weighed; communications, software, advertising, financial services. They trade, write, design, talk, spin and create: rarely do they make anything (Leadbeater, 1999, p. 18).

Drucker argued that information and knowledge have overtaken traditional capital, natural resources and labor in their roles as basic means of production.

> The basic economic resource – "the means of production," to use the economist's term – is no longer capital, nor natural resources (the economist's "land"), nor "labor". *It is and will be knowledge*. The central wealth-creating activities will be neither the allocation of capital to productive uses, nor "labor" – the two poles of nineteenth – and twentieth century economic theory (...) Value is now created by "productivity" and "innovation", both applications of knowledge to work (Drucker, 1993, p. 8, Italics in original).

In observing major occupational groups in the US from 1964 to 1975, Daniel Bell identified a shift from a goods-producing economy to a service economy, including personal (retail stores, laundries, garages, beauty shops), business (banking and finance, real estate, insurance), and public (transportation, communication and utilities, health, education, research and government), and the pre-eminence of the professional and technical class (white-collar workers) over blue-collar workers (Bell, 1973).

The emergence of information as a new major source of production, "informationalism", caught the attention of scholars and other audiences in the late 1970s (Nora & Minc, 1978). Castells argued that the new ICTs enable people to manipulate knowledge or information, increasing the productivity and competitiveness of entities in the economy by their capacity to generate, process and apply knowledge-based information efficiently; economic activities of production, consumption and circulation as well as their components can now be organized on a global scale in a global network (Castells, 2000, p. 77).

2.3.3 Changes in production processes

Sayer (1991, p. 28), along the lines of an argument of Marx about capitalist production[2], argued that in the early industrial age, individuals in a division of labor produced for unknown markets, and did not know whether or not their labor met a consumption need or not.

The transition from industrialism to informationalism was not only in the means of production, but also in the processes of production. For example, there have been major changes to production processes from the Taylorism or Fordism styles of mass production to the Toyotism of flexible production approach in the post-war era. The Toyotism that originated from Japanese automobile firms was "just in time (…) management-worker cooperation, multifunctional labor, total quality control, and reduction of uncertainty" (Castells, 2000, p. 169).

Enabled by ICTs, producers who work online either individually or for a company know almost in real time what and how customers want a product. For example, the core system operation of the company, Cisco is its website. Customers can select from a number of options on different product lines to specify their preference. They can seek online advice and support from Cisco engineers. Once specified, the order is automatically transferred to the network of suppliers, and manufacturers ship the requested product to the customers (Castells, 2000, pp. 181-182).

Peter Golding (2000) divided technologies into two types. "Technology One" makes existing social actions and processes occur more speedily, more efficiently, or more conveniently, for example in management processes or in communication. "Technology Two" impacts comprise new forms of activity which were previously impracticable, or even inconceivable. In the author's view however, Golding's classification is not much different from Marx's distinction between "a new use for well-known use-values" and "discovery of new use-values" (Marx, 1969, p. 440).

May claimed that ICTs are mistakenly seen as a Technology Two (May, 2002, p. 153) because ICTs "lack of a manifest revolutionary effect requires the identification of a truly transformative information age to be constantly presented as a forthcoming development, as it frequently is". In other words, ICTs do not have the capacity to enable new forms of activity which never existed before.

However, it is argued here that ICTs have the characteristics of both types of technology. This can be explained using two examples: the impact of mobile communication on field workers, and e-business – the creation of a new form of business using the Internet such as Amazon.com.

The impact of mobile communication on field workers

Today, wireless devices like the mobile phone and personal digital assistants (PDAs) are more personal, portable and pedestrian (Ito, Okabe, & Matsuda, 2005). They are becoming essential equipment for many people, especially those whose work relies heavily on travel, like truck drivers, construction workers, maintenance engineers and the police (Agar, 2003; Group, 2005; Lundin, 2005).

In Scandinavian countries mobile instruments permit truck drivers to communicate and coordinate with one another and with their headquarters on the move. They

---

[2] "The essence of bourgeois society consists precisely in this, that a priori there is no conscious social regulation of production. The rational and naturally necessary asserts itself only as a blindly working average" (Marx & Engels, 1975).

can seek assistance for unpredictable situations or receive new assignments, thus being very flexible and responsive to circumstances in the field. Workers can work on other jobs in the time between two pending activities (Laurier, 2002; Perry, O'Hara, Sellen, Brown, & Harper, 2001).

On the one hand, from a productivity viewpoint, the twenty-four hours a day, seven days a week availability allowed by mobile technology is beneficial. On the other hand, the new technology arguably blurs the boundaries of social space and time, and between work and individual hours. Workers nowadays can be permanently monitored, controlled and timed by their supervisors (Laurier, 2002, p. 50). Castells et al. refers to mobile communication as "a wireless leash" (Castells et al., 2007, p. 80).

E-business: the creation of a new form of business using the Internet – the case of Amazon.com

The Internet is transforming business practices in a fundamental way in relation to customers, suppliers and production processes. The emergence of e-commerce firms like Amazon.com is one of many examples of the Technology Two type that only the Internet can support. Launched as an online bookstore in July 1995 by Jebb Bezos, an Electrical Engineering & Computer Science graduate from Princeton University, Amazon.com Inc. soon become one of the leading e-business companies selling goods to the world via the Web. Despite the dot.com bubble in the late 1990s, its revenue kept growing, from $3.9 billion in 2002 to $8.7 billion in 2005, though the profit recently dropped. While a traditional bookstore might sell up to 200,000 titles, the online catalogue offers millions of items, including books and music, DVDs, videos, electronics, computers, software, apparel & accessories and more (Hof, 2005).

According to Saunders (1999), Bezos saw an opportunity that others, even those in the book industry failed to see: the potential of the Web for e-commerce. Amazon was a vanguard in using the Internet to achieve syndicated selling by linkage with more than 60,000 sites, including the top websites in the world such as AOL.com, Yahoo!, Netscape, Geocities and Excite whom they call "Associates" (Saunders, 1999, pp. 17-19). Among one of billions of sites on the Web, Amazon is successful because it makes it easy and enjoyable for the customer, and diminishes any fear and reluctance in buying things on the Net.

Spector summarized the simple procedures needed to make an online order on Amazon.com. Returning customers need only to enter their password. New customers are required to provide their email address and method of payment. Several minutes after an order is placed, Amazon.com sends an email to confirm the order. After the order has been shipped, another email is sent to notify the customer, and if some products are not yet available, the customer will be told about this as well (Spector, 2000, pp. 141-143).

Another step Bezos undertook was to put the entire product line into the cyberstore. A customer who might be uncertain when selecting a book in a large conventional bookstore receives a list of items retrieved by the Amazon search engine which have been purchased by others who have ordered the item. Each item is given a page with a photo of the cover of the package, a description of the contents, a rating score by customers and other information (Saunders, 1999, p. 37).

As remarked by Mellisa Bane, a senior analyst of Yankee Groups, "[t]he Internet creates novel advantages and opportunities for those like Bezos who can master and control the new technology" (Saunders, 1999, pp. 22-23).

Saunders wrote of Internet-based businesses:

> They work hard to keep their customers. They have the advantage that they work on the Net all the time, whereas their competitors are brick-and-mortar stores that are still trying to figure the Net out (Saunders, 1999, p. 22).

2.3.4 Conclusions about changes to the social forces of production

There have indeed been revolutionary changes in the social forces of production, at least in the developed countries and in many emerging economies. These changes may be continual and not visible by daily measurement in the developed countries. However such changes can be seen as a clear shift in some countries, for example in the Asian Pacific Rim, depending on where and when the researcher makes observations, and more importantly, what system of political-economy is used as a point of reference.

However it is argued here that ICTs were not the sole cause of the knowledge economy, if it ever existed.

In considering the changes to labor and production processes in the analysis above, the expansion of the service sector in the US took place in the 1960s and Toyotism after 1948.

The explosion of ICTs occurred much later in the 1990s. The huge investment in the ICT industries was driven by economic profit and potential profits. The affordability, speed and reliability of the new technologies have led to its use in meeting social and individual needs and to its rapid take-up in many parts of the world.

The examples above illustrate the typical characteristics of ICT technologies today. They are commonly dynamic, interactive, globalized, individualized, customized, competitive, networked and risky.

In the developing world, ICTs offer enormous potential if people can harness the new ICT applications to manipulate space and time for their development purposes.

But the capacity to make use of the new productive powers that ICTs brings to people requires appropriate social and economic conditions to be in place. For example, a farmer in Vietnam can search for a book on Amazon.com from Seattle with just a few clicks. But he cannot make that order until he gets enough money, say by selling his (low-priced) rice and having enough resources to obtain a credit card.

Drachmann pointed out that the first steam turbine was invented not by James Watts, but by Heron of Alexandria, an ancient Greek engineer about 150 BC, two millennia before Watt modified the Newcomen's machine (Drachmann, 1963). Why did the industrial revolution not happen at that time? Why was the steam engine not widely applied in non-western countries such as eighteenth century China where labor was so cheap and feudal monarchy ruled?

In other words, societal changes such as the industrial revolution and the knowledge "revolution" depended also on a specific social context, or what Marx called the "social relations of capitalist production".

The next section will examine what social and economic conditions facilitate the development of a knowledge economy – if that exists; and how capitalism rules the world economy and the way that wealth is made.

2.4 Knowledge/material exchange and the evolution of the social relations of production

2.4.1 Knowledge and productivity

Creating wealth is both the means and ends of capitalism, and knowledge and ICTs become new keys for capitalists to appropriate wealth. Alstyne and Buckley (2004) used two models, "homo economicus" and "homo computicus" to explain how knowledge and ICTs create new value. The homo economicus model deals with the concepts of value and information defined as facts. The homo computicus model applies rule-based logic to explain the concepts of efficiency and information defined as instructions. They hypothesized that (1) more precise information improves decisions by reducing waste, risk and duplication and (2) increasing the control of undervalued assets creates new options for workers and promotes economies of scope and scale.

Drucker assumed that knowledge creates value by increasing productivity (Drucker, 1993, p. 167). However, in checking the data of productivity growth rate for the 1870-1969 period in the G7 and OECD countries, Castells found no evidence for the hypobook that the application of ICTs leads to a significant growth in productivity. In contrast, data from the US Bureau of Labor Statistics showed only an extraordinary growth in the computing industry of the US in the 1972-1999 period (Castells, 2000, pp. 80-93). As there was no evidence of productivity gain, Castells assumed that the application of technology was for the sake of profitability, competitiveness and "*the growth of value of their stocks*" rather than for productivity (Castells, 2000, p. 94, Italic in original).

What does this mean? It can be argued that the uses of ICTs come first in the industries which are at their source. The ICTs then spread to other industries. It can be said that it is the ICT industry in the US that enjoys the most profitability from the ICT revolution, because other industries and countries are unable to compete locally and globally without the application of ICTs (Castells, 2000, p. 94).

Boisot and the Principle of Least Action

Boisot proposed a theory to explain that everything can be collapsed around data. To him, "[i]n *any physical system, learning processes bring about the substitution of data for energy, space, and time, as the short term, of the Principle of Least Action*" (1995, p. 31, Italic in original). Boisot argued that as with biological systems, humans can minimize entropy production along the Principle of Least Action by learning. Codification and abstraction minimize the volume of time-space and energy to contain data within in very small quantities, thus physical factors, such as energy, time and space can be saved by learning a new technology, from high physical input to high data input. In other words, the latter progressively substitutes the former in the learning curve. When there is a shift of the core technologies, i.e. industrial and information, there are discontinuities in the history with a respective revolution, as shown in Figure 2.2 below. Therefore, talking about the ecological crises, Boisot believed they:

> are certainly not imaginary; they have to be taken seriously. But they are far from constituting the absolute limitations on growth that a conventional eco-

> nomic perspective makes them out to be. They can be circumvented by our proven ability to make novel and effective use of data and information; that is to say, by our ability to *learn* (Boisot, 1995, p. 432, Italic in original).

However, Boisot's theory arguably faces several major practical problems.

Firstly, human beings cannot follow the Principle of Least Action in consumption. People cannot live with only data and information. Physical needs for human existence, such as food and clothing, could hardly be reduced by learning process (though learning can save the amount of raw materials and energy in the production of goods and services for these physical needs). With increases in population and consumption resulted from the increase of economic earnings, global consumption, at least in the foreseeable future, will rise dramatically. Matos and Wagner, amongst other authors noted that the consumption of raw materials in the US has increased significantly throughout the 20th century, and most of these raw materials are non-renewable (Matos & Wagner, 1998). Foster and Clark (2004, pp. 182-200) wrote:

Extraction of raw materials for commodity production is organized around meeting the demands of the countries of the North, where approximately 25 per cent of the world's population lives but which consumes 75 per cent of global resources.

Secondly, if learning can save the production costs per production unit, for the sake of maximization of profitability, capitalists cannot stop the expansion of production scale. Thus, the cost of production eventually increases in total.

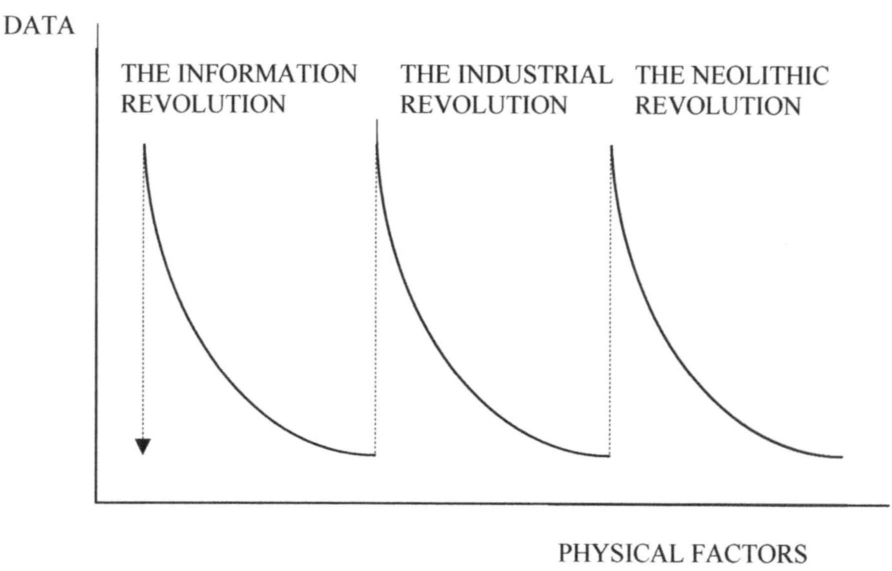

Figure 2.2 Collapsing physical factors into data - an evolution by technology. Source: Adapted from Boisot (1995, p. 38)

Thirdly, human beings have a bottomless cupidity and never stop the demand. For example, ICT capacity grows with an exponential pace in the recent decades by Moore's law, human demand always exceed that pace. Gantz from IDC, the premier global provider of market intelligence, advisory services, and events for ICTs, tallied up all the digital information the world generated in 2006, including photos, videos, e-mails, Web pages, instant messages, phone calls and the likes. He estimated that about 161 exabytes of digital information were created and stored in 2006 and that number would jump up to 988 exabytes (or 1 zettabyte) in 2010. This volume of data if stored in hard drives requires about 644 million 250-GB-hard-drives and would be "equivalent to 12 stacks of the latest Harry Potter novel, each reaching from the earth to the sun" (Bergstein, 2007). This is a gigantic rise from 5 exabytes, the figure that Berkeley estimated in 2003 (Bergstein, 2007; Kessler, 2007). Although the methods used in the two studies were not identical, the case suggests that humankind tends to maximize the benefits of new technology, requires and consumes more than the capacity that technological progress can afford. By doing this, learning process demands even more energy and space, too.

Laozi, the ancient Chinese philosopher living in the 6th century BC had found out the contradictory movements between humans and Nature when he wrote about humans' desire in his famous Tao Te Ching or the SAGE. He pointed out that Nature follows the principles of harmonization and least of action while men act on the revease (Laozi & Henricks, 1989, p. 77; Laozi & Merel, 1995).

Boisot suggested that by learning we can survive. But human beings, as material creatures, are living in a material world, and *cannot exist without materials*. Even we have learned the SAGE of Laozi for 2,600 years, we have not changed much our behavior to save the nature. Therefore, learning is necessary but not adequate.

So what is the secret of a knowledge economy? James Heskett observed that "[m]anufacturing is substituting information for assets" (Heskett, 1986, p. 160). Gilder, in "Microcosm: the quantum revolution in economics and technology", saw the information revolution as essentially one of "mind over matter" (Gilder, 1989).

In the author's view, a knowledge economy can only exist if it can swap its non-physical products for materials demanded by consumption. Therefore, a knowledge economy requires at least three conditions:

(1) The existence of economies which produce physical goods to meet physical needs in contrast with developed economies which create and accumulate knowledge capital, typically the U.S. This will be examined further by reviewing the division of labor in the global economy;

(2) An international mechanism that allows the selling of "knowledge goods and services" with high price and also the purchase of physical goods with low price. This is discussed later in an examination of the accumulation of knowledge in developed countries and of how knowledge properties are protected in international trade by regimes like the Trade Related Aspects of IPRs of the World Trade Organization; and

(3) The inequality of the exchange of mind over matter. This is further examined in a discussion of the concept of Terms of Trade.

Figure 2.3 below illustrates these relationships. They are examined in more detail in the next sections.

2.4.2 The division of labor in the global economy

Three points are argued here by the author:

1. The division of labor has existed at least in the iron age;
2. Today the division of labor operates as a global level;

3. The Third World, and especially emerging economies, in coping with this challenge, is trying to move to the so-called knowledge economy to diminish the divide.

These points are now examined in more detail.

There has always been a division of labor

Firstly, a division of labor has been a fact in economic activities since the ancient age. This author disagrees with Marx's book that there was no class in the hunter-gather society. Even in simple societies of ants, bees or monkeys, a certain hiararchy and division of labor always exists. In ancient Greece, Plato saw it as a source of inequality in the society as well as the basis of the state in his famous philosophy and political work The Republic in 360 BC,

> Well then, how will our state supply these needs? It will need a farmer, a builder, and a weaver, and also, I think, a shoemaker and one or two others to provide for our bodily needs. So that the minimum state would consist of four or five men....(Plato, 1955, p. 103).

And in the Orient, two centuries before Plato, Laozi had argued "the master does nothing, yet he leaves nothing undone. The ordinary man is always doing things, yet many more are left to be done" (Laozi & Henricks, 1989).

The division of labor is operating today at a global level

Adam Smith in the early industrial age saw specialization and the division of labor within industrialism, on the one hand, and the dynamic engine of economic development on the other as causing a "mental mutilation" of workers; they become ignorant and insular as their working lives are confined to a single repetitive task (A. Smith & Seligman, 1970).

Marx distinguished between the political division and the economic division of labor. That is, some forms of labor co-operation are due to economic necessity, but others are purely a result of a political/social control function related to a class and status hierarchy. If these two divisions were merged, the existing division of labor would appear to be economically inevitable and immutable, rather than politically constructed and influenced by power relationships (Marx & Engels, 1975, Preface). In recent times the division of labor has evolved to maximize the value of surplus in the world's capitalist system. Today the division of labor no longer occurs within the boundary of a nation but on an international scale, a world economic system. Before the 20th century, due to limitations in transportation and communication and the absence of an international regime to legalize free trade across borders, the division of labor remained primarily within the boundary of nations, and was a key factor within class stratification and struggle though activities like colonization, spice trade, discovery or shipping of slaves Today however, the division of labor is evolving at a global level. Developed economies are able to specialize in high-value industries, such as services and ICTs, and are leaving the production of low-value industries to the rest of the world. A common view is that a new international division of labor has emerged within the world economy since the 1960s. Some authors have raised the question of legitimacy of Drucker's term "knowledge worker", and the statisti-cal data based on that (Bologna, 2004). Huws (2003, p. 130), for instance, claimed that the decline of employment in agriculture only reflects the numbers of people who directly work on the land, and does not include those who engage in agriculture-related industries, such as making tractors, fertilizers, packing and preparing foods, etc. In the author's view, though, the concept of "knowledge workers" has some grounds as the people in such jobs are in fact no longer working as unskilled labor but are trained, and therefore, in many cases are skilled and knowledgeable workers.

Jenkins noted such an increase within the service sector in five developed countries. The economies in the developed countries where the growing service and non-goods-producing sectors have actually replaced the material sectors, as shown in Figure 2.4, are typified by four main trends in the composition of employment and GDP:

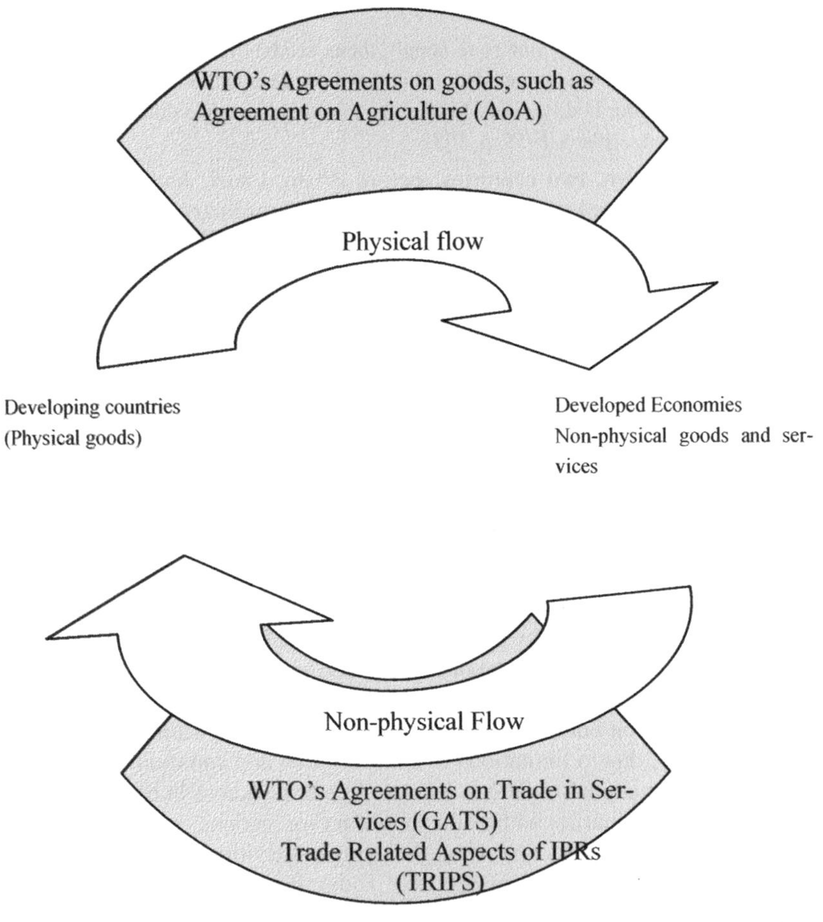

Figure 2.3 Knowledge and physical capitals exchange in the global market

1. a reduction in employment and GDP proportion within the agricultural sector;

2. an increase in employment and GDP proportion within the industrial sectors from 1890 to peak in the post war, then a decline afterwards; and
3. an increase in employment and GDP proportion within the service sector (Jenkins, 1992a).

Service work does not necessarily include knowledge work. A seller of Coca-cola soft drink provides a service, but requires little knowledge though somewhere in the

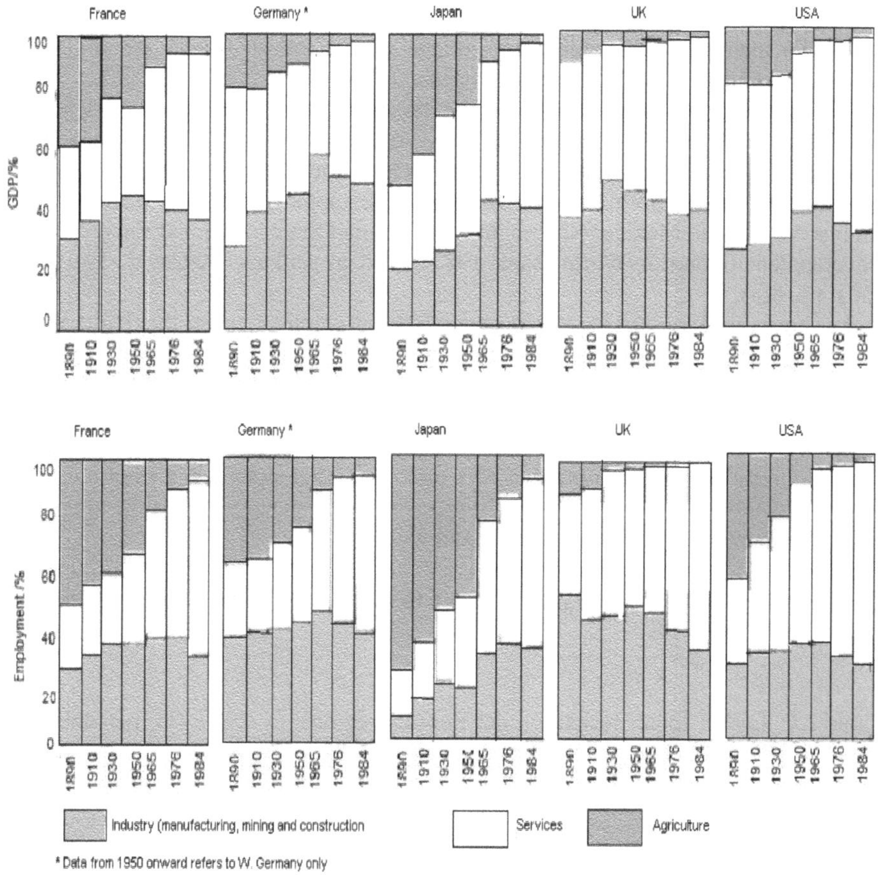

Figure 2.4 Sectoral shares in employment and GDP in developed countries from the 19th century to 1984. Source: adapted from World Bank 1987 (cited in Jenkins, 1992a, p. 30)

production line computer and other machines have been used to make the drink.

In defining knowledge work, Neef said it includes:

> the use of knowledge in the form of complex problem solving, technological innovation, creative exploitation of new markets, and the development of new products and service offerings (…) (Neef, 1998, p. 2).

The non-goods producing economies of developed countries rely both on the relocation of manufacturing employment to other parts of the world (even though such employment might still be conducted by the same companies), and on the supply of resources from developing countries.

Already in 1998 Neef observed that about 20% of production by US companies took place abroad. Multinational companies (MNCs) such as Nestle (headquartered in Switzerland) had 98% of their production capacity outside their host nation (Neef, 1998, p. 6).

Many economies in the developing world are major suppliers of resources and goods, for example within agriculture or the mining industry. Such resources and goods can only be supplied by depleting natural resources which are finite. While the production of goods is declining in the developed world, the maintenance of that production in developing countries plays the key role in feeding the world economy, including the "weightless economy". For instance, the majority of gross US oil imports in 2000, apart from Canada, came from developing countries, namely Saudi Arabia, Venezuela, Mexico, Nigeria and Iraq (United States. National Energy Policy Development Group & World Bank 1987 (cited in Jenkins, 1992a, p. 30) Bush, 2001, pp. 8-4).

As shown in Table 2.1, by 1980 about 80% of employment in the Third World remained in the goods-producing sectors and these sectors were the main source of physical goods and raw materials of the world (International Labour Office, 1984). Even service or knowledge work is never really "immaterial". On the contrary, all services and any human activity in some way must rely on certain physical resources. From thinking as a product of a brain to digital signals in a machine, none can exist without physical attributes. In fact, simple existence requires the alienation of physical resources such as air, water and food.

Developing economies are trying to move to the so-called knowledge economy to diminish the divide

Table 2.1 Labor force structure: developing regions in 1980 (in percentages). Source: (International Labour Office, 1984, p. 7)

Group	Agriculture	Industry	Service
All developing countries	59.1	19.9	21.0
Latin America & Caribbean (low income countries)	49.3	19.8	30.9
Latin America & Caribbean (middle income countries)	31.8	25.8	42.4
China	60.0	25.8	14.2
Asia (other low-income countries)	65.5	11.8	22.7
Asia (middle-income countries)	52.5	19.4	28.1

Africa and Middle East (low-income countries)	80.0	8.6	11.4
Africa and Middle East (middle-income countries)	55.6	19.0	25.4
Africa and Middle East (capital surplus oil producers)	51.8	18.3	29.9

To cope with the comparative disadvantage in pricing systems, many developing countries, especially the Asian Rim Pacific, have been undertaking dynamic strategies to shift from a resource-based economy toward a knowledge-based economy, for example, Singapore's strategy to be an "Intelligent Island" (Droege, 1997); South Korea's "e-Korea" (Dahlman, Andersson, Organisation for Economic Co-operation and Development, & World Bank Institute, 2000), and recently China's "Seizing the 21st century through knowledge" (Dahlman et al., 2001). These countries are robustly transforming their economies from goods-production to services and knowledge production sectors. In China, the share of the primary sector in employment declined from more than 70% in 1978 to about 50% in the late 1990s and the secondary sector increased respectively (Dahlman et al., 2001, p. 12). The change is also reflected in the composition of manufactured exports from these countries. As shown in Figure 2.5, resource-based exports of China, Malaysia, the Philippines, Thailand and Indonesia dropped dramatically, and medium and high-technology sectors gained considerable growth from 1985 to 1995, though low-technology production remains high in some countries like China and Hongkong.

2.4.3 Economic/political rules that enable wealth appropriation through the exchange of nonmaterial goods with material goods

In the next two sections this book looks at two methods employed by developed countries to appropriate wealth through exchanging nonmaterial goods with material goods: (1) The protection of Intellectual Property Rights (IPR) and (2) Terms of Trade.

Protection of IPRs, a major tool of wealth appropriation

Arrighi and Silver, in reviewing the birth, growth and decay of the Dutch and British hegemonies in the 17th to the 19th centuries discovered that during the growing phase (or the phase of material expansion), financial capital was accumulated. When these hegemonies went down (the phase of financial expansion) the accumulated high finance was expanded into markets and colonies with the protection of political and military power (Arrighi & Silver, 1999, see Chapter 14).

In recent times knowledge and non-physical capital accumulated in the US requires political and commercial protection regimes to expand to the non-US market. The protection of intellectual property rights (IPRs) are one of these regimes. The alienation of the physical resources of developing countries is realized using the protection of IPRs by international jurisdiction, typically the Berne Convention for the Protection of Literary Artistic Works of the World Intellectual Property Organization, and the Trade Related Aspects of Intellectual Property Rights (TRIPS) of the World Trade Organization (WTO).

The original motive of IPRs arose in the 17th century from John Locke's suggestion that only through rewarding intellectual labor would human beings try to improve the world. By withholding the use of intellectual properties for a temporary period, IPRs create a short-term monopoly over certain uses of a work and thereby, a limited scarcity in relation to the properties. The limited time of protection of IPRs was to ensure that on the one hand those who could use knowledge objects most efficiently could secure an appropriate reward for such usage, on the other hand by ensuring the transfer of knowledge to the most efficient users, the public good was maximized along with the totality of social welfare.

In reality the right of the direct creator of intellectual property is seldom highlighted in IPRs. The TRIPs agreement favours the right of knowledge of owners over the right of knowledge of creators (May, 2000, p. 73). In employment contracts, the ownership of the employers is well established so that "first to convey to the employer any rights the employee or independent contractor may have in specific copyrights, patents, trade secrets or trademarks" (Little & Trepanie, 1997). In other words, the ownership and control relations pertaining to society's productive assets and key resources, a major component of Marx's relations of production, remain in the hands of the capitalist.

The emergence of IPRs is firstly in the interest of specific groups in society: those who possess such resources can utilize it to accumulate more resources and the dominant discourse of IPRs is defined by the dominant actors. Within WTO where the power is largerly in the hands of the two leading players – the US and the European Union (EU), the talk on IPRs was initiated in the Uruguay Round by multinational corporations (MNCs) in the pharmaceutical and IT industries which claimed huge losses due to inadequate protection of their property overseas (Jawara & Kwa, 2003, pp. 36-57).

Maskus (2000) argued that the globalization of intellectual property really only benefited the US, the world's biggest net IPR exporter, and to a lesser extent the EU. It is the US and the EU that have the world's dominant software, pharmaceutical, chemical and entertainment industries and the world's most important trademarks. When TRIPS was negotiated, only 1% of 3.5 million patents belonged to the developing nations who were in the position of being importers of intellectual goods and services.

The IPRs regime makes ideas artificially scarce so that they can be given a price as material goods. Publishers and manufacturers prevent knowledge objects from unlicensed copy in order to reinforce the scarcity and maximize revenues from monopoly rights. TRIPS artificially protects the scarcity and maximizes returns by a constructed monopoly (May, 2000, p. 42). However, intellectual properties are non-rivalry capitals. Goods are called "non-rival" if their consumption by one person does not diminish its availability for use by any other person (Romer, 1990). In other words, an intellectual property can produce infinite copies. The real trick is here, with a marginal cost of reproduction almost as low as non-existent, an intellectual property, such as a software package or an information service, can be sold for as long as, and at as high a price as the IPRs jurisdictions can extend and permit.

That is the way the developed states try to nourish the knowledge economy: by extending and expanding the term of IPRs protection. The US House of Representatives, for instance, extended the term of protection of copyright from 17 to 85 years after the death of an author. The EU similarly in the Directive of 29 October 1993 in "harmonizing the term of protection of copyright and certain related rights" increased the term from 50 years to 70 years. The scope of IPRs protection is also expanded from artistic and literature works to many areas, such as education, biotechnology and to software programs (N. T. Nguyen, 2001).

Not only protecting knowledge created within the developed countries, the protection spread out to common knowledge and properties which had previously been available for public use. The patenting of life forms and genetic resources, including plants and their medicinal functions under the TRIPS Agreement (Article 27.3b) allows biotechnology MNCs such as Monsanto to slightly change seeds, and patent them for twenty years. These patented seeds are then sold to farmers worldwide who are not allowed to use seeds from the previous harvest for the next year's planting, even though they may have been applying that practice for hundreds of years. With the control and monopoly of genetic resources, MNCs gain huge profits from TRIPS protection (Jawara & Kwa, 2003, p. 36).

By signing TRIPS, agricultural exporters find that they have to pay more for the patented agricultural and chemical inputs from the developed countries. Thus an institution like TRIPS has actually become a vehicle for the transformation from knowledge power to physical capital, thereby helping the advanced countries to appropriate natural resources from the less advanced countries (Deardorff, 1990).

2.4.4 Terms of trade – the invisible alienation of nature

Along with the protection of IPRs, the terms of trade reflect the trend of giving a high price to non-goods-products compared with that of goods-products. Referred to as the relative prices of a country's exports and imports calculated by dividing the index of export prices by the index of import prices, Presbisch and Singer independently found that there is a tendency for the terms of trade of developing countries to decline over time in dealing with developed countries (Prebisch, 1962; Singer, 1950). In 1998 Neef remarked that service prices had increased more than three times as rapidly as industrial prices since the 1960s and the percentage of the American GNP dedicated to services had risen from 50% to more than 80%; 64% of the world's wealth then consisted of "human capital" (Neef, 1998, p. 2).

Dowlah indicated that there was a precipitous decline in the prices of primary commoditie:

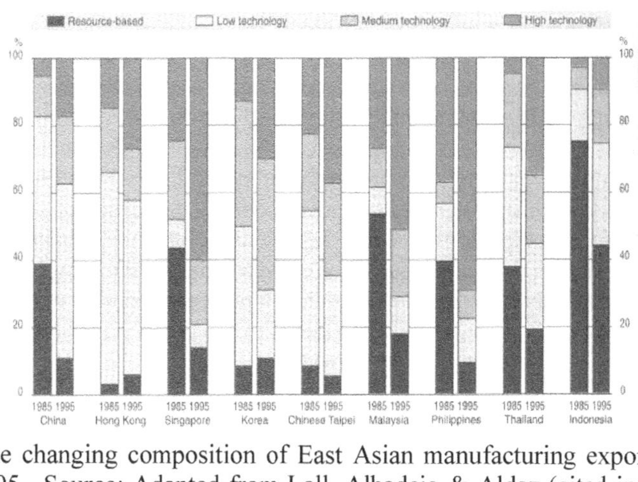

Figure 2.5 The changing composition of East Asian manufacturing exports, 1985-1995. Source: Adapted from Lall, Albadejo & Aldaz (cited in Dahlman et al., 2000, p. 30)

> Between 1997 and 2001, international price indices of food fell by more than 30 percent and that of agricultural raw materials, minerals, ores, and metals fell by 20% (Dowlah, 2004, p. 16).

Recalling that material goods and natural resources like coal, are finite and rival, one person's use of coal reduces the amount available to others. In most cases, nevertheless, natural and human-made capitals are complements, not substitutes. Such unequal pricing mechanisms, the "double standard" allows more flow of physical resources from the South to the North.

Costanza and Daly (1992) asserted:

> It should be obvious that the human-made capital of fishing nets, refineries, saw mills, and the human capital skill to run them does not substitute for, and would in fact be worthless without, the natural capital of fish populations, petroleum deposits, and forests (Costanza & Daly, 1992).

The economy and human beings operate only on the basis of extraction of resources from the environment and the disposal of waste back to the environment. In other words, the physical economy is vital to human existence.

The process of development as promoting structural relationships asymmetrically benefits the North at the expense of the South. Theorists including Futado and Presbisch argued that pervasive inequality and unfavourable terms of trade would lead to a state of persistent "dependency", in which wealth would flow from the "periphery" of the developing economies to the "core" of dominant nations (Harris, Wise, Gallagher, & Goodwin, 2001). This perspective is associated with an advocacy of strong state intervention to promote import-substituting industrialization, using tariffs and state direction of economic development. In reality the net flow of financial resources from the developing to the developed countries is increasing. In the 1960s the ratio was three to one (that is, for every one dollar that went to the developing countries, three went from South to North through debt servicing and other financial flows). By 1998, it was seven to one (J. W. Smith, 2000, p. 27). Thereby, J. W. Smith concluded that the new world economic order has been purposefully structured largely to benefit the already wealthy.

Similarly, Nobel Winner Stiglitz (2002, p. 61) also asserted that the "double standard" globalization agenda that the North erects is to ensure that it earns a disproportional share of the benefits. Advanced countries are able to refuse to open up their markets to the goods of developing countries, for instance, by setting quotas on numerous goods from textiles to agriculture, while insisting that the developing countries open up their markets to the manufactured and knowledge goods of the developed countries. Whilst developed countries continue to subsidize their own farmers, they advocate that developing countries should eliminate subsidies for the poor farmers, and accuse poor countries, such as Vietnam, of "dumping". The elimination of subsidies floods the markets of developed countries with cheap products (Mathew, 2004). As a result, the net effect is to lower the prices of natural capital of some of the poorest countries in the world.

2.5 Main insights from Chapter 2

Insight 2A: While early Utopian claims for the knowledge economy were exaggerated, ICTs have enabled a shift in what Marx termed the social forces of production, towards much higher knowledge intensity, initially in developed countries but spreading to the developing world.

Insight 2B: With knowledge resources, humankind can minimize the use of physical resources. ICTs enable more efficient monitoring and control of production processes, reducing wastage in the conversion of nature-made to human-made resources. This is one side of the so-called "substituting data resources for physical resources" (Boisot's Principle of Least Action) through changing the social forces of production.

Insight 2C: While knowledge resources are endlessly renewable, many natural resources are finite and exhaustible. Since knowledge production is renewable, it has an advantage over non-renewable natural-resource production. Within the international division of labor, ICTs confer an advantage on countries with more developed knowledge-producing economies compared to countries whose economies depend on the sale of natural resources. That is the other side of "substituting data resources for physical resources" through changing the social relations of production.

Insight 2D: Increasing knowledge-intensity gives an impression of "weightless" production in developed countries, but this is largely achieved by shifting an accelerating burden of environmental resource exhaustion to developing countries under terms of trade and double standards unfavorable to the latter. On a global scale, ecosystem losses from increasing physical resource production in developing countries largely offset ecosystem gains from "weightless" knowledge production in more developed countries.

Chapter 3

Knowledge economy for sustainable development: reality or utopia?

3.1 Scope of the chapter

- Presents an international "state of the art" account of sustainable economic development and its relationship to knowledge and ICTs.
- Focuses on:
  - The global environmental crisis
  - The tensions between "sustainability" and "development"
  - Efforts to achieve international co-operation on environmental issues
  - Potentialities for, and limitations of, knowledge strategies and ICTs in helping deal with degradation to the world's ecosystem
  - Relevance of resistance by developing countries to global ICT hegemonies.
- Introduces basic equations foreshadowing theorizations in Chapters 14 and 15.

3.2 The seriousness of the world's environmental problems

This section describes the extreme seriousness of the world's environmental problems. It highlights two inter-connected areas of concern: the ever-increasing demand for non-renewable resources and the degradation of the ecosystem.

3.2.1 Ever-increasing demand for non-renewable resources

Pearce et al. showed that as the world's population grows, the demand for non-renewable resources rises. They also argued that the consumption of materials per capita also increases with the increase of living standards. Between 1960 and 1984 the world's total consumption of energy had increased by 46 percent due to population growth and 54 percent through increased per capita consumption levels (Pearce et al., 1991, p. 123). The increase was remarkably high in Asia where densely populated centres like China and India are making considerable strides in economic development.

According to Aplin, in addition to the growth of the world's population level (Pn), and consumption level (C), the level of technological development (T) is another factor impacting on the environment. Some authors have suggested that the

Impact (I) caused by humans on the environment can be roughly measured by the formula I = Pn x C x T (Aplin, 2002, p. 22)[3].

Human activities are changing the environment in a way unlike that of any other era. At the beginning of the twenty-first century, extensive and excessive resource use, energy-inefficient lifestyles, industrialization and the pursuit of economic growth have been inextricably linked to environmental degradation, within and across state borders. The global economy has expanded "five-fold in the past half-century, three-fold since 1980 alone" (Rees, 2002, p. 24). World exports have increased to be worth more than USD 7 trillion, over 20% of the world GDP (World Bank, 2003b, p. 25). Since the 1950s, world industrial production has increased fourfold (Sitarz, 1993, p. 38).

In the Roma Club report, Meadows et al. (1972, p. 56) calculated that with the exponential growth rate, much of the world's important non-renewable minerals, including aluminium, copper, gold, lead, manganese and mercury could be used up within this century. Many have criticized the validity of the 1972 Roma Club report (see, Hammond, 1998), but on the basis of the evidence considered to date, the present author contends that the irrevocable degradation of the earth is merely a question of time. Energy, the lifeblood of the world's economy, for instance, was consumed at an average growing rate of about 2% per year between 1972 and 1999, leading to a total growth of energy consumption of almost 70% over that time (UNEP, 2002, p. 35).

Some authors have estimated that the world has consumed about half of its estimated economically recoverable oil reserves – the oil and gas reasonably certain to be producible using current technology at current prices, with current commercial terms and government consent (James, 2000). Appenzeiller quite bluntly referred to the situation as being "the end of cheap oil" in an article of the same title (Appenzeiller, 2004, p. 84).

Klare (2004, p. 169), interpreting a chart from the US National Energy Policy 2001, estimated that domestic US oil production would decline from about 8.5 million barrels per day (mbd) in 2002 to 7.0 mbd in 2020, while oil consumption would jump from 19.5 mbd to 25.5 mbd. Klare thus concluded that US oil consumption would continue to exceed domestic production from 11 mbd in 2002 to 18.5 mbd in 2020 which could only be met by foreign sources. The shortage of energy occurs not only in the developed world but also in other parts of the world, especially emergent economies like China and India. The two countries have actually become major rivals of the US in the importation of energy. Net oil imports by China are expected to rise from 1 to about 8 million mbd in 2020, with 70% of this amount from the Middle East (United States. National Energy Policy Development Group & Bush, 2001, pp. 8-4, 8-14).

3.2.2 The degradation of the ecosystem including environmental habitat, raw materials, biodiversity, the atmosphere, rivers and oceans

In addition to the ever-increasing exploitation of nature as a source of raw materials, pollution from solid, liquid and gas wastes, and chemical discharge from production

[3] Aplin used the symbol P for population level. The present author has substituted Pn to facilitate later use.

and consumption severely threaten the ecosystem. It is a serious problem that waste is thrown into the commons, such as the ocean, atmosphere, land and rivers as an unlimited dumping ground. As a result, the concentrations of carbon dioxide emissions in the air, mainly emitted by developed countries, have increased exponentially since the Industrial Revolution, and now stand at 370 parts per million (ppm). Computer models, such as those of Huang, Dickinson and Chameides (2006) suggest that microscopic aerosol particles are powerful agents which could severely impact climate change over East Asia in the 21st century. Between 70 to 80 percent of marine pollution is from land-based sources which include agricultural run-off (pesticides and fertilizers), sewage, oil and hydrocarbons, synthetic compounds and a range of heavy metals (Sitarz, 1993, p. 145).

The new global division of labor described in the previous chapter, by means of which manufactured goods are increasingly produced in the developing countries, has led to a situation in which high pollution and energy and resource consumption levels are moving to large parts of the developing world, allowing the most developed economies, such as the US, to keep their environment relatively clean.

According to the UN's "Millennium Ecosystem Assessment Report" (Millenniumassessment.org, 2005), over the past 50 years, humans have changed ecosystems more rapidly and extensively than in any comparable period of time in human history, largely to meet rapidly growing demands for food, fresh water, timber, fiber and fuel. This has resulted in a substantial and largely irreversible loss in the diversity of life on Earth. Sixty percent of the ecosystem services examined during the assessment have been degraded.

The changes that have been made to ecosystems have contributed to substantial net gains in human well-being and economic development, but these gains have been achieved at growing costs in the form of the degradation of many ecosystem services, increased risks of nonlinear changes and the exacerbation of poverty for some groups of people. An example of nonlinear changes or regime shifts in an ecosystem is "the case of tropical coral reefs, where nutri¬ent loading, declines in herbivorous fish, and reef degradation collectively trigger shifts to algal-dominated systems" (Millenniumassessment.org, 2005, p. 6). These problems, unless addressed, will substantially diminish the benefits that future generations obtain from ecosystems.

Pearce et al. suggested that the "bearing capacity" of a given area is "the maximum number of people that can be sustained by the resources on that land" (Pearce et al., 1991, pp. 114-115). Sachs expressed deep concern that human activities may have "exceeded the biosphere's capacity since the mid-1970s" (Sachs, 2002, p. 13). Together with non-renewable resources, the existing "renewable" resources, such as land, water, air, and biodiversity are being used beyond the capacity from which they can be recycled naturally, and thus, may no longer be seen as renewable.

Water provides a useful illustration. Many parts of the developed world are using more water than can be replenished. Citizens in the world's thirstiest cities such as Houston and Sydney, and many large areas in the United States are using substantially more water than nature can replenish (WWF, 2006a). Major constructions in developing countries, such as dams are built without adequate consideration of environmental costs. In India, much of its agriculture is under threat from rampant overexploitation of water resources. International concerns were raised over the scale and the ecological and human costs of China's massive water constructions. A report from the Mekong River Committee showed that the water level of the lower reaches of Mekong River reached its lowest level in 1993 right after China had built the Manwan Dam at Yunnan province, and the country planned to build another six dams on the river. Those constructions endanger the life of 60 million people living in the delta area, a region as large as the land surfaces of France and Germany put together (Minh Sơn, 2004).

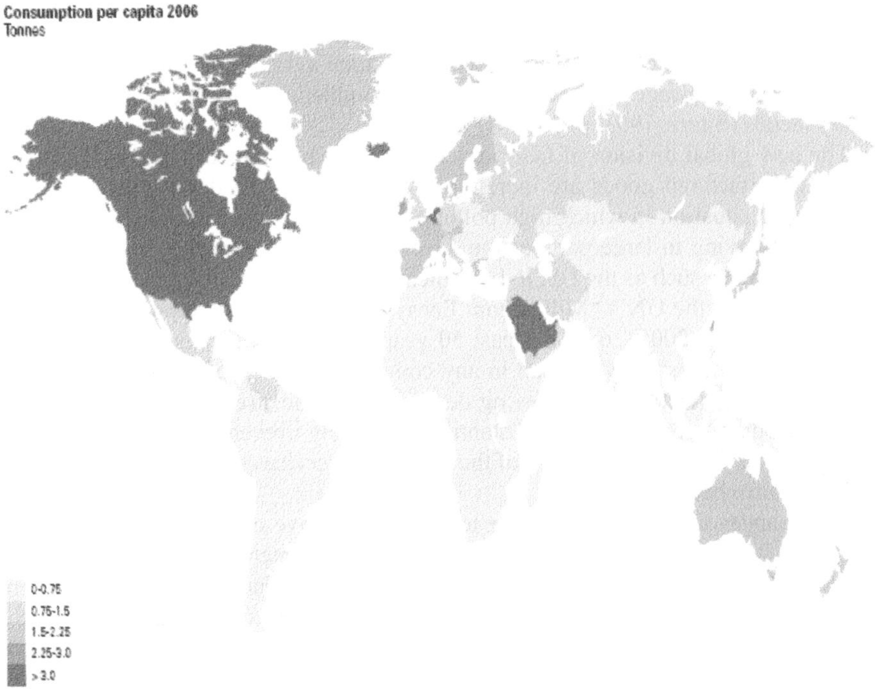

Figure 3.1 Energy consumption per capita 2006. Source: Adapted from BP (2007a, p. 13)

Barlow and Clarke noted that classical scholars like Malthus and John Stuart Mill might fleetingly have considered the impacts of possible scarcities of food and raw materials, but these scholars would never have envisaged a shortage of fresh water so dire that no remedy is possible. In recent years fresh water is no longer considered to be an infinite resource, but is in fact "blue gold" (Barlow & Clarke, 2002).

Less than one per cent of all the water on the earth is available for human beings. By the mid-1990s the water supply per capita in developing countries was only one-third of its 1970 level (UNDP, 1996). Over one billion people do not have access to safe drinking water, and about two billion are without adequate sanitation.

The water crisis has affected not only the poor, but also the rich. In its report "Rich Countries, Poor Water" (WWF, 2006a) the WWF argued that the water crisis has become a key problem on a global scale for wealthy nations. European countries along the Atlantic have been suffering recurring droughts. In the Mediterranean areas, water-intensive tourism and the expansion of irrigated agriculture are threatening the resource. Contamination of water supplies is a serious issue in Japan. Salinity and evaporation are major threats to the dry continent of Australia. Jamie Pittock, Director of WWF's Global Freshwater Programme argued:

> Economic riches don't translate to plentiful water (....) The crisis in rich nations is proof that wealth and infrastructure are no substitute for protecting rivers and wetlands, and restoring floodplain areas (WWF, 2006b).

There is considerable consensus that the burning of fossil fuels has led to climate change resulting in global warming. In a 1990 publication, Houghton, Jenkins and Ephraums predicted that the growth of emissions would interfere with the climate system at a rate of change in average global temperatures that would be faster than at any time in the last 10,000 years, and sea levels would rise about three to six times faster than in the last 100 years (J. T. Houghton, Jenkins, & Ephraums, 1990, p. xxviii).

Garrett Hardin (1968) likened the unregulated use of the oceans and atmosphere to the pre-industrial overuse of the English agricultural commons, and argued that such unregulated use would be a "tragedy of the commons" because "freedom in a commons brings ruin to all".

3.2.3 Impacts of shortages of physical resources and energy

Shortages of physical resources and energy inevitably lead to market conflicts. Such conflicts can be classical causes of war. In the late 20th and early 21st centuries, the demand for petroleum to meet US needs was stated in the US policy on energy:

> A fundamental imbalance between supply and demand defines our nation's energy crisis .... With forward-looking leadership and sensible policies, we can meet our future energy demands and promote energy conservation, and do so in environmentally responsible ways that set a standard for the world (United States. National Energy Policy Development Group & Bush, 2001, p. viii).

Its own thirst for oil is seen by many to be the real motive of the US Administration in invading Iraq to seize hegemonic control over the main oil wells in accordance with the US "standard for the world". Klare explained this standard in "Blood for Oil: the Bush-Cheney Energy Strategy", stating: "The only way to ensure the continued flow of energy is to guard the oil fields and pipelines with American soldiers" (Klare, 2004, p. 170).

Hirst believed that if energy scarcity through economic consumption has already been a main cause of war, the impacts of global warming, such as water shortages, the sinking of coastal land due to the melting of the Polar ice caps and the rise of sea levels could threaten the entire globe, from Bangladesh to the Maldives, and from southern Europe to North America. Hirst made the point that a huge wave of migration could provoke fighting and environmental war in the second half of the 21st century:

> In such an extended crisis we must assume violent conflicts between nations and the probable limited use of weapons of mass destruction. The nature and outcome of such conflicts is beyond calculation (Hirst, 2001, pp. 102-103).

3.3 The challenge of sustainable economic development

3.3.1 Views on sustainable development

The concept of sustainable development has been examined by authors working within different disciplinary frameworks.

Three key interacting dimensions of sustainable development have been recognized: economic, environmental and social. Harris and Goodwin summarized these as follows:

> An economically sustainable system must be able to produce goods and services on a continuing basis, to maintain manageable levels of government and external debt, and to avoid extreme sectorial imbalances that damage agricultural or industrial production.
>
> An environmentally sustainable system must maintain a stable resource base, avoiding overexploitation of renewable resource systems or environmental sink functions and depleting non-renewable resources only to the extent that investment is made in adequate substitutes. This includes the maintenance of biodiversity, atmospheric stability, and other ecosystem functions not ordinarily classed as economic resources.
>
> A socially sustainable system must achieve fairness in distribution and opportunity, adequate provision of social services, including health and education, gender equity, and political accountability and participation (Harris & Goodwin, 2001, p. xxix).

"Soft approaches" towards sustainable development were considered in the 1980s by policy makers and environmentalists who were concerned about the capacity of the natural resource base to maintain and increase human welfare. Soft approaches peaked with the famous definition of "sustainable development" in the Brundtland Report of the World Commission on Environment and Development (WCED): "development that meets the needs of the present without compromising the ability of future generations to meet their own needs" (World Commission on Environment and Development, 1987, p. 43).

This definition became popular, but some felt that "no one could really explain what was meant in practice" (Grainger, 1997, p. 64). The concept of soft approaches to sustainability was also criticized as focusing too much on intra-generational equity rather than on a North/South imbalance (Wise, 2001). A number of authors believed that not only must the value of the global stock of natural resources be conserved, but these resources must be redistributed to assure a certain level of consumption for all people in the world (L. M. Elliott, 2004, p. 3). Harder approaches to sustainability have been formulated, originally by Pearce and his colleagues in "Blueprint for a Green Economy" (Pearce, Markandya, & Barbier, 1989) which have attempted to develop theoretical bases that can be expressed as mathematical formulae – these are mentioned later in this chapter.

Pearce and his colleagues also differentiated between different conditions of sustainable development, namely weak and strong sustainability (Pearce et al., 1989, p. 127). "Weak sustainability" requires that the total value of both natural resources and human-made capital should not decline over time: "the wellbeing of a defined population should be at least constant over time and, preferably, increasing" (Pearce et al., 1989, p. 32). This definition implies that for the sake of people's happiness some level of substitutability between natural and human-made resources is acceptable. "Strong sustainability", however, argues for limits to be set on economic development involving human made resources in order to preserve the environment, even if some hardship for people is a result. "[T]he economic system cannot grow beyond the limitations set by the regeneration and waste-absorption capacities of the ecosystem" (Harris, 2001, p. 6).

Strong sustainability is often expressed in terms of the "precautionary principles" (United Nations, European Commission, International Monetary Fund, Organization for Economic Co-operation and Development, & World Bank, 2003, p. 7) which are expressed below:

- Renewable resources should not be used in excess of their natural regeneration;
- Non-renewable resources should be used prudently and efficiently with care that the same function is available to future generations, say by technological development or shift to use of renewable resources;
- Sink functions should not be used beyond their assimilative capacities;
- Activities which cause degradation in (environmental) service functions should be avoided or at least minimized.

Bartelmus and van Tongeren proposed the use of a new indicator, "Eco Domestic Product (EDP)" by which to assess economic development. The indicator includes a quantification of natural resource depletion. EDP aims at:

> Replacing conventional growth indicators, notably GDP or NDP, by EDP and expanding the scope of key variables such as capital and capital formation to include natural capital [use] in dynamic growth models (...) [EDP] could thus provide early warning-signals about the trends and limits of sustainable economic growth (Bartelmus & van Tongeren, 1994, p. 21).

UNESCO's Universal Declaration on Cultural Diversity (2001) stated that it was not only the conservation of natural values which was important, but also human values, such as cultural heritage:

> Cultural diversity is as necessary for humankind as biodiversity is for nature (...) [I]t is one of the roots of development understood not simply in terms of economic growth, but also as a means to achieve a more satisfactory intellectual, emotional, moral and spiritual existence (UNESCO, 2001).

3.3.2 The distinction between human-made capital and natural capital, and the possibility of substituting human-made for natural capital

This book argues that the distinction between human-made and natural capital is fundamental for reaching an understanding of sustainable development. Pearce et al. (1989, pp. 42-43) coined the terms of human-made capital (or manufactured capital) and natural capital for the first time. They later argued:

> The genetic information, the stock and variety of species, and the stock and variety of ecosystems and their function are all "natural capital". Economists are used to analysing economic processes that include man-made capital – machines, roads, buildings – and human capital – the stock of knowledge and capabilities (Pearce et al., 1991, pp. 22-23).

The "Handbook of National Accounting: Integrated Environmental and Economic Accounting 2003" defined natural capital as comprising three main types of functions, which should be included in national accounting systems:

> Resource functions cover natural resources drawn into the economy to be converted into goods and services for the benefit of mankind. Examples are mineral deposits, timber from natural forests, and deep sea fish;
>
> Sink functions absorb the unwanted by-products of production and consumption (…) These three destinations [air, water and landfill sites] are often referred to as "sinks";
>
> Services functions provide the habitat for all living beings including mankind. Some aspects of habitat are essential, such as air to breathe and water to drink. These are called survival functions (…) (United Nations et al., 2003, p. 5)

The "environmental accounting" or "green accounting" above also emerged from Pearce et al.'s suggestion about a "green economy" (Pearce et al., 1989). They claimed that the transformation of natural capital into human-made capital is an inevitable consequence of development, and in that process, society in general, and also economists often give priority – within the constraints of monetary policy – to promote the outputs of goods and services (measured by GDP) rather than paying the costs of environmental impacts. They suggested that environmental assets or natural capital must be valued by putting a monetary value on the services generated by nature. For sustainable development, they argued, it is very important to "ensure that environmental values are integrated into economic decision-making" (Pearce et al., 1989, p. 122) by giving them a "proper pricing" (Pearce et al., 1991, p. 167). However, the method of "discounting the future" of Pearce et al. was criticized for undervaluing natural assets (Pearce et al., 1991, p. 137). The author of this book argues that in accounting terms, natural assets are not a redundant item to be discounted; there must be practical accounting methods for quantifying their value this notion will be developed in terms of "absolute value" as the theorization strand of the book develops in later chapters.

Costanza and Daly wrote that human-made capital refers to the things "produced by humans" (Costanza & Daly, 2001, p. 14). Costanza & Daly (1992) divided natural capital into renewable natural capital and non-renewable natural capital.

Grainger suggested that human-made capital (HC) includes "the total stock of material and intellectual capital accumulated by the growth of civilization" which consists of productive capital (PC) like industrial capital and farmland; cultural capital (CC) such as education services; and infrastructure capital (IC), for instance, the provision of roads and sewers (Grainger, 1997, p. 64). Grainger included renewable and non-renewable resources (RRC and NRC) in his concept of resources capital (RC), and added environmental quality (EQ) as a new component. He thus defined human made capital as a total derived from the equation: $HC = PC + CC + IC$. And the total natural capital (NC) is: $NC = RC + EQ = (RRC + NRC) + EQ$ (Grainger, 1992, pp. 64-65).

Authors like Meadows et al. (1972), Pearce, Markandya, & Barbier (1989) and Costanza & Folke (1997, p. 67) have put forward other equations for modelling the condition of sustainable development within the context of environmental economics theory. Grainger (Grainger, 1992), for example, wrote that: "Hyper-strong" sustainable development which results in "no decline at all in natural capital [NC]", can be expressed as $\partial NC \geq 0$. In this expression no substitutability between human capital (HC) and natural capital (NC) is allowed. In the "weak condition" there is no decline in total capital and substitution of the two kinds of capital is possible, so that a rise in human capital (∂HC) can at least balance a decline in natural capital (∂NC), that is $\partial HC \geq |\partial NC|$."[4] Grainger pointed out (p. 67) that such calculations would be unlikely to be applied in a developing country, and that "questions remain about how changes in natural capital and human capital can be compared directly".

Grainger went further in initiating a national index of sustainability development, S, for weak sustainable development. This index was defined as, $S = f(\partial NC + \partial HC) = f(\partial RC + EQ) + \partial HQ$ (Grainger, 1992). Grainger argued:

> For development to be sustainable in a given year, S must be greater than, or equal to zero, since the fall in natural capital must at least be balanced by the rise in human capital" (Grainger, 1997, p. 71).

Grainger pointed out that the formula poses difficulties in that "S is a pure number with no relation to any point on a scale of development" (p.71).

Serafy (1997, p. 220) believed that weak sustainability is a "positive" (that is, a statement about what is, containing no indication of approval or disapproval), rather than "a normative concept" (that is, a judgment about what ought to be). Furthermore since it focuses on a yearly basis, it does not guarantee long-term sustainability.

3.3.3 Sustainable development as a contradiction in terms

Suspicion that there are limits to possible economic growth has existed for a long time. Adam Smith for instance raised the question of the long-term future of economic development in Britain in his Wealth of Nations in 1776 (1798). Thomas Malthus (Munda, 2001) was pessimistic about prospects for long-term economic growth, despite a surge in industrial productivity beginning in the late eighteen century and the increase in agricultural land available in the New World. The pessimism came from the scarcity of natural resources and the absence of mechanisms other than limiting goods supplies to restrain the growing population. However, at the end of the nineteenth century, Victorians were convinced that economic growth in Britain, Western Europe and North America was inevitable since most of the rest of the world consisted of colonies which supplied raw materials to the economic centers. In recent times there has again been pessimism about the feasibility of infinite economic growth because of concerns about whether both sustainability and development can be achievable goals (Mikesell, 1993, pp. 1,31).

Meadows et al., in the report "The Limits to Growth", argued that the developed world will need to stop growing if there are not to be enough natural resources for it to survive (Meadows et al., 1972).

---

[4] The formula $\partial HC \geq |\partial NC|$ is proposed by the present author. Originally Grainger wrote $\partial NC \geq \partial HC$. This appears to be a mistake because often in practice $\partial HC \geq 0$ and $\partial NC \leq 0$. If $\partial NC \geq \partial HC \geq 0$, that would be more than the condition of hyper-strong sustainability, that is $\partial NC \geq 0$.

Some neoclassical economists denied any special standing to natural capital; others recognized an obligation to keep the value of natural capital at a constant level for strong sustainability, because, as Costanza and Folke pointed out, if natural and manufactured capital can be fully substituted, there is no particular reason to conserve natural resources. For example, if forest products are used for industrial development, the depletion of the forests is then justified (Costanza & Folke, 1997).

The human species, in using nature for its survival, depletes it. In "The Entropy Law and Economic Process", Georgescu-Roegen (1971) argued that it is ultimately impossible to maintain a constant stock of natural capital, since all planetary resources will eventually be degraded or used up according to the Second Law of Thermodynamics, the Entropy Law. The author agrees with Georgescu-Roegen that the world's physical resources will eventually be completely depleted, but does not agree with him that the thermodynamics law is the main cause of the depletion. The author sees social laws and behaviors, in the way they are enacted by humans, as being the main agent for the frightening imperilment of civilization, humanity and the planet Earth.

Ayres and Kneese believed that all social and economic (i.e. human-related) activities extract resources from the global environment and dispose of waste back into it (Munda, 2001). The question of sustainable development is only applicable in the global context and all human activities should take into account the limits of this one global environment.

As indicated earlier, there is serious concern about the effects of the deterioration of the global environment on the future growth and liveability of the earth. There is great concern regarding the issue of whether it is possible to have both sustainability and development as realistic goals. The concept of sustainable development is thus a contradiction in terms (Redclift, 2006). Pairs of opposing concepts are, for example: today's growth versus tomorrow's needs, infinite demand versus a finite ecosystem, present development versus future survival. These pairs all emerge from the needs of our species Homo Sapiens to survive through the use and management of nature. Certainly Shakespeare's question of "to be or not to be" is relevant today – current behavior patterns of the human species are, intentionally or not, suicidal in terms of environmental depletion. The question is acute because humans are approaching the limits of what nature can offer.

Efforts to reconcile these contradictions are faced with major social and political challenges: ignorance versus knowledge, knowledge versus behavior, and local versus global arenas. It is complex because either people unintentionally do not understand the crisis of sustainability, or they may intentionally ignore it as they are too busy with the daily concerns of personal existence. In addition, even if the problem is understood, the perceptions and interests of different groups of people are unlikely to be the same, for example, the different countries' attitudes towards the Kyoto Protocol. Additionally, the distance between knowledge and behavior or action can be very large. These issues are dealt with further below.

3.4 Difficulties in achieving international co-operation to heal the environment

Given the global nature of sustainability, what are the conditions for international sustainable development? Concern about sustainability is not at the same stage in every country of the world.

In Vietnam, a country where more than 80 percent of people live on less than one dollar per day, people are unlikely to have the time and strength to worry about the survival of future generations when they themselves are struggling to survive in the short term. A poor country with plentiful forests may have considerable justification for chopping down ancient trees to support its economy.

But as economies grow and become "informationalized", it is necessary for people to devote attention to the balance between the present and future generations.

Even if all national governments wanted to achieve global sustainability, national economic policies would be likely to fall short of achieving the same collective result that could be realized by a world economic authority. Mikesell argued that each government could not adopt the goal of maximizing world net social output, but they could rather seek to maximize national welfare (Mikesell, 1993, p. 17).

It is difficult to negotiate and implement agreements on matters such as ocean fisheries, greenhouse gases and CFCs. The problem is often complicated by disagreements among governments as to the perception and extent of global warming that will occur, and how its harm will be distributed to each nation.

A national commitment to sustainable development will be largely limited to the resources of the country or to regions within countries. If one country, for instance, the US, seeks to preserve its forests, it will import more timber from other countries. This is thus a notion of "importing sustainability". Poorer countries exporting sustainability would risk non-sustainability for the benefit of the importing nations (L. M. Elliott, 2004, p. 12).

Bottlenecks in achieving sustainable development also lie in the difficulties of setting up international regulations and actual enforcement of these regulations by disparate stakeholders of the international community.

3.4.1 Conferences and summits

Elliot noted that in the decade before the 1972 Stockholm Conference on the Human Environment there was growing scientific and public concern about environmental degradation, yet the development of international environment law and the democratization of international environmental policy making were initiatives that developed at a very slow pace. At the 1972 conference, developed countries were cautious about setting up any institution which would require substantial funding, and developing countries were reluctant to support an institution whose decisions might restrict their own development. Also existing UN agencies were jealously guarding their own environmental related activities that they felt might be taken over by United Nations Environment Program (UNEP) (L. M. Elliott, 2004, p. 12).

Positive international efforts could be seen, such as the Summit on Environment and Development held in Rio De Janeiro in 1992, in which 179 participating countries passed the "Rio Declaration on Environment and Development" which was composed of 27 basic principles and "Agenda 21" solutions for sustainable development for the world in the 21st century. Ten years later, at the World Summit on

Sustainable Development held in 2002 in Johannesburg, 166 participating members passed the Johannesburg Declaration and the Plan of Implementation on Sustainable Development. By 2004, 113 countries and regions had developed national sustainable agendas and set up respective implementing agencies (Government of Vietnam, 2004d).

However, Elliott noted that the optimism engendered by the 1992 UN's Conference on Environment and Development in Rio did not last long. By 1997, the multilateral consensus was fragile, with disputes about how to manage the process and the consequences of globalization, issues which were not resolved at the World Summit on Sustainability Development in 2002 (L. M. Elliott, 2004, pp. 7-8).

3.4.2 Kyoto Protocol

The difficulties in enforcing the 1997 Kyoto Protocol of the Convention on Climate Change (United Nations, 1998) is an example of the failure to date of universal co-operation in tackling climate change. The aim of the Protocol was to stabilize atmospheric concentrations of greenhouse gases at levels that will avoid major damage to economies and ecosystems by defining, assigning and enforcing property rights in greenhouse gas emission from developed and transitional economies. Developing countries view this failure as an irresponsibility of the developed economies. According to Barrett, the US did not ratify the Protocol because 1) the reduction of the emissions of developed and transitional countries to 95 per cent of their 1990 levels can at best, only slow the rate of growth of atmospheric greenhouse gas stocks because much of the greenhouse gases emitted between 1997 and 2012, the end of the agreed period, will still be in the atmosphere in 2050; 2) the Protocol places no restrictions on the emissions of developing countries such as China and India, that will soon overtake developed countries' emissions; 3) there has been no commitment on greenhouse emissions after 2012 (Barrett, 2003).

As a symbol of international effort in cutting greenhouse emissions the Kyoto Protocol has also become a symbol of international failure in cooperation against global warming. Wills blamed the refusal of the USA as well as developing countries to adopt emission targets, and the absence of effective enforcement measures in adoption of the Protocol, as one of the biggest obstacles to the efforts of the international community (Wills, 2006, pp. 255-258).

On the positive side, the Kyoto Protocol continues to be a global attempt to take action towards sustainable development. Though it was accepted in reality by some dozens of developed countries, it raised the need for global coordination in keeping a clean environment. The theme caught the attention of many nations, firstly big pollu-tion emitters. This even became a hot topic in G-8 meetings, although the US always stayed outside the talks (Eilperin, 2005).

3.5 The potential of knowledge and new technologies in facilitating adaptation to new situations

3.5.1 Knowledge and new technologies

Many authors believe that knowledge, in the form of both appropriate education and the development of human resources, does not only contribute to current economic growth but in the long term, is also likely to contribute solutions to the global problems relating to sustainability.

Thurow suggested that knowledge is the ability to learn to adapt to new situations and change (Thurow, 1996). Individuals become more socially responsible towards the future of the environment if they have knowledge about the threats to their future and about ways to reduce such threats (Steiner & Steiner, 2001). Environmental economists regard expenditure on human resources as a greater contribution to social output than investment in human-made capital.

One example is the role of knowledge in stemming population growth in poor developing countries with very limited land and resources. Birdsall noted that for sub-Saharan countries with spiraling population growth, foreign capital and the application of technology by themselves cannot prevent a rapid decline in soil productivity and the spread of desertification unless populations are brought under control. He suggested that the dissemination of information and literacy programs can play a key role in reducing both fertility and child mortality rates (2001, p. 130).

Landes stated:

> Many Third World countries, moved more by ideology than reason, have so multiplied the obstacles (…) as to force them into the kind of high-effort, low-profit situation where they prefer to leave or stay away. In general, attitudes are changing, and the most resolute developers among Third World countries have been trying to shift from ideology to reality. What is a basis for sustainable growth is knowledge, because knowledge can yield a continuing, growing payoff (Landes, 1998, p. 67).

Knowledge also has a key role to play in the affluent developed world. In the 1960s the US contained approximately one fifteenth of the world's population, but was using about one third of the world's resources. The self-regulation of consumption can contribute much to the sustainability of the environment, but this can only take place with increasing awareness and understanding about what is at stake. Arguably people with a high standard of living have more capacity to gain awareness of environmental concerns than those who are struggling daily for their basic survival.

In 1982 K.E. Watt estimated the carrying capacity of the world to be 600 million people, assuming an American standard of living (Watt, 1982). In 2007, the world's population is 11 times above that figure. In a 2006 Oscar winning movie, "An Inconvenient Truth", Al Gore told the audience how to change their lives to curb their carbon-gobbling ways: switch to compact fluorescent light bulbs, use a clothesline, drive a hybrid, use renewable energy, dramatically cut back on con-sumption. Yet critics pointed out that the former US vice president was a hypocrite as he used more than 20 times the national average of gas and electricity at home (BBC, 2007). Post-environmentalists called on the North rather than the South to follow asceticism as way of living in the post-modern world (P. J. Taylor, 1996). However, without the intervention of governments, for example through taxing con-sumption or pollution, such calls have little chance of being heeded.

Among, and within developing communities it would be desirable to share information about, for example, successful environmentally responsible cultivation or production projects. West Malaysia has enjoyed mostly sustainable development for nearly three decades because it successfully capitalized upon its initial natural resource base virtually banishing rural poverty and environmental degradation. At the same time, in the two eastern states of Sabah and Sarawak, which have had richer natural resources, natural forests have been consumed by unsustainable practices, largely as a result of serious policy failure (Gillis, 2005, p. 29). At a regional level, it can be asked whether West Malaysia's knowledge and experience might be shared with the Eastern region, or whether such knowledge might not be applicable to that region.

3.5.2 The discovery or creation of new resources

Furthermore, new knowledge and technologies can lead to the discovery or creation of new resources, and ways to use resources.

New reserves of oil and minerals have been found recently through advances in technology making it possible for fossil fuels and minerals to be taken from deeper layers than before, both on land and in the ocean. New knowledge can also include new ways to use traditional resources more economically. For example, much larger quantities of ore can now be handled which (together with new metallurgical methodologies) allows metals to be extracted from much lower grades of ore than in the past. In addition, new manufacturing and energy production technologies allow lower grades of minarals or fuels to be used directly (Aplin, 2002, p. 115).

The most promising application of the power of knowledge is the development of forms of energy other than the traditional fossil fuels: solar, nuclear, geothermal, wind, tidal and fuel cells. Fuel cells are a promising technology which can produce electricity by direct conversion from bio-fuels such as alcohol, ethanol or hydrogen. In 2007, Shelley Minteer, at Saint Louis University developed fuel cells which can run on sugar (Choi, 2007). The use of fuel cells as a source of energy could revitalize agriculture, and developing countries could share in the wealth created by a new fuel cell industry. However concerns have already been expressed that mass production of fuel cells may lead to a new form of eco-crisis: the overuse of water for irrigation to produce sugar or other organic compounds as energy sources for fuel cells.

Nature with its biodiversity is not only the major source of physical resources but is also the source of knowledge for bionics or "bionical creativity engineering". Frederic Goettmann, a chemist at the Max Planck Institute for Colloids and Interfaces in Potsdam, Germany, has managed to produce energy by simulating the way in which plants break very stable bonds in CO2. With a new catalyst, mpg-C3N4, he can split carbon dioxide gas into CO from which liquid fuel then could be made (Goettmann, Thomas, & Antonietti, 2007; Simonite, 2007). It is very promising if scientists can improve the efficiency and scale so that fuel can be made directly from CO2. With this plant-imitating energy, we might also reduce the accumulation of carbon dioxide from the atmosphere (Simonite, 2007). Unfortunately, solutions from the sciences are not yet developed enough to offer economic substitution of fossil fuels.

By 2000 only two percent of electricity of the US was generated by renewable sources, and 20 percent by nuclear power (United States. National Energy Policy Development Group & Bush, 2001, p. xiii). Bauer, writing in 1995, noted that there is resistance to new technologies and even technophobia. Investment in scientific and technological solutions has been very modest (Bauer, 1995).

In summary, knowledge-based solutions do offer hope, but without major effort and investment, solutions to the problem of sustainability are still far off.

3.5.3 The beneficial uses of ICTs in the linking of minds and wills

Sustainable development requires the participation of all people on the earth, from individual citizens to non-governmental organizations, from family enterprises to transnational corporations, and from local government to the United Nations. Disinterest or ignorance, conflicts of interest or miscommunications - any of these will negate the chances of gaining success in turning around the environmental problems facing the world.

ICTs can be used and managed to play a crucial role in building up the motivation and the cooperation needed – the achievement of a vast network of minds and wills. Efforts such as those expressed in the principles of building the Information Society that the United Nations called for in the World Summit on Information Society are clearly critical (United Nations. National Assembly, 2001; World Summit on Information Society, 2003).

Welsh described steps taken by the Brazilian national government in March 2007 which offers free satellite Internet connections to indigenous tribes in the Amazon to crack down on illegal logging in the world's largest tropical rain forest. In Brazil, as a result of growing beef, chicken and soy for ever-expanding export markets, deforestation peaked in 2004. The Internet is expected to be a way to open communications between 150 small communities in the Amazon and other remote areas including the Pantanal wetlands and its arid Northeast. The tribes use the Internet to spread the message to each other, the police, or even the country's president. Indigenous leaders support the program but worry that computers might erode native cultures in the country (Welsh, 2007).

Technologies in the areas of information and communication have the capacity to facilitate bringing the voices of the people to policy makers and the capacity on the one hand to enhance the partnership between citizens and the government, and on the other hand to improve the validity and credibility of policy implementation. According to Tapscott et al. (1998, p. 352),

> Grass-roots programs tailored to unique local needs will replace top-down national programs. Information technology will enable local organizations (both governmental and nongovernmental) to obtain expertise and information previously restricted to national agencies. They will combine this with knowledge of community needs to find effective solutions to problems.

Rather than expecting top-down solutions, citizens will look to their national governments to supply them with the information they can then use to leverage resources, organize themselves, and solve their own problems. On information equity, the poor and disadvantaged people should be encouraged more by government in participation in the network economy. "Market mechanisms can work magic, but only if everyone can participate" (Tapscott et al., 1998, pp. 352-353). (Kennedy, 2004)

In dealing with divergence among stakeholders about approaches to, and responsibilities of sustainability, ICTs can help people to gather and disseminate information about threats and risks, for example the non-compliance to agreements of an individual member which might endanger the whole world's chance of survival.

3.5.4 Skepticism about the use of knowledge and ICTs

Technologies – in this case, in the areas of information and communications, can be used for both good and bad purposes. Knowledge can be ignored, hidden, or manipulated by those in control. Robert F. Kennedy Jr., in his book "Crimes against Nature", accused Bush and his administration of encouraging giant conglomerates to abandon any notion of civic responsibility about climate change.

> You're talking about a president who says that the jury is out on evolution, so what possible evidence would you need to muster to prove the existence of global warming. We've got polar ice caps melting, glaciers disappearing all over the world, ocean levels rising, coral reefs dying. But these people are flat-earthers (Kennedy, 2004).

In the hands of those in power, knowledge and information, and the technologies managing them can be distorted or suppressed to serve personal motives which are not in the public interest.

For example, one could have a skeptical view of the World Summit on Information Society. Rather than interpreting the event as a worthy attempt at global cooperation, one could see it as a mechanism whereby developed, capitalist countries promote a new market, the market for knowledge services and ICT commodities in developing countries. Poor countries need to be watchful and cautious about the motives of different initiatives from the developed countries to ensure that they do not find themselves with even greater debts and further depletion of their natural resources. It could be achieved if poor countries better understood how wealth is made in the global digital economy – a "secret" upon which this book seeks to shed more light. Mechanisms with the potential for such learning are exemplified by the WSIS process, which was co-sponsored by developing countries, notably Tunisia which hosted Part 2 of WSIS in 2005. Perhaps unfortunately WSIS was more pre-occupied with the governance of the Internet than the economics (e.g. at Tunis control of the root servers of the Internet by the US Department of Commerce was challenged), but at least WSIS demonstrated that developing countries can and must seriously engage with inequities in the global information society.

It is all too easy to incur mounting debt for ICT imports. Even Australia has a revealing "growing annual ICT trade deficit of $21 billion". In 2006 Australia's ICT exports were worth AUD5.7 billion, with significant gains in ICT services exports, while ICT imports cost AUD26.6 billion (author's italics) (Houghton, 2007). To place these figures in perspective, the total income of Australia was AUD23.3 billion, from coal exports and AUD14.5 billion from iron ore exports in the same year (Department of Foreign Affairs and Trade, 2007).

As noted previously, the imbalance between the price of material resource based commodities, the main economic outputs of the South, and that of knowledge and ICT commodities is remarkable. How many hours of labour and resources does a farmer in a poor country need to spend to buy a computer with its many versions of software? It is argued in this book that barriers erected by the prices of hardware and copyright restrictions prevent poor people from the chance to use these new forces of production. This gives extra meaning to the often-used phrase, "the digital divide".

3.5.5 ICT responses by the developing world

In relating to the ICT market hegemony of major corporations in the developed countries, the developing world responds with two kinds of practices - piracy and copyleft or the use of open software sources.

Piracy

In reality, it is not uncommon for conformity with IPRs to be violated. According to Business Software Alliance (BSA) and IDC, the US anti-piracy organization, globally 35% of the software used in 2005 was illegal. The rate was relatively high in the Asia Pacific region (54%), Central/Eastern Europe (69%) and Latin America (68%).

The rate of piracy in the European Union, United States and Canada may be relatively low in proportion to the total volume of sales, but in large markets even relatively small piracy rates can add up to large absolute losses in revenue (BSA & IDC, 2006).

Piracy is sometimes regarded pragmatically as a way for the poor to gain access to the networked world. Romanian President Traian Basescu told Bill Gates in the opening of a Microsoft global technical center in Bucharest:

> Piracy helped the young generation discover computers. It set off the development of the IT industry in Romania. It helped Romanians improve their creative capacity in the IT industry, which has become famous around the world (Reuters, 2007).

The Romanian case shows that poor countries can benefit much from piracy for their local IT industry.

The copyleft movement and open source software and platforms

The copyleft movement or the GNU (Gnu's Not Unix) project, initiated by Richard Stallman, is to give all users the freedom to redistribute and change GNU software. Instead of putting GNU software in the public domain, GNU allows recipients to "copyleft" it. Copyleft licences can be given from the website "Creative Commons" which says,

> anyone who redistributes the software, with or without changes, must pass along the freedom to further copy and change it. Copyleft guarantees that every user has freedom. A widely used and originating copyleft license is the GNU General Public License (GPL) (Stallman, 2006) .

The open source principle gives alternative software not only to individual users, but also to big platform and "silo" companies, IBM, Novell, HP, Sun. Leading cyber providers such as Amazon and Google also build their platforms on cheap and free open source materials. A series of governments, including France, United Kingdom, Germany, Italy and Finland, developed regulations on the use and acquisition of Libre software for customers based on an open source platform (DiBona, Cooper, & Stone, 2006). In India, which is an emerging software developer, the Government sponsors open source software (OSS) initiatives. In 2003, a memorandum of understanding between the state of Karnataka and IBM to build an OSS center of excellence in the government-built Hubli software park was signed. In particular, OSS is viewed by many Indian companies and by government as a key for improving the quality of education, though educational applications are still at an early stage (Sharma & Adkins, 2006).

However, GPL is not completely free. The first court case in relation to a GPL dispute was the Dutch Court upholding a Creative Commons license. Adam Curry, a Netherlander, posted his photographs on flickr.com under a Creative Commons Attribution-Non-commercial-Share-alike license. The Dutch weekly, "Weekend", a gossip magazine, reproduced four photos in a story on Curry's children without seeking Curry's prior permission. Curry sued the magazine for copyright and privacy infringement. As to the copyright claim, "Weekend" argued that it was misled by the notice "this photo is public", and that the link to the CC license was not obvious. On March 9, 2006 the District Court of Amsterdam rejected the magazine's defense, and held as follows:

> All four photos that were taken from www.flickr.com were made by Curry and posted by him on that website. In principle, Curry owns the copyright in the four photos, and the photos, by posting them on that website, are subject to the [Creative Commons] License. Therefore Audax should observe the conditions that control the use by third parties of the photos as stated in the License. The Court understands that Audax was misled by the notice 'This photo is public' (Hugenholtz, 2006).

In this case, the distinction between GPL versus copyright seemed blurred.

Can open source products challenge closed sources, the mainstream commercial products? Benkler (2002) called open source a community-based peer production which lacks motivation of intrinsic benefits of participation. Furthermore, all of the OSS projects are concerned mostly with the improvement of the products, and attention may not be given to interoperability with other products. At this stage, owing to a lack of strategic planning and coordination, OSS still comprises mostly fragmented projects which are not a serious rival to commercialized products.

3.6 Main insights from Chapter 3

Insight 3A: The global environmental crisis is an urgent reality. The health of the global ecosystem affects all countries and needs to be dealt with globally through international co-operation. Individual countries are constrained in how far they can deal with ecosystem degradation on their own.

Insight 3B: There is a tension between sustainability and development that must be taken into account. In many ways the term "sustainable development" and thus also "sustainable economic development" is self-contradictory.

Insight 3C: Progress made at the interface between economics and environmental studies have yielded rich concepts and formulations, for example about relationships between human-made and natural capital. These provide a foundation for the present author's attempt, in later chapters, to extend such formal reasoning to include a finer-grained account of information and knowledge as factors in sustainable economic development.

Chapter 4
The cultural context of Vietnam

4.1 Scope of the chapter

- Using literature and fieldwork data, presents cultural and societal dimensions that may affect the failure or success of the emergence of a knowledge society and sustainable development in Vietnam.
- Outlines positive and negative effects of economic progress and cultural change.
- Opens up the theme of cultural heritage and indigenous knowledge in economic development.
- Presents a diagrammatic overview of "bamboo fence" cultural fractures between the Vietnamese rural and urban populations and the outside world.

4.2 Defining culture

National culture can be defined in different ways, subject to the perspectives of different disciplines. It could be 'shared values' (Triandis, 1972) or "the collective programming of the mind acquired by growing up in a particular country" (Hofstede, 1997, p. 262). UNESCO's Universal Declaration on Cultural Diversity argued

> … culture should be regarded as the set of distinctive spiritual, material, intellectual and emotional features of society or a social group, and that it encompasses, in addition to art and literature, lifestyles, ways of living together, value systems, traditions and beliefs (UNESCO, 2001).

Ho Chi Minh, the founder of Vietnam, suggested that culture comprises the creations and inventions of people which enable them to live and survive:

> For survival and for the purpose of life, mankind created and invented languages, script, ethics, laws, science, religion, literature, art, daily tools for clothing, food, dwelling, and made use of them. All of those creations and inventions constitute culture (X. N. Pham, Galla, Logan, & Slatis, 2003).

Even within a small group of people there are likely to be differences in beliefs, attitudes, and behaviors. Corbitt, Peszynski, and Intranond (2004) argued that general structural frameworks of national culture are simplistic descriptions that may overlook minorities within a nation. According to Fukuyama (1995) and Groeschl and Doherty (2000), subcultures may exist within one border, and a nation does not necessarily consist of one culture.

Culture is not fixed and rigid, but evolves, along with economic, political, technological and social changes. Under external pressures, especially growing globalization, members of different geographical and ethnic groups might work and live together. In an evolving culture, people are not as constrained as they were traditionally, but have opportunities to partake in dynamic events (Giddens, 1984), and creatively reconstruct the social systems in which they live recursively (Walsham, 2002).

Hofstede (1988, pp. 4-21) defined five attributes of national culture: individualism/collectivism, masculinity/femininity, power of distance, uncertainty avoidance, and long-term vs. short-term orientation. Rossen (2003) and Ford and Chan (2003, pp. 11-27) suggested that the cultural dimensions in Hofstede's model are likely to have an impact on knowledge management processes. This chapter includes an analysis of the negative effects of economic progress and cultural change which inevitably hinder the advancement of a knowledge economy.

4.3 Vietnam: brief historical overview

Vietnam is a country with a long, proud, independent history and a rich culture with much diversity in terms of geography, ethnicity and religion. For thousands of years Vietnam built an extensive agrarian culture during uninterrupted patriotic wars. Nguyen Trai, the genius strategist of the Vietnamese victory over the Wu to gain the independence for the country, proclaimed in 1428:

> Our Great Viet
>
> such a cultural and constitutional nation in the South
>
> that has long been independent
>
> in land, mountains and rivers
>
> and different in its custom from the North [i.e., China] (T. Nguyen, 1993).[5]

In Vietnamese culture, the natural and constructed human environments coalesce. According to historian Tran Quoc Vuong (Q. V. Tran, 2001) this culture originated from a drawn-out struggle for autonomy, and also from continuous water-rice cultivation. The irrigated rice culture gave birth to the "Red River civilization", the cradle of the Vietnamese culture, which distinguishes it from the "Yellow River civilization" created by the dry millet and wheat cultivation of the Chinese.

By legend, the very embryonic states of Vietnam, Van Lang (Village of Literature) and then Au Lac, existing in the first millennium before Christ in the early Bronze Age with 18 Hung kings, already manifested characteristics of a national culture. Archeological evidence shows that the Dong Son civilization from around that time consisted of inter- and super-village communities, for the purpose of resisting invaders and building dykes for rice paddies (Langlet, 2004, p. 35).

For a millennium after the Au Lac, the country was trampled under the domination of Chinese dynasties, which were finally thrown off by Ngo Quyen, the first king of the independent Dai Viet (Great Viet) in 938 BC. The post-Chinese domina-

---

[5] The quotation is translated by the author.

tion period was characterized by two parallel trends of Han acculturation and anti-Han acculturation. The Dai Viet period was the second peak of Vietnamese culture, featuring the Ly, Tran and Le dynasties from the 10th to 16th centuries when Vietnamese culture underwent a rapid restoration under the dominance of Buddhism and Taoism (Langlet, 2004).

The period from 1802, when the Nguyen dynasty tried but failed to restore Confucian culture, to 1945, when Ho Chi Minh declared the independence of Vietnam from the French imperialists, was marked by a cultural mix of westernization and anti-westernization. The struggle was between patriotic culture and colonial, paternalistic culture. Vietnam in the 20th century manifested a strong patriotism, allied with Marxism and Ho Chi Minh ideology (Q. V. Tran, 2001). The terrible impacts of the US-Vietnam conflict in the period 1959 to 1975 and the ensuing embargo are summarized in the next section.

Since 1986, when Doimoi was introduced, the country has increasingly become an integrated, post-modern civilization, but has also maintained efforts to ensure the preservation and the enhancement of its national identity (X. N. Pham et al., 2003).

4.3.1 The impacts of the US war

The US-Vietnam conflict occurred from 1959 to April 30, 1975. This was followed by a 20 year embargo placed on Vietnam by the US which continued until February 3, 1994 when President Bill Clinton lifted the embargo to lay the foundation for the normalization of relations between the two nations. During the war, 14 million metric tons of bombs, napalm and landmines, as well as 72 million liters of toxic chemicals were dropped by the US army on to Vietnamese land (Westing, Phuong, Vo, Bui, & Dwernychuk, 2002).

About five million Vietnamese persons, four million of whom were civilians, lost their lives during the war (Agence France Presse, 1995). Ten million people were injured by explosive weapons, or contaminated by the chemical, Agent Orange (containing dioxin). More than thirty years after that war, innocent persons are still being killed or injured by landmines. It is claimed that serious health problems have been associated with Agent Orange, such as cancers, chloracne and birth defects (T. Q. Hoang, Constable, Noller, Dao, & Hay, 2002).

The damage to the environment destroyed about ten percent of the forests in the southern areas, which included almost a third of Vietnam's coastal mangroves. It was so severe and widespread that it was termed "ecocide", as Westing et al. described:

> On an area less than eight percent that of the United States, the amount of high explosives employed was almost double the amount expended by the USA during World War II. Left in the earth were many millions of large bomb craters, bomblets, and other ordnance which continue to take a heavy toll of live and limb.
>
> Over 72 million liters of herbicides destroyed roughly ten percent of southern Vietnam's valuable forests, including nearly one-third of the coastal mangroves which play vital roles in coastal ecology and in sustaining fish stocks. Toxic chemicals contained in the herbicides, arsenic and dioxin in particular, are expected to continue posing a significant health threat long into the future (Westing et al., 2002).

In brief, the thirty-five years of war and embargo had destructive impacts on the development of Vietnam in every dimension: environmental, social and economic, and made Vietnam one of the poorest countries in the world in the 1990s:

> Vietnam is one of the poorest countries in the world. . . . Under the embargo, Vietnam's GNP [Gross National Product] has grown at much lower rates than its neighbors: In the late 1970s, Vietnam's GNP growth averaged 0.4%, while that of [neighboring] ASEAN countries was over 15%. (Burke, Franklin, Urata, Searcy, & Cutting, 2002, p. 10)

Another ongoing negative impact of the war is the hostility of remaining pressure groups of war refugees, mainly in the US, Canada, Western Europe and Australia who are virulently anti-Communist. In other words, the US war still divides the Vietnamese people into two fronts, though the Government of Vietnam has been trying, and continues to try to heal that conflict (Ho Tai, 2001).

With Doimoi, the Government of Vietnam initiated a spirit of "closing the past and looking to the future" to give the US the opportunity to overcome its own inferiority complex about its experience in relation to the war, and also to engage Vietnamese overseas to join in the construction of the nation. That policy is proving to be very successful in normalization of the relationship between the two ex-enemies and is attracting investment in Vietnam from the US, Viet Kieu and many other external parties. Thus many "Viet Kieu", a Vietnamese term to refer to Vietnamese people living overseas, no longer regard themselves as refugees who cannot return to their country of origin, but rather as international citizens of Vietnamese ethnicity who freely travel to and from Vietnam for family or business reasons.

4.4 Religion and philosophy

The emergence of Vietnamese nationality over millennia was influenced by internal and external pressures. The Vietnamese culture was influenced by philosophies and ideologies, mixing ancient Vietnamese animism with Confucianism, Taoism, Buddhism, and Christianity. Confucianism had a strong influence in the mandarin class which emphasized royal and familial duties. Following this philosophy, many Vietnamese youth were eager to pass the imperial mandarin exams to advance themselves in social status. At the same time, some famous Vietnamese scholars did not adhere unwaveringly to Confucianism, but rather flexibly explored the spirit of Buddhism and Taoism in order to become more open, closer to the people, and more harmonious with nature. For them, loyalty to a monarch included following Confucianism, but their commitment was conditional upon royal success at defending national territory (Langlet, 2004).

4.5 Community life

The village as a community continues to be a special characteristic of Vietnamese agricultural society, in association with many primitive remnants, such as old festivals, Taoist dualism, and a mindset that prefers emotional experiences to rational logic, and concrete images to abstract concepts. Tran spoke of traditional culture as being "conservative, parochial and excessively egalitarian" (Q. V. Tran, 2001, pp.

200-239). As mentioned later, these traits can be both strengths and weaknesses in attempts to modernize the country. To an outsider, the culture of Vietnam has perhaps seemed to be conservative and unchangeable, formed for thousands of years behind the protection of the village bamboo fence and the national borders. However to an insider, the features of Vietnamese national and cultural identity are by no means immutable, according to two eminent historians Cao Xuan Huy and Tran Quoc Vuong, who saw the Vietnamese people as being as "adaptable and flexible as water" (Q. V. Tran, 2001, p. 33).

4.6 Ethnic groups and languages

According to data from the 1999 Vietnam Census, 54 ethnic groups in Vietnam give the country the richest and most complex ethnic make-up in Southeast Asia. The Kinh ethnic group, who speak Vietnamese as a native tongue, account for 86.2% of the total population, but live in less than half (yet the most fertile part) of the country. Among 53 ethnic minorities who inhabit the majority of the land (albeit the least fertile), Tay (Tho), Tai (Thai), Hmong (Meo or Miao), Muong (Mol) and Nung are the largest groups, with more than half a million members each. Some other groups, like Romam or O-du, are believed to have diminished to only a few hundred. Ethnic minorities have their own languages and cultures, but few have written texts. Language is a big communication barrier for ethnic people (N. V. Dang & D. D. Le, 2004).

4.7 Family

Marriage and family play a big part in Vietnamese life. Ancestor worship profoundly influences Vietnamese thinking. Children are taught from an early age that they owe everything to their parents and ancestors, thus they have to do well in school and work hard to respect their parents and the family reputation. Traditionally, the male was primarily responsible for the economic well-being of the family, while the woman, the "noi tuong" (the general-in-house) looked after her husband, children and her in-laws. Now there are big changes in family life. The August revolution in 1945 gave equal economic and political rights to Vietnamese women. In the long ensuing wars, women also had to take on many traditionally male duties (Thai, 2004).

4.8 Positive effects of economic progress and cultural change

Vietnam is no longer wedded to the goals of national defense or patriotic war, but rather to constructing a socialist-oriented market economy to build a country with a new motto of "... affluent people, strong state, equal, democratic and civilized society" as stated in the country's Constitution Amendment in 2001 (National Assembly of Vietnam, 2001).

Doimoi, the recent successful strategy of Vietnam, aims to cope with globalization and

> ...synergize reforms in all socio-economic activities, including policy planning, public administration, government structure, the economy, privatization, research and development, and international co-operation (Ministry of Agriculture and Rural Development, 2004, pp. 5-22).

For 20 years the Doimoi process has brought many opportunities and challenges to the economy, society and culture across the country, and to rural areas in particular.

In an interview with a senior official of the Ministry of Agriculture and Rural Development the point was made that under the new economy, farmers are given full rights to land use, and trade liberalization optimizes their farm production. This new approach avoids the problems associated with the saying, "everybody's business is nobody's business" that is, nobody is motivated to take real responsibility for the success of the cooperative. Such an attitude has long discouraged creative and competitive production under a co-operative regime (S39, interview, Dec 1, 2004).

S36, another senior officer of the Ministry of Agriculture and Rural Development added that in the agricultural sector, the steady high growth (5.5% p.a.) is securing the country against fluctuations in other economic sectors. This was very obvious during the global economic crisis of 1997 when neighboring countries suffered severe recession. Vietnam has jumped from being a country of chronic food shortage to being the world's second largest exporter of rice. Furthermore, primary products have been diversified. They are not only confined to rice, but include higher value plants and livestock, such as coffee, tea, fruits, flower, rubber, catfish and shrimp. New policies liberate farmers and give them autonomy in regard to land and products, and encourage the application of technology to breeding, cultivating and processing. The agriculture of Vietnam has become a new competitor in the world market. Living standards have been improving in both urban and rural areas. The progress made in the elimination of hunger and poverty is remarkable. The part of the population living in poverty (less than $1/day) had reduced by half in less than a decade (from 58% in 1993 to 29% in 2002) (S36, interview, Nov 30, 2004).

A World Bank report noted that in 2000 Vietnam successfully completed the popularization of primary education with illiteracy elimination, and more than 90% of the population now has access to health services. The country also achieved considerable improvement in the Human Development Index, rising from 0.11 in 1992 to 0.682 in 1999 (World Bank, 2003). A recent CIA report has stated that the overall adult literacy rate is very high, with little disparity between males and females, 93.6% and 86.9% in 2002, respectively (Central Intelligence Agency, 2007).

According to an expert in Development Studies, Vietnam is becoming a learning society where people typically make every effort to learn in order to have a better job, to work for a foreign company, or to study overseas. He said:

> While in some parts of the world like Seattle or Cancun, anti-globalization protesters march down the street shouting 'no-WTO', it is more likely that Vietnamese people favor integration and globalization, especially when it concerns music, imported culture, and the Internet (S43, interview, Dec 07, 2004).

More open, democratic, transparent, free speech in the Vietnamese media is apparent. A journalist from Vietnam Post and another from VnExpress.net remarked that the Vietnamese press appears more modern, critical, professional and proactive in processing information than it was. The mass media plays a revolutionary role in fighting social evils and helps to build a fairer society. The corruption of many officials has been disclosed by paper-based and online newspapers (S11, interview, Nov 2, 2004; S49, interview, Dec 14, 2004).

Since the first official e-magazine "Que Huong" (Motherland), launched in the lunar new year 1997 (Commission for Vietnamese Overseas, n.d.), there have been numerous Vietnamese online publications from government, organizations, communities and single persons, booming in cyberspace. An example was given by the interviewee from Vnexpress.net. Vnexpress.net, "the first Vietnamese website to be positioned in the top 500 global busiest websites is a wonderful directory for sharing new knowledge and good practices for successful farmers" (S49, interview, Dec 14, 2004). As S11 pointed out, information about Vietnam is open and accessible to the rest of the world via such instant media (S11, interview, Nov 11, 2004),.

Another VnExpress.net staff member said that countless virtual communities are forming electronic forums.

Table 4.1 Rank of selected Vietnamese home pages. Source: Axela.com, viewed August 1st, 2004[6]

Website	**Host**	**Area**	**Global Rank**	**rpmu ***
Vnexpress.net	FPT Communications	Newspaper, forum	448	1650
vn99.net	blog**, student-developed	Music, entertainment, forum	9,596	165
sextvvvvvv.com	Blog***	Pornography, forum	9,846	50
Laodong.com.vn	Vietnam Trade Union	Online newspaper	17,508	70
Mofa.gov.vn	Ministry of Foreign Affairs	foreign policy, press release and news	59,379	20

*reach per million users.

** blog: personal web log.

*** This anonymized blog in fact is hosted and operated by Vietnamese overseas.

> E-communities are new playgrounds for Vietnamese youth to meet, chat and share information. In cases where government education is inadequate, students learn by themselves necessary English and IT skills, and exchange new technologies with virtual peers, to create their own hobby homepages and forums, (S48, interview, Dec 9, 2004).

Two students interviewed described Vn99.net, a voluntary website, which was developed and maintained by moderators who are first or second year university

[6] Alexa.com was a website information company that provided information about page views (number of reach per million users -rpmu) and traffic rankings (based on three months of aggregated historical traffic data from Internet users) of frequently accessed websites. The statistics in Table 4.1 were seen at Alexa.com on August 1st, 2004 by the author.

students. It was a forum to share music, stories, paintings, and opinion, a site where people could express “love”, “hate”, “anger”, as they wished. The website moderators pointed to its sudden success: it attracted 143,000 young members (99% are students), with 829,785 threads by August 1, 2005, just two years after launching (S69 and S70, online interviews, May 19, 2005). In 2005, this amateur forum ranked much higher than many professional websites (see Table 4.1 below). However this community-based work was not sustainable but died in 2006. (Vn99.net and e-community in Vietnam are dealt with again in a later chapter).

4.9 Negative effects of economic progress and cultural change: obstacles to the advancement of a knowledge economy

Together with positive trends, the new economy, population growth, urbanization and globalization inevitably also have negative effects on livelihoods and quality of life.

When traditional village culture is confronted with changing values, social structure, behavior and attitudes, and technologies, it is not unexpected that problems will arise. The author argues here that these need to be addressed if a knowledge economy or networked society is to be achieved in Vietnam.

4.9.1 The gap between rich and poor

Globalization weakens national integrity in many respects, and promotes global capitalism (The Group of Lisbon, 1995). A recent dispute between Vietnam and the USA on seafood was perhaps the first both sweet and bitter experience of the Vietnamese farmers in the wake of globalization. In 2002, Vietnam produced 2.4 million tons of shrimp, of which 250,000 tons valued at $2 billion were exported to the US market (213% more than in 1995). Aquaculture sustained the livelihoods of thousands of farming communities living along Mekong River. Suddenly, Vietnam became the second largest exporter of shrimp and Basa fish to the US market (Ministry of Agriculture and Rural Development, 2004), thereby becoming an emerging competitor to American producers. US consumers have demonstrated a preference for Vietnamese catfish over US varieties. US catfish producers, led by the American Catfish Farmers Association (CFA), lobbied the US International Trade Commission, saying that Vietnamese farmers were dumping the price. A punitive 64 per cent tariff was then levied on Vietnamese catfish, threatening the livelihoods of more than 100,000 catfish farmers (Oxfam International, 2005).

Urbanization is occurring very quickly, mainly along land beside roads that used to be fertile rice fields (S4, interviews, Oct 28 and Nov 10, 2004). After farmers receive some compensation for losing their inherited land, they have to live in multistorey buildings. The Vietnam Farmer Association estimated that each year, about 200,000 hectares of land are lost due to urbanization, the equivalent to 1.5 million peasants becoming jobless (Vietnam Farmer Association, 2005). The peasants are very vulnerable in a strange environment where they are not equipped with the adaptive knowledge to survive.

Numerous farmers are trapped in a cycle of poverty: Urbanization and industrialization devour more and more farm land, which leads to pressure on jobs, low skill levels, and a soul-destroying quality of life. Jamieson, C.T. Le and Rambo (1998) remarked that "poverty is characterized not only by low incomes, but also by lack of access to basic services, such as medicine, education, information and entertainment".

Rural and mountainous areas accommodate 90% of the poor, where the gap between rich and poor is expanding. On the one hand, the World Bank (World Bank, 2003a, p. 11) acknowledged that "the poverty rate was literally halved in less than a decade, and very few countries in the world can report a comparable success" as an outstanding achievement of the Hunger Eradication and Poverty Reduction Program of Vietnam. Poverty fell from 58% in 1993 to 29% in 2002. On the other hand, arguably by way of backlash, an assessment of the socioeconomic status of Vietnam (Government of Vietnam, 2004d, p. 19) recognized that "the gap between the poor and the rich and social stratification tend to be on a rapid increase in the market economy".

The 2002 Vietnam Household Living Standards Survey (cited in the World Bank, 2003, p.25) revealed that "the biggest gap of all is between urban and rural areas. Other things equal, an urban household spends 78 percent more than a rural one". The World Bank (2003a, p. 6) predicted a greater widening of the gap:

> The next decade may well be characterized by increasing inequality in Vietnam. (….) further integration with the world economy will primarily benefit the economic hubs of the country. At least for some time, the gap in earnings between urban and rural areas will increase. And the earnings gap between skilled and unskilled workers could increase as well.

Two interviewees commented on the plight of farmers and the impact on the environment: "In many cases farmers are the ultimate arbiters of whether to keep or destroy forest, or high-quality arable land. The poor tend to think of nothing more than what is necessary for the daily survival of the family" (S52, interview, Nov 17 and 23, 2004).

Interviewee S01 said:

> Poverty today takes on the form of mass imitation, under pressure of the market economy and globalization. Farmers are pragmatically minded. Many mortgage all their land and valuable properties, to compete with their successful neighbors by planting coffee, cashew, sugarcane or shrimp, then they have to destroy them all, because they do not have timely information about world markets (S01, interview, Oct, 28, 2004).

Ha and Phan in 2005 wrote:

> The commune today is a very different place from in the past. The old commune may have been poorer but at least safer for ages with its self-reliant economy. Now it only takes one day for three American states -- namely Louisiana, Alabama and Mississippi – to ban Vietnamese seafood in their shops, and promptly many Vietnamese families suffer. There are other risks; for instance, if Vietnamese farmers do not have the know-how to prevent antibiotics use in their aquaculture, they may find out that the recently-opened American market could soon refused their contaminated products (Ha & Phan, 2005).

As noted previously, a drawback of a market economy is that the gap between the poor and rich continues to extend and social disadvantage multiplies quickly.

4.9.2 Changing social structures and values

Today money resumes its commanding influence in Vietnam, after four decades of quiescence under a command economy. In the 17th century, Locke asserted that "money is the mother of civil government' in a laissez-faire economy" (cited in Caffentzis, 1989, p. 70). In a market economy, everything can be measured and converted into financial value.

Money has a devastating power to tragically change social values and structures, as Nguyen Du (1983, pp. 36-37), author of the Vietnam's greatest masterpiece "Truyện Kiều" (Tale of Kiều) in the 18th century, noticed:

> When money's held in hand it's no great trick,
> swaying men's hearts and turning black to white.

The government laments that serious social evils such as drug addiction, prostitution, HIV, smuggling, and corruption are shaking social stability and culture (Government of Vietnam, 2004d).

There has been radical change in attitudes to personal wealth. Excessive wealth and money-making were for decades perceived as "evil". Now the norm is reversed – "making money is not bad." Changes in perception and values, without proper control, can be potential causes of social chaos, as S23 pointed out in interview.

> With regard to a socialist structure, Vietnam possessed a simple egalitarian society where citizens did not need to extend their reach; personal creativity was not encouraged. People now have more incentives to work at whatever they want. A drawback is that rapid change in social structure, co-existing still with old structures, creates an uncertain, undirected society and vulnerable families (S23, interview, Nov 16, 2004).

Increased mobility means that transport accidents involve tens of thousands offenders and immense human loss. Obviously, the use of a car is neither bad nor good in itself, but the social consequences of widespread adoption have to be assessed. In interview, S23, regretted that "social evils sometimes appear to be beyond the control of any effective social intervention" (S23, interview, Nov 11, 2004).

4.9.3 Parochialism and negative attitudes to laws and regulations

Two interviewees who are experts in the field of socio-economics used the metaphor of "farmer frame of mind" to describe Vietnamese people entering the new high risk economy, arguing that even though they may have become richer their behavior remains little changed.

> Like paternalistic landlords in the past, nouveau riche villagers strive for the trappings of wealth first and foremost. Thus, some villagers who have received large compensation from land sales or suddenly become wealthy by selling materials for recycling waste, now splurge their new-found wealth on luxurious Mercedes cars, and care nothing of tomorrow (S55, interview, Dec 21, 2004).

An example of an excessively local mentality is to be found in scant regard for the rule of law. Widespread unlawful behavior is observed when few Vietnamese people stop for a red traffic light unless they are watched by police nearby. Similarly, though many laws exist, they are seldom enforced. S41, a senior academic made the observation that agricultural life involves an element of 'law negation' because it militates against the power of "external" law. This may be explained by a folk proverb, "The king's proclamation is not as important as the village's norm." It is not a problem of lack of awareness of broad moral standards, but of resistant parochial attitudes and behaviors (S41, interview, Dec 2, 2004).

The author suggests that negative attitudes, like these towards laws and regulations, can impact badly on progress towards the building of a knowledge society.

4.9.4 Attitudes to knowledge sharing

A senior government staff member said in an interview that though Vietnamese people may be clever, they are clever with short-term goals and in a reactive way[7], (S18, interview, Nov 3, 2004). In the author's thinking, savvy suited farming or guerrilla war may not be relevant in long-term and large-scale projects that require the co-operation of many people and all the forms of resources necessary to a networked knowledge society.

Along similar lines, short-sightedness and superficiality were raised as negative traits. Two economics experts (S58, interview, Dec 22, 2004) and (S05, interview, Oct 21, 2004), noted these characteristics in connection with the planning and designing of new urban centers recently that were fragmented and tattered and often became backward after use.

One interviewee felt that envy may be nominated as a negative characteristic of the Vietnamese people (S59, interview, Jan 6, 2005).

Although Vietnamese are hardworking and canny, the named weaknesses are obstacles to co-operation and sharing knowledge, or to the building up of large corporations, in the opinion of a government official (S03, interview, Oct 19, 2004).

4.9.5 Difficulties experienced by women farmers and by ethnic minority groups

A World Bank report revealed that

> ... in all rural sites except Nghe An, women reported that despite performing much of the agricultural labor, they were likely to be overlooked when extension training took place (World Bank, 2003a, p. 35).

Consequently, it is obviously harder for women farmers to find new jobs and to join the new economy, especially when they have been removed from their land. On the other hand, in the study, several villagers were found to be very successful and knowledgeable about the new economy. However their knowledge could not be transferred to others due to the lack of useful channels of communication.

Interviewee S43 believed that

> The new social and cultural challenges are even more serious for some groups in the population, especially children, women and ethnic groups. Children are forced to study much harder than before. Women - who account for 60% of farmers - have limited education or understanding of technology (S43, interview, Dec 7, 2004).

A article published in 1998 entitled "The development crisis in Vietnam's mountain", identified the following difficulties experienced by ethnic minority people: i)

---

[7] S18 used a French word "malin" which means cunning, crafty or petty clever in English.

isolation physically and in terms of communication: they have the least information; ii) language barriers: the minorities often speak their own languages, not Vietnamese; iii) fast population growth: family planning methods are not widely appreciated; iv) technology deficit: technicians from the lowlands often do not know languages to communicate with local people; v) marginalization in society and the economy: they are the poorest members in the society, their indigenous culture is under threat; vi) political-administrative hindrance: lack of fluency in the national language and a low level of literacy among village administration puts them at a gross disadvantage within the new, larger cultural system; and vii) lack of social services: health, education and entertainment (Jamienson et al., 1998).

Challenges such as these make it harder to develop a knowledge society in mountainous areas.

4.9.6 The education system

The education system contains many "abnormalities", to use a word of interviewee S43 (interview, Dec 7, 2004). Interviewee S02 said:

> Many graduates with many degrees are not qualified enough to find a modern job; for example, only 7 or 8 out of 1,000 applicants were satisfactory in a recruitment of Nortel Corp (interview, Oct 27, 2004).

A company training manager said that his company once had 30 job vacancies but none of the new IT graduates could be accepted (I41, interview, Mar 21, 2006). A CEO of another company estimated that about 10 percent of the newly graduated students could match the market's criteria (I42, interview, Mar 24, 2006). An entrenched bookish cultural and educational system merely teaches people learning by rote, and encourages neither creativity nor the means of self-retrieving and processing information, especially from the Internet (S41, Dec 2, 2004; or S23, Nov 16, 2004).

S41 and S23 spoke of an "ossified sham examination system" and pointed out that though Vietnam has plenty of labor, the quality of the skills of human resources is often too poor to meet the demands of the market (S41, Dec 2, 2004; or S23, Nov 16, 2004).

In Vietnam, the average age of scientists is exceptionally high (60 for professors and 56 for associate professors) and there are no resources to replace them because the salary offered only meets up to one third of their minimum needs (V. D. Nguyen, 2000).

"Corruption occurs extensively in Vietnam and it seems to be that the more the government fights against it, the more corruption appears" (Thanh, 2005). In the author's opinion, the basic reason for corruption is that people first and foremost have to exist and help their families to survive. If the salary is too low, corruption, either in money, time or in knowledge, is inevitable. And that is also the reason why an "overwhelming majority of Vietnamese youth do not intend to follow a career in either education or research" (V. D. Nguyen, 2000).

In another interview, a scientist then working as a government official, when asked about the government's talent treatment policy, answered by telling the story of his experience when he studied in France. He met an old overseas Vietnamese in Paris. Though the old man loved the motherland so much that he still kept his Vietnamese citizenship for three decades, he did not want to return home. The younger scientist asked the old man why he did not return home when the government had offered open policy to attract overseas Vietnamese scientists. The old man answered that "I don't look at the policy. All policies are meaningless. But I look at the way the government actually treats scientists like you" (S04, interview, Oct 28, 2004).

An interviewee, in recalling the 2000 Conference of Knowledge Economy and its issues to Vietnam (see Chapter 6), told the author:

> If Vietnam does not have enough money, it can borrow. But we cannot borrow humans. IT specialists, IT users, and the general IT capacity of our society are weak. At the moment, the government cries out about outsourcing to Japan or exporting our labor to other countries. However little attention is paid to develop indigenous capacity to create adequate domestic demand for IT ... IT demand is the biggest challenge to develop an IT industry and knowledge economy. Let's take e-commerce as an example. The problem is neither money nor infrastructure (that we can borrow by loan to build). The problem is that we do not have people (with sufficient knowledge of IT and understanding of international standards) to build up e-commerce and people who can use e-commerce (S02, interview, Oct 27, 2004).

4.9.7 Telecommunications difficulties

The telecommunications network in communal and remote areas is not reliable, affordable or comprehensive. Although there is major progress towards telecommunications and Internet development in Vietnam, the heavy domination by the state-owned corporation, Vietnam Posts and Telecom, is widely regarded as a restraint on development. Accessibility is mostly good in urban areas, yet in villages, even on the outskirts of Hanoi, such as in Bat Trang village, local people experience poor quality connections (no broadband) and exorbitant costs (S37 and S38, interviews, Nov 30, 2004).

Reliability and security of telecommunications is also worrisome. In a UNDP report (United Nations Development Programme, 2003, p. 14), Nam Roi Pomelo, a successful application of ICTs by farmers was quoted: a 22-year-old company director described how she used a local website to promote her famous Nam Roi pomelo fruits. The author however notes that the recommended website (www.5roi.com) does not exist on the Internet now (perhaps being hacked, or now owned by others). The story illustrates the difficulties of achieving a stable networked environment.

4.9.8 Pornographic websites

Other social values are violated online. Privacy is ignored by the networks. Pornography websites (most operated by Vietnamese overseas; see Table 4.1) publicize stolen intimate pictures or video clips of Vietnamese film actresses or models. They attract high traffic from Vietnamese youngsters. As a consequence, controversial policies have been developed by the government, such as a 2005's circular No.02 of the Government on tighter control of Internet cafes, which have had the effect of narrowing potential extended access, and banning both bad and good knowledge in unison (B. Dang, 2005).

4.9.9 Adaptive capacity of the government

It is not surprising that the pace of socio-economic and technological change is much faster than the adaptive capacity of the government. Many interviewees raised this concern. These points demand further thorough investigation, but at this stage some notable examples have been collated. The unfamiliarity with a market economy among many leaders is a big concern. Here are some quotations from interviews:

> The knowledge economy is said to be important because it has been asserted as such worldwide. In fact there is no actual pressure or motivation to join it.... Low knowledge management awareness and understanding among leadership results in half hearted determination, thus assigning an incapable person who is not managing properly (S04, interview, Oct 28, 2004).

> Few leaders go to a site to meet farmers. How can they really understand farmers' problems or help them? (S59, interview, Dec 5&6, 2004).

> Many middle managers are presumably opportunists and lack IT skills and knowledge, thus giving wrong advice to decision-makers. For example, the failure of the ambitious software plan 2005, which targeted $US500 million for domestic software production, is blamed on men who never touched a keyboard, yet determined that Vietnam should export software and software experts (S04, interview, Oct 28, 2004).

> Many civil servants do not make much effort to study, and are inflexible. They have nominal degrees, but no actual qualifications, either because they avoid study, or study a lot of useless topics, e.g., political dogma. Many government staff work in a specialty, but study in a different specialty, because the latter is easier to get a degree in than the former. Corruption and excessive bureaucracy are the most criticized problems of the government. IT investment is in dribs and drabs. Interdepartmental management and policy is lacking. Government systems are slow, separate from each other, and ineffective (S04, interview, Oct 28, 2004).

> Investment in research and development declines. Application of science and technology is very slow and the lowest in the region (S59, interview, Dec 5&6, 2004).

A senior government official expressed his concern about the slavish imitation of fashion in the government sector:

> Too many fashionable priorities make a jackfruit policy without fully understanding the feasibility or usefulness of each new fashion. Consequently investment and resources are scattered too thinly. For example, when information technology emerged in the policies of other countries, it was soon made as a high priority in the Vietnam agenda too, but then it was side-stepped to make way for new hot topics, such as bio-tech, nano-tech, the knowledge economy, or sustainability when these are canvassed as priorities elsewhere (S07, interview, Oct 14, 2004).

4.10 Cultural heritage for a sustained development

Giddens (1999) claimed " ... traditional cultures didn't have a concept of risk because they didn't need one". However, "in advanced modernity the social production of wealth is systematically accompanied by the social production of risks" (Beck, 1992). Amongst those, traditional culture is more vulnerable with risk.

In the next chapter examples are given of the vulnerability of traditional cultural endeavors under the impacts of modernization, and the creative strategies being employed to counter such impacts.

Enterprises such as Dong Ho folk printing, Quan Ho folksong and Bat Trang ceramics (discussed in Chapter 9) are parts of Vietnam's rich cultural heritage, and a source of valuable indigenous knowledge and intellectual property that can bring sustained wealth to Vietnam in the form of cultural tourism and related cultural goods and services.

Cultural services are already important to both regional and local economies in terms of income and employment generation. Vietnam was host to 2.63 million foreign and 13 million domestic tourists in year 2002, with total industry revenues amounting to VND 23.5 trillion (US$1.5 billion) (X. N. Pham et al., 2003).

The tourist trade, and especially cultural tourism based on Vietnam's cultural assets have been highly successful during the 1990s. Tapping into the global heritage conservation system, Ha Long Bay, Hue, Hoi An and the Cham remains at My Son have been successfully recognized as World Heritage assets. The inclusion of these areas on the World Heritage List has given further impetus to cultural tourism, to associated arts and crafts, and to local development. The proposal to list Ha Noi's Ancient Quarter and citadel, together with the nearby Co Loa citadel, will provide much new publicity for Vietnam as a tourist destination. In addition, there are currently more than 2,700 cultural heritage sites and monuments inscribed on the National Heritage List (Viet Nam National Administration of Tourism, UNDP, & World Tourism Organization, 2001).

Clearly, cultural tourism expansion, if well managed, can lead to growth in areas such as arts and crafts, food, beverages and transportation, which will then have beneficial social impacts in other sectors such as employment and health. The development of tourism that is friendly to the environment, "Sustainable tourism" can be beneficial in contributing to the long-term survival of both the environment and the industry.

Pham et al. noted that:

> A shift in the tourism paradigm towards sustainable tourism development requires a community development process by which communities can recognize, interpret and manage their own cultural and environmental wealth, and actively participate in the protection and enhancement of these resources. In order to achieve this shift, there is a need for demonstration projects, for example, through volunteerism in integrated heritage preservation and tourism development (X. N. Pham et al., 2003).

4.11 Management of risk

Vietnamese people look to the future with contrasting expectations. Ngo Duc Thinh (2002) wrote that the disruption of traditional culture, norms and values in rural areas has exacerbated poverty and created all sorts of unwelcome interference. However three experts with science backgrounds in separate interviews (S56, interview, Dec 21, 2004; S12, interview, Nov 4, 2004; S23, interview, Nov 16, 20004) took a more optimistic long-term view, arguing that although the disruption may be worrying, in general the evolution of Vietnamese society and the economy offers a promising future.

This chapter has argued that aspects of the knowledge society have been both destructive and creative for Vietnam. The author acknowledges that there is a need for a positive national approach to managing the risk of departing from centuries-old ways and as a result, breaking through the bamboo fences and national borders that have been historic shields of Vietnamese traditional culture, and believes that traditional knowledge and new knowledge will form a synergy leading to positive outcomes for a sustainable knowledge economy in Vietnam.

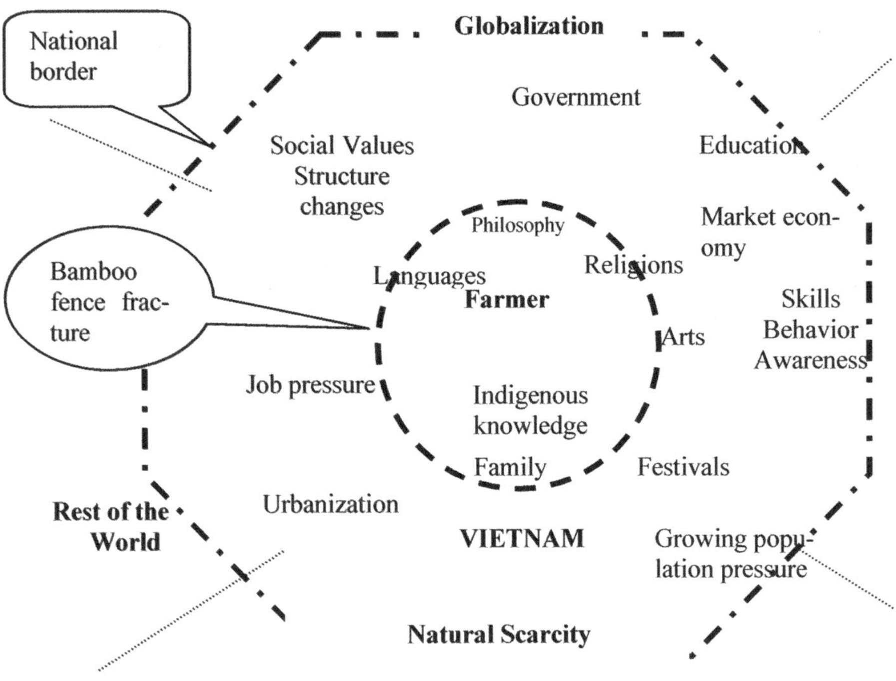

Figure 4.1 Key elements affected by economic growth, featuring bamboo fence and national border "fractures" between rural and urban Vietnam and the outside world

4.12 Key elements affected by economic growth

Figure 4.1 provides a diagrammatic representation of the elements affected by economic growth and the fracture between old and current culture, and some key relationships within conflicting forces. The diagram aims to summarize the discussion above about primary social and cultural factors that can affect the development of a knowledge economy in Vietnam. In the center is the Vietnamese farmer who is no longer protected by the traditional bamboo fence. Together with the disappearance of that fence or metaphoric shield is the deterioration of traditional values, languages, religion, culture, philosophy, family and indigenous knowledge. In a larger context, the blurring of national borders under the impact of globalization results in new opportunities and also challenges for the domestic economy and the society: the changes in social values, urbanization, population increase, the quality of education versus the demand of economy growth, the depletion of the ecosystem, and finally the capacity of the government to respond to these challenges.

4.13 Main insights from Chapter 4

Insight 4A: Vietnam is a society which has experienced and withstood internal and external tensions for centuries, sheltering itself behind the mental construct of the bamboo fence. It demonstrates a great desire for progress in a globalized economy provided that core aspects of traditional culture are maintained.

Insight 4B: Doimoi and other reform measures have helped Vietnam to recover from the US War and become a competitor in the world market, especially in commodities such as coffee and rice, and aquaculture products.

Insight 4C: Increased prosperity has been two-edged. Average literacy rates and incomes have risen, but key social inequities (rich/poor, rural male/female, rural/urban, Vietnamese/ethnic minorities) persist or worsen. Accelerated degradation of the ecosystem has been a major concomitant of economic development.

Insight 4D: With high population pressure and governmental structures still transitioning from a command to a market economy, there is a widely acknowledged need for rapid change, but with risk minimization.

Chapter 5
Policy achievements and challenges

5.1 Scope of the chapter

Based on literature analysis and the author's experience as an officer in the Office of the Steering Committee of the National Program on Information Technology 1996-2000, Ministry of Science, Technology and Environment, Ministry of Posts and Telematics, and lastly the Ministry of Information and Communications) this chapter:

- Provides background data on Vietnam demographics, natural resources and economic growth
- Presents an overview of Vietnam's policy-led achievements and challenges in relation to the following, with special attention to government policy development and implementation:
  - Sustainable economic development
  - ICT use and development.
- Explains the positioning of government and business relevant to the deployment of ICTs in all sectors of the economy, and across the rural/urban divide
- Also focuses on the:
  - The progress of Internet connectivity
  - The nationwide provision of CPCPs (Communal Posts and Culture Points),
  - The building of ICT skills across the population.

5.2 Vietnam fact sheet – location, population, natural resources, biodiversity and economic growth

5.2.1 Location

Vietnam, or the Socialist Republic of Vietnam is situated in the Indochina Peninsula, bordered by China to the north, Laos and Cambodia to the west, and the South China Sea to the east. At the national level, the Government in Vietnam consists of a dual structure comprising the Communist Party and the National Government. At the level of local government, the country is organized into 64 provinces. The provinces are divided into 588 districts, which are further subdivided into 9,069 communes (General Statistics Of-fice of Vietnam, 2006b).

5.2.2 Population

According to the CIA World Factbook (2007) the area of Vietnam is 329,560 sq km, about half the size of Thailand but twice more crowded, with a total population of 85.26 million (July 2007 est.) (see Table 5.1). It is the 13th most populous country in the world.

The population is unusual in that a majority of the population is younger than 30 years old. The median age is 26.4 years, comprising 26.3% of 0-14 years, 67.9% of 15-64 years, and only 5.8% of 65 years and over (Central Intelligence Agency, 2007). Approximately 60 million people live in a rural setting with farming as the main occupation (General Statistics Office of Vietnam, 2006a).

5.2.3 Natural resources

Vietnam's natural resources include minerals, water and biodiversity.

Minerals

According to the US "Geological Survey Minerals Yearbook 2005", important minerals of Vietnam are antimony, bauxite, chromite, coal, copper, natural gas, gold, iron ore, nickel, crude petroleum, phosphate rock, rare earths, tin, titanium, tungsten, barite, carbonate rocks (limestone and marble), gemstones (ruby and sapphire), graphite, lead, manganese, silica sand (Wu, 2006). The Yearbook noted that except for carbonate rocks, coal and hydrocarbons, most mineral resources remain largely unexploited owing to Vietnam's outdated mining equipment and technology, poor infrastructure, and uncompetitive Government policy to attract foreign investors in mining. Exploitation of the country's ferrous and nonferrous metal resources is still at an early stage of development (Wu, 2006).

Water

Before 2001, Vietnam was regarded as a country rich in fresh water with a high rainfall (an average annual rainfall of 1800-2000 mm) usually replenishing the supplies of the surface and ground water and a relatively dense network of rivers with the system of two rivers of Red and Thai Binh in the North; the systems of Ca and Ma rivers and of the Han, Thach Han, and Thu Bon rivers in the Centre; and the system of the Mekong and Dong Nai rivers in the South. At present, these river systems still exist, but the country is suffering more severe shortages of water in many months of a year because 1) the dry season is longer than before, which is believed to be the result of global warming, and 2) the degradation of water quality from rivers and groundwater because of human pollution (see the section of Challenges to sustainable development below).

Biodiversity

Vietnam is one of the 16 countries of highest biodiversity in the world in terms of species composition and diversity, both in landscapes and ecosystems. According to the National Environmental Agency of Vietnam, now the Vietnam Environment Protection Agency (VEPA), in the 4 major biodiversity centers in Vietnam, namely Hoang Lien Son, Northern Truong Son, Central Highland and South Eastern regions, even recently several new mammals, birds and plants have been discovered, such as the Sao La, also known as the Vu Quang ox, a species of bovid, discovered in 1993 in the Phu Quang nature reserve. Forests which covered 28.8% of the total land area of the country in 1999 were home to a diverse and abundant flora and fauna. Inland freshwaters of Vietnam were endowed with a rich diversity of flora and fauna including algae, aquatic macrophytes, aquatic invertebrates, insects and fishes. There is also a rich biodiversity of native plants in Vietnam (National Environmental Agency, 2002).

Recent studies reported 13,766 species of plants in Vietnam of which 2,393 were lower plant species and 11,373 vascular plant species, and 10 percent of the plant species were endemic. Of the 21 primate species occurring in the Indochinese subregion, 15 species were recorded in Vietnam, of which 7 species and subspecies were endemic. The biodiversity of cultivated plants of Vietnam was also high with 734 species of popular plants belonging to 79 families (National Environmental Agency, 2002).

For a knowledge economy, that biodiversity is an enormous source of genes - crucial raw materials for the potential biotechnology and pharmaceutical industries, important knowledge based sectors which Vietnam can develop.

5.2.4 Economic growth

Wars of national identity: damage and recovery

In the decade after the end of the US war in 1975, Vietnam was isolated from the world capitalist system as a result of the damage of the war itself and the US embargo which followed the end of the war. The uninterrupted century-long series of wars for identity against France, Japan, China and the US, and especially US sanctions devastated the economy, especially after the unification of the country in 1976. Vietnam became one of the most impoverished countries in the world with an annual per capita GNP of about USD150.

By the mid-1980s, at the time of the collapse of the socialist bloc in Eastern Europe, the country faced "serious food shortages, large trade and fiscal deficits, and hyper-inflation". Fortunately, the social welfare network developed evenly in the country to achieve relatively good indicators: life expectancy was over 60 and adult literacy reached 87 percent (World Bank, 2001, p. 1).

The era of Doimoi and the socialist-oriented market economy

Since 1986 Vietnam has undertaken a far-reaching process of social-economic reform known as "Doimoi" (innovation), a major reform program. The Government of Vietnam committed to increased economic liberalization and enacted structural reforms needed to modernize the economy and to produce more competitive, export-driven industries. It

> devalued the exchange rate to equal the parallel rate; decontrolled most prices to allow market forces to work; relaxed official marketing constraints; introduced fiscal and tax reform, formalized and lengthened land tenure for family farms; re-estabilshed state owned enterprises as financially independent units; and passed regulations to encourage the non-state economic sector, i.e., to promote private sector development (World Bank, 2001, p. 1).

Through the introduction of a socialist-oriented market economy, the private economy, including non-state sectors and family business has become a new major engine of the country's development.

Achievements of Doimoi have been spectacular. According to a World Bank report in 2003, the progress made in Vietnam in alleviating poverty has been one of the most successful stories in world economic development in recent decades. Poverty rates measured at international level (one dollar a day) have halved from 58% in 1992 to 37% in 1997 and 29% in 2002 (World Bank, 2003a) and further down to 24.1% in 2004 (Asian Development Bank, 2005).

Engagement with trade liberalization

Jenkins noted that the trade liberalization that began at the end of the 1980s in Vietnam was characterized by the liberalization of entry into international trading activities; the removal of most export taxes and non-tariff barriers; a reduction in tariff levels and bands, from 200% to 120% in tariff, the signing of various international agreements – ASEAN Free Trade Area (AFTA), agreement with the EU in 1992, a Bilateral Trade Agreement with the US in 2000 and with the WTO in 2006; and the establishment of export processing zones, industrial zones and high technology parts (Jenkins, 2004).

Scale of economic growth

Together with poverty reduction, the country had very successful economic growth since 1990 with around 6.8% annual GDP growth from 1997 to 2004, 8% in 2005 and 8.2% in 2006, making it the world's second-fastest growing economy (Central Intelligence Agency, 2007). Export increased from USD9.1 billion in 2001 (Dapice & Fellow, 2003) to USD 39.92 billion in 2006 (Central Intelligence Agency, 2007).

Legalization of private business, foreign investment and WTO membership

In 2000 the Government officially allowed the establishment of private businesses by the introduction of the Law of Enterprise. After that, private investment increased quickly with the number of private industry (excluding household business) increasing nearly 20% per year. This created 1.75 million new jobs from 2000 to 2002, compared to almost zero growth in the public sector (Dapice & Fellow, 2003). In 2006, nominal GDP and per capita GDP were USD 48.43 billion and USD 576, re-spectively. Estimated on purchasing power parity, these figures were USD 262.8 bil-lion and USD 3,100 (Central Intelligence Agency, 2007).

Foreign direct investment (FDI) grew twofold, from USD3.2 billion in 2001 to USD6.8 billion in 2005 (Tuoi Tre, 2007). The online newspaper Tuoitre (Youth) estimated that FDI reached USD10.2 billion in 2006 (Baker, Tumbarello, & Ahmed, 2006). After joining the WTO Vietnam had an opportunity to receive a new wave of FDI, about USD15 billion have been invested in Vietnam by November 2007 (Chau, 2007).

5.3 Challenges to sustainable development in Vietnam

The quick growth of the economy and the population, and also global climate change have had extremely negative impacts on the sustainable future of Vietnam.

The quantity and quality of its natural resources, such as petroleum, water, biodiversity and forests are severely threatened.

Table 5.1 Vietnam and others - Selected social economic data. Source: tabulated by the author from data supplied in the World Fact Book (Central Intelligence Agency, 2007)

Country/ Group of countries	Population 2007 est. ('000)	Population density per km²	Population growth rate 2007 est. (%)	Sex rationmale/ female 2007 est.	Literacy rate (%)	GDP in official exchange rate 2006 (USD billion)	GDP growth rate 2006 est. (%)	GDP in PPP 2006 (USD billion)	2006 GDP in PPP per capita (USD)
Vietnam	**85,262**	**258**	**1.00**	**0.98**	**90.3**	**48.43**	**8.2**	**262.8**	**3,100**
Indonesia	234,693	122	1.21	1.00	90.4	264.7	5.5	948.3	3,900
Philippines	91,077	303	1.76	0.99	92.6	116.9	5.4	449.8	5,000
Thailand	65.068	130	0.66	0.98	92.6	197.7	4.8	596.5	9,200
Malaysia	24,821	75	1.76	1.01	88.7	132.3	5.9	313.8	12,900
China	1,321,851	137	0.60	1.06	90.9	10,170	10.7	2,518	7,700
Singapore	4,553	6,579	1.27	0.95	92.5	122.1	7.9	141.2	31,400
South Korea	49,044	50	0.39	1.01	97.9	897.4	4.8	1,196	24,500
Australia	20,434	2.6	0.82	0.99	99.0	644.7	2.7	674.6	33,300
France	63,716	97	0.59	0.96	99.0	2,149	2.1	1,891	31,100
The whole world	6,602,224	44.3	1.16	1.01	82.0	65,950	5.3	46,760	10,200

As noted earlier, global warming is believed to have had a deadly influence on the river systems in Vietnam. In recent years the dry season extends longer - about 8 months a year – causing serious droughts and badly affecting cultivation and life. The 2007 hydrographical data showed a lower rainfall in the country compared to the averages of previous years, and the rainy season ended one month earlier than normal. Vietnam, from being a country in the past with abundant water resources, has been added to the list of countries that are poor in water resources (Tien Phong, 2007). [8]

Due to the extension of the dry season, the flow of rivers, streams, lakes and dams fell drastically. On January 23, 2007, the water level of the Red River in Hanoi, for instance, was only 1.13 meters, the lowest level in the past 100 years (Khanh, 2007).

[8] The International Water Resources Association regards a country as water poor if the per capita water level is below 4000 m³ a year.

In addition 60 percent of the rivers – the source of ground water, originating from neighbor countries like China, Myanmar, Laos, Thailand or Cambodia - can be seriously depleted or used up before reaching Vietnam (National Environmental Agency, 2002; T. X. Tran, 2004).

Serious droughts have occurred more frequently than before, especially in the southern part of the central region. The 2007 flows of rivers were around 50—90% lower than average. In the central provinces from Quang Ngai to Khanh Hoa many water reservoirs were well below their designed capacity. In the south the water level at the riverhead of the Mekong River fell very low (Tien Phong, 2007).

Domestic economic and demographic growth has also severely impacted on the water system. Pollution is making water from the rivers near Hanoi and Ho Chi Minh City practically unusable. The Vietnam Environment Protection Agency has reported that the system of three major rivers of Cau, Nhue and Day near Hanoi and Dongnai River near Hochiminh city which supplies water to valleys consisting of 37,400 km2 for 16 million people in 11 provinces and cities has now become sources of dead water. Liquid and solid wastes from industrial and medical centres and household activities in Hanoi and Thainguyen province discharged directly into the Cau, Nhue and Day Rivers. Pesticides, heavy metals, oil and organic pollutants have flowed from factories in Hochiminh city into the Dongnai River. The surface water of these rivers is so polluted that it is not at all suitable for agricultural uses, and absolutely impossible for living purpose (Vietnam Environment Protection Agency, 2007).

Through intensive mining, the reserves of major natural resources, such as mineral and marine products are being depleted. With limited proven reserves of oil and coal (400 million tons and 150 million tons, respectively) (BP, 2007b), in 2006 Vietnam was nevertheless the world's third leading producer of anthracite coal, the sixth largest producer of crude petroleum in the Asia and the Pacific region, and one of the major producers of ilmenite and phosphate rock in Asia. In 2005, Vietnam produced 19.1 tons of crude oil and 18.3 million tons of coal (BP, 2007b). The country is likely to run out of its own fossil fuels in the next two decades. Since 2005, under a regulation published by the Ministry of Industry, Vietnamese enterprises are only allowed to export processed minerals that meet the regulations of the Vietnam Laboratory Accreditation Scheme, but are not allowed to export raw materials, including ilmenite, rutile and zircon (Wu, 2006). In a 2007 annual plan launching ceremony held in PetroVietnam, the Prime Minister, Nguyen Tan Dung, suggested that Vietnam should not export crude oil, but should rather promote processing and services to earn added value (Vietnam Government Portal, 2007).

5.4 Action planning towards sustainability, Vietnam Agenda 21 (VA21), and the implementation of sustainability measures

Vietnam viewed environmental protection as a high priority in the "National Plan on Environment and Development in the period 1991—2000" dated June 12, 1991.

A Directive No. 36-CT/TW dated June 25, 1998 of the Politburo about strengthening environmental protection in the period of industrialization and modernization stated:

Environmental protection is inseparable from the lines, directions, and socio-economic plans of authorities at all levels and in all sectors and is a significant basis for ensuring sustainable development and successful implementation of the cause of industrializing and modernizing the country (cited in Government of Vietnam, 2004d, p. 7).

In August 2000, the Prime Minister assigned the Ministry of Planning and Investment and the Ministry of Science, Technology and Environment to work out a national program on sustainable development which would serve as a foundation for sustainable development in the country. With assistance from the United Nations Development Program (UNDP), the VIE/01/021 project operated by the Ministry of Planning and Investment (GTZ & MPI) conducted a series of conferences in Haiphong, Hochiminh city and Cantho (in May 2002) and Hue and Sonla (in June 2002) to share international and national experience on sustainable development and prepare inputs for the Vietnam Agenda 21 (Ministry of Planning and Investment of Vietnam, UNDP, & Danish International Development Agency, 2003).

The "Strategy for Socio-economic Development in the period 2001-2010" reaffirmed the role of sustainable development as:

> … fast, effective and sustainable development, economic growth should occur in parallel with the implementation of social progress and equality and environmental protection (…) socio economic development is closely tied to environmental protection and improvement, ensuring harmony between the artificial and natural environment and preserving bio-diversity (cited in Government of Vietnam, 2004d, p. 7).

In 2002 the international "Rio Earth Summit" drew up an action plan for sustainable development for the world in the 21st century, often referred to as "Agenda 21".

A Government of Vietnam report entitled "The Strategic Orientation for Sustainable Development (Vietnam Agenda 21)" noted that Vietnam's system of policies and legal instruments were not uniform enough to ensure the effective combination of three key elements of development, namely economic and social development, and environmental protection (Government of Vietnam, 2004d, p. 8).

As a result of the work, on August 17, 2004, the Government of Vietnam issued the decision No. 153/2004/QD-TTg to pass the "Strategic Orientation for Sustainable Development in Vietnam" or "Vietnam Agenda 21", as a Vietnamese version of the Rio Declaration. According to this plan, the overall objective of sustainable development in Vietnam was to

> create a materially, culturally and spiritually plentiful life for the people, quality of life for all citizens, the consensus of society and harmony between the people and the nature (Government of Vietnam, 2004d).

To achieve that goal, eight main principles were defined. They are summarized briefly below:

> First, human beings are the centre of sustainable development. (…)
>
> Second, consider economic development as the central task in the coming development period (…) thriftily and effectively utilize natural resources within the permitted limits in conformity with sustainable eco-system and environment protection. (…)
>
> Third, protection and improvement of environment quality are to be considered as an inseparable factor from the development process. (…)

> Fourth, the development process must equally satisfy the needs of the current generations without causing obstacles for the life of future generations. (…)
>
> Fifth, science and technology is the foundation and momentum for the country's industrialization, modernization, quick, strong and sustainable development.(…)
>
> Sixth, sustainable development requires the cause of the whole Party, authorities at all levels, the ministries, sectors and localities, agencies, businesses, social organizations, population communities and the whole people. (…)
>
> Seventh, the establishment of an independent and autonomous economy is tightly attached to the international economic integration on the basis of self initiative. (…)
>
> Eighth, social-economic development should be closely combined with guarantee of national and security as well as social safety and order (Government of Vietnam, 2004d, pp. 22-25).

The Vietnam Agenda 21 viewed ICTs as facilitating the monitoring of sustainable development through the establishment of information systems and networks in the following important areas: environment, land and natural resources, population, the labor market and the scientific and research experience of other countries. The Agenda also set out the need for regulations requesting large and medium sized enterprises and businesses to establish their own environmental monitoring systems which would provide information about waste and pollution resulting from their production activities (Government of Vietnam, 2004d).

A National Council of Sustainable Development was set up by the Prime Minister in 2005. A Vice Prime Minister is chairman, the Minister of Planning and Investment is a standing vice chairman, and a leader of the Ministry of Natural Resources and Environment is another vice chairman. Forty one heads of national ministries are Council members. They include a leader of the Ministry of Posts and Telematics (Government of Vietnam, 2005a).

The Sustainable Development Office of Vietnam (or VA21 Office), based at the Ministry of Planning and Investment, is the coordinating agency on sustainable issues in Vietnam. VA21 organized 2 national conferences on sustainable development, the first in 2004 and the second in 2006 to increase awareness, share experiences and review the coordination of related agencies and provinces. The author of this book attended the 2004 conference during his first fieldwork period through which he was able to gain access to many relevant documents and interview key speakers of the conference. Many of these important references are included in the book.

The VA21 Office does not have representatives from the provinces or sectors. However the Ministry of Natural Resources and Environment (MONRE) and its affiliate, Vietnam Environment Protection Agency (VEPA) have branches at all provinces and districts and have representatives in some major national agencies. The mandate of VEPA, the focal point of environment protection, is to

> Manage the environment protection by inspecting, supervising, preventing, repairing pollution, deterioration, and problem of environment; improving the environment quality, preserving biodiversity, measuring environment; applying technology, building database, statistics, information and reports on envi-

> ronment, educating and improving the community awareness about environment (Ministry of Natural Resources and Environment, 2005).

With this mandate, VEPA is responsible for many applications of ICTs in environmental protection, for instance, VEPA hosts http://www.nea.gov.vn/English /index.asp as the homepage.

Another agency, the Ministry of Agriculture and Rural Development (MARD) is responsible for forest protection and water management. MARD played a crucial role in increasing forest coverage from 33.2% in 1999 to 36.1% in 2003 as a result of the 327 Project on rehabilitation of degraded forest lands (Ministry of Agriculture and Rural Development, 2004, p. 26).

Despite the excellent intentions of the VEPA mandate, the results of environment protection is as yet very limited because

> There have remained overlapping of functions, duties among the management levels and sectors in environmental protection. The state management of the environment has been implemented at only the central, sectoral and provincial levels, but not at the district level and no management at commune/precincts level (...) there has been an absence of legally binding provisions that localities and sectors must participate in the formulation process of such kinds of planning and comply with them (Government of Vietnam, 2004d, p. 20).

The intervention of governmental agencies with respect to environment pollution is not yet strong enough to reduce or stop bad practices. Sources of pollution, such as from the Thainguyen Steel Corporation into the Cau River, or from the industrial zone into the Dongnai River, received mild or even no penalties from environment protection authorities. In many cases, the measures of those agencies are recommended, but not mandatory.

The establishment of a new police department specializing in environmental protection within the Department of Environment and its systems in the provinces in March 2007, and also the tightening of environmental protection laws are expected to strengthen the enforcement of law and regulations relating to sustainability in Vietnam (Ministry of Natural Resources and Environment, 2007).

Although strategies and policies about sustainable development are prevailing in Vietnam, the implementation of these policies so far is not adequate to maintain a sustained future for the country. The burden of economic growth is so severe upon the environment in all areas that some scholars, such as professor Le Quy An, have asked whether there is a threshold of economic development for Vietnam (Q. A. Le, 2002).

5.5 ICT use and development

5.5.1 ICT phases

The history of ICT in Vietnam can be divided into three big phases: late 1960s to 1993, 1993 to 2000, and 2000 up to the present.

The first phase of IT in Vietnam started in the 1960s with the introduction of large computers and mini computers of the second generation from the former

USSR in the North, and IBM large computers in the South (Tran Luu Chuong, interview, Oct 27, 2004).

The second phase started with the introduction of the Government's Resolution No. 49/CP dated 04/08/1993 on "Information technology development in Vietnam in the 1990s" (Government of Vietnam, 2004d) which set up the first "National Program on IT to 2000" (often called NPIT2000). The Program's steering committee members were heads of related ministries and agencies, such as Science, Technology and Environment, the Government Office. The NPIT2000 provided guidelines and funding to all ministries and provinces/cities in the country to implement their own IT projects. Unfortunately, the on-going NPIT was suddenly disbanded at the end of 1998, only 3 years after its commencement.

The third phase followed the termination of NPIT 2000 when the focus of IT was placed upon the IT products research and development section in the framework of an IT Techno-economic program led by the Ministry of Science and Technology. Several groups of ICT experts, with backing from the Vietnam-Canada Information Technology Project (VCIT) of which the book author was a vice director, developed an important document that then was issued by the Politburo of Vietnam's Communist Party in late 2000 - the "Directive No. 58/CT – TW". This Directive 58 was a key milestone in the history of ICT development in Vietnam. (It is discussed in more detail later in the book). From then on, IT and telecommunications were combined into one sector in order to form "one of the most important motivations to the development [of Vietnam]" (Vietnam Communist Party Poliburo, 2000).

After the Directive 58, a series of strategic policies were prepared and issued to form a clear legal corridor for ICT use and development in Vietnam. The most important of these documents are listed in Appendix 7.

Apart from the Directive 58, many of these were direct or indirect outcomes of the VCIT project in which the book author actively participated as a key designer and manager. The 2005 law on Electronic Transaction originated from a sub-project titled "E-commerce". During 1998-2000 VCIT had assisted the Ministry of Trade in developing a framework plan for the acceptance and application of electronic commerce in Vietnam (GAIA Corporation, 2001, p. 56). Policies, such as the Government's "Decree on Development of the Software Industry in 2000-2005" and the "Master Plan on Use and Development of IT to 2005", "Open Source Software Policy", "Plan of Telecommunication and Internet Development in Vietnam to 2010" were all developed from other original drafts of VCIT projects (GAIA Corporation, 2001, pp. 16-18).

Other documents were outcomes of the second phase of the VCIT project, the "IT Policy Implementation Assistance Project" or IT/PIAP, funded by the Canadian International Development Agency (CIDA). The book author was also a vice director of the IT/PIAP project, and participated in the preparation of the "Regulation of IT Procurement by Government Budget in Agencies and Organizations" and the "Law of Information Technology" in 2006 (T. T. Nguyen, 2002).

In Annex 7, the author has grouped the titles of documents relevant to IT development in Vietnam into 3 groups:

1. Laws and ordinances on IT, e-commerce, posts and telecommunications, and IPR related subjects;
2. Strategies and plans including the 2020 vision, 10 year strategies and 5 year plans (to 2005 and 2006-2010) on ICT, Internet, telecommunications and software industry;

3. Governmental regulations on concrete matters, such as digital signature, IT procurement, funding, and utilization by the government sector.

5.5.2 The position of ICTs in supporting an economy that is environmentally sustainable in government policies

Elmer and Nguyen Thanh Tuyen claimed that because of inadequate integration of ICT policies into Vietnam's broader national, social, and economic development strategies, particularly for the sustainability of the economy, the role of ICTs as a possible enabler in development strategies and development had been underplayed (Elmer, 2002; T. T. Nguyen, 2003a). On the one hand, ICT policies primarily concentrated on short-term and medium term targets and paid little attention to the long-term target of economic sustainability; on the other hand, national strategies on sustainable development, such as the "Strategic Orientation for Sustainable Development in Vietnam", took insufficient account of the potential contribution of ICTs.

Only Directive 58 and the "Strategy of Information and Telecommunications to 2010 and Vision to 2020" make a connection between ICTs and sustainable development. The Strategy states:

> ICTs are the most important tool to realize the millennium development goals, to build up an information society and shorten the course of industrialization and modernization of the country (Government of Vietnam, 2005d).

The only project relating to sustainable development –"Building an information system on land, the environment and natural resources" is given little emphasis within a long list of 74 ICT projects of the above Strategy.

Similarly, other documents, such as the "Plan of Telecommunication and Internet Development in Vietnam to 2010" and the "Program of software industry in Vietnam to 2010" (both of which consist of practical guidelines to realize the Strategy) make no specific reference to sustainable development, and there is no specific financial commitment to realize these activities.

5.5.3 ICT structure

Before July 2002, Telecommunications and IT were two separate fields. Telecommunications were managed by the General Department of Posts and Telecommunications, and IT was not an economic sector. Even the Steering Committee of the NPIT 2000 was not a permanent organization but an ad-hoc apparatus. When the NPIT 2000 was dismissed, the management of IT was transferred to the Ministry of Science and Technology.

The establishment of a ministry in charge of both IT and communication, the Ministry of Posts and Telematics (MPT) in July 2002, and a National ICT Steering Committee (SC58) in the same year, marked a significant turning point in ICT development in Vietnam. From then on, IT and telecommunications merged into one single sector in Vietnam.

In terms of organization, the MPT consisted of general and specialized agencies. General agencies included departments of planning, international relations, personnel and organization and the ministry's office. Specialized agencies included departments of ICT industry, IT applications promotion, telecommunications, posts, IT utilization, radio frequency, ICT quality control Vietnam, the Vietnam Internet Center and the National Institute of ICT policies and strategies. The Steering Committee on Implementation of the Directive Number 58-CT/TW (SC58), an upgrade of SC-NPIT, was chaired by a Vice Prime minister. Its members include the Minister of MPT, the heads of related ministries and 3 major municipalities and representatives of ICT associations. MPT worked as the focal point of SC58 whose office was located at the Agency of IT Utilization. In each province a post and telecommunications department, managed administratively by the provincial authority but guided technically by the MPT, provided state management in the locality. There was no lower branch of post and telecommunications at the district and commune level.

On July 31, 2007, the new National Assembly of Vietnam merged the Ministry of Posts and Telematics with the mass media and publication sections of the Ministry of Information and Culture into the Ministry of Information and Communications.

Besides the governmental system, ICT industry associations operating at the national level are the Vietnam Information Processing Association (VAIP), Vietnam Software Association (VINASA), Vietnam Electronic Industries Association (VEIA), and Vietnam Radio & Electronics Association. These associations have provincial branches, such as the Hochiminh Computer Association and the Hanoi IT Association which are members of VAIP. These organizations are a channel through which the voices of ICT businesses can be transmitted to the government – heads of VEIA and VINASA are members of SC58. Currently there is no association of ICT users in Vietnam.

5.5.4 ICT investment

The ICT industries are given much encouragement by Government. As a percentage of GDP, Vietnam's ICT spending is competitive with many other Asian countries. Vietnam spends 7.3% of its GDP on ICT, while Japan spends 7.4%, Hong Kong 8.4%, and South Korea 6.7%. Indeed, with the exception of Singapore (10.5%), Hong Kong and Japan, Vietnam devotes a higher share of GDP to technology than any other Asian nation (VNE, 2006). Though the absolute figure is much lower than that of the counterparts, this investment provides a good basis for long-term development and encourages money from other investors.

In the past two years, there was a big shift of foreign direct investment in the ICT sector in Vietnam. Nidec, a Japanese chip and component manufacturer, increased its capital by USD 500 million more to expand its production lines for chips and fluid dynamic bearing disk drive motors in the Saigon Hi-Tech Park (SHTP), bringing its total investment in Vietnam to USD1 billion and making Vietnam the second largest production base after China (D. Tran, 2006).

Similarly, Intel, the world's largest chip producer, decided to triple its investment to USD1 billion in its chip building plant in Hochiminh City (Hochiminh Computer Association, 2006). Canon Vietnam committed to invest an additional USD110 million, adding up to a total of USD370 million of investment in the country by late 2007 and turning Vietnam the largest base for manufacturing Canon printers outside Japan (Hochiminh Computer Association, 2006). Taiwanese companies, such as Foxconn Honhai, intended to invest USD5 billion in a multi-phase project in Bacgiang and Bacninh provinces. Compal has a USD500 million project to produce Compal laptop computers in Vinhphuc province (Tuoi Tre, 2007).

5.5.5 IT Industry

The average growth rate of the IT industry in the period from 2002 to 2005 was relatively high at 31.03% per year. In particular, the IT output in 2005 increased to USD1.4 billion at 50.54%. The value of the hardware products often accounted for over 80% of IT industry revenue.

Table 5.2 IT production in 2000-2005. Adapted from Hochiminh Computer Association (2006)

Year	Software, Content and IT Service					Hardware		Total	
	Local (1)	Growth rate %	Export (2)	Growth rate %	Growth rate %	Revenue (3)	Growth rate %	Total (1+2+3)	Growth rate %
2002	65		20			550		635	
2003	90	38.4	30	50.0	41.18	700	27.27	820	29.1
2004	125	38.8	45	50.0	41.67	760	8.57	930	13.4
2005	180	44.0	70	55.5	47.06	1,150	51.32	1,400	50.5
2008[9]	1,120					4,100			

In 2008, the hardware industry hit a record of over USD1 billion for the first time, of which the export and domestic value were USD1.042 billion and USD108 million, respectively. With these figures the IT industry became one of the 7 industries with export value over USD1 billion. However, the highest contribution was from foreign companies led by Fujtsu producing hard-disks valued at USD515 million and Canon which making printers valued at USD450 million (Hochiminh Computer Association, 2006). The rest of the revenue was shared among local companies, reflecting the fact that the ICT industry of Vietnam was characterized by small/medium size companies. In addition, computers were often assembled by hand and simple in-struments, except for a few large corporations like FPT Leads and CMS which had assembly lines. These were major disadvantages of domestic hardware enterprises compared with foreign competitors.

The software industry was expected to be a leading sector and favored as taking advantage of the intellectual potential and manpower resources of Vietnam. That industry was also seen as providing possible impetus to economic development in the information era. In recent years many software parks and enterprises were established, of which the most successful is the Quangtrung Software Park in Hochiminh City. By the middle of 2005, there were around 600 enterprises of ICT services and software operated in the park, approximately 5 times the figure in 1997. Twenty to

[9] Data from Ministry of Information and Communications (2009).

thirty services-software enterprises were newly formed each year during 1996 - 1999 on average, and up to 75-80 in 2000 – 2005.

It was estimated that about 15,000 workers were working in the software industry by mid 2005, with an average of 25 laborers per enterprise. This figure suggests that Vietnamese software enterprises were still young and characterized by small numbers of workers.

Table 5.3 Number of services and software enterprises (1996-2005). Adapted from Hochiminh Computer Association (2006)

	1997	**2000**	**2001**	**2002**	**2003**	**2004**	**2005**
Number of companies (by end term)	115	229	304	328	370	470	600
Number of companies increased during term	20	59	75	24	71	100	130
Number of software programmers	2,300	4,580	6,080	6,560	7,400	12,000	15,000

The Vietnamese software and services industry grew at an average rate of 43.30% per year. In 2005, this segment saw its revenue growing at 47% to USD250 million of which USD180 million from the local market accounted for 61.1% and USD70 million came from software outsourcing contracts with foreign companies (38.9%).

In 2000 - 2005, the IT local market[10] expanded gradually by 22.72% per year. In 2002, the revenue made a breakthrough to over 28.83% from only 17.6% in 2001, and continued the development in the next year. In 2005, the growth rate decreased sharply to 20.9%. However, it is still an impressive growth rate given that global IT spending was at 7%, and that of the Asian-Pacific region was around half of the Vietnamese figure.

The hardware market increased 2.5 times from USD250 million in 2000 to USD630 million in 2005. The percentage of spending on hardware over software was 83.37%. However, as it can be seen from Table 5.4, there is a slight trend of a reduction in hardware spending from 83.35% to 76.09%, and an increase in software spending from 16.67 to 23.91%.

The Vietnam Customs Office reported that in 2005 USD 522 million out of USD 630 million (86.57%) consisted of revenue from the sale of imported products, mainly from Japan. Thus most of the hardware sold in the Vietnam market was made from outside, while domestic enterprises supply only 15% of local demand. In other words, the products made by domestic enterprises, mainly spare-parts, accounted for a very small market share in comparison with that of foreign enterprises.

In contrast with the hardware market, domestic products dominated at around 90% of the software market share. There was a significant change in the software

[10] IT Market is the sum of domestic production and importing for domestic consumption.

market, marked by spending an amount in 2005 which was 4 times higher than in 2000. However, its percentage was relatively small compared to hardware.

Table 5.4 IT spending in Vietnam market 2000-2005 (Unit: Million USD). Adapted from Hochiminh Computer Association (2006)

Year	Software Market	Software/ Total %	Hardware Market	Hardware / Total %	Total	Growth Rate %
2000	50	16.67	250	83.35	300	n/a
2001	60	17.65	280	82.35	340	13.3
2002	75	18.75	325	81.25	400	17.6
2003	105	20.39	410	79.61	515	28.8
2004	140	20.44	545	79.56	685	33.0
2005	198	23.91	630	76.09	828	20.9

In 2001, the percentage of software and services out of the total spent on IT was around 17.65%, far lower than the average rate in the world of 49%. The percentage gradually increased over time and stood at 23.91% in 2005. In fact, these software expenditure figures do not adequately reflect software usage due to high software privacy rates in Vietnam. Copyright violation was estimated by the Business Software Alliance to be at 90% in 2005 (Bergstein, 2007; BSA & IDC, 2006). The Government committed to bringing down the piracy rate by the use of open source software and by strictly abiding by software copyright in procurement of IT products in the government sector (Government of Vietnam, 2007a). Besides, the introduction of low cost alternatives to Microsoft products, such as Kingsoft Office 2007 from China, is expected to improve the situation greatly (H. Hung, 2007).

5.5.6 Telecommunication industry

According to Prof Do Trung Ta, the Minister of Posts and Telematics, the telecommunication sector appeared in Vietnam much earlier than the IT industry, a few days before the birth of the country in September 1945.

In the Doimoi period from 1993 to 2000, the telecommunication industry followed an "Acceleration Strategy" which resulted in the digitalization and automation of all the national communication networks. In the current 10 year period of 2001-2010, the strategy of the sector is "Integration and Development" which

> upholds all internal force, creates a wide and deep competitive environment in order to innovate technology, expand the network capability, diversify service modes, accelerate the use and development of IT, reduce cost and proactively integrate to the global economy (Do, 2006).

As a result of the openness for competition in the telecommunications market since 2000, domestic telecommunications has been growing to become an emerging market in the ASEAN region. Over the past five years, telecommunications tariffs have been dropping quickly from being one of the world's highest down to a competitive level in the region.

Statistics of the International Telecommunication Union indicate that between 2002 and 2006 fixed telephones sustained a growth of 44.1% - the world fastest speed, compared with the global average of 5.3% and the Asian of 11.9%. The number of mobile cellular and Internet users increased at an even higher rate at 62.7% and 86% per year, respectively. By June 2006, there were 20.1 million fixed telephones, about 22 million mobile phones and 13.4 million Internet users. The densities per 100 inhabitants were 24.2, 26.2 and 16.1 respectively (Do, 2006; Hồng Anh, 2006; International Telecommunication Union, 2006). Historical data are given in Table 5.5.

Table 5.5 Telecommunication and Internet in Vietnam. Source: synthesized by the author from data of (*) ITU (2007b), (**) Vietnam Internet Center (2006), (+) Do (2006), and (++) estimated by Ministry of Posts and Telematics (Hồng Anh, 2006)

Indicator	2002	2003	2004	2005
Main telephone line (thousand)	3,929.1	4,402	10,124.9	15,780
Subscribers per 100 people (*)	4.48	5.41	12.28	18.73
Mobile cellular (thousand)	1,902.4	2,742.0	4,960.0	9,000.0
Subscribers per 100 people (*)	2.34	3.37	6.01	10.68
Internet subscriber (thousand)		450	1,123.8	1,899
Internet users (thousand)	1,300	1,709.5	4,311.3	7,185
User per 100 inhabitants (**)	0.01	2.14	5.29	8.71

In relation to the questions raised at the beginning of the chapter, here below are the author's remarks drawn from the statistical data above:

- Vietnam ICT businesses can partly meet growing local market needs given that ICT services and localized software being the main requirements.
- Vietnam ICT industry possesses a potential capacity to produce a large amount of human-made capital in the form of services, software and the value of information that can be supplied to the local people. Software production and value-added hardware can then be exported to the overseas market as human made capital. In the global market this human made asset in the short and medium term can replace traditional industries based on natural capital, such as coal-mining.
- However, most of the hardware and software comes from overseas. If the contribution from foreign invested companies in Vietnam such as Fujitsu (USD515 million) and Canon (USD450 million) were excluded, a modest number of 300 local companies remain. The majority of the local ICT enterprises of Vietnam are small and medium in size.
- In addition, since most of the exported hardware components are assembled from imported parts, the value actually gained in Vietnam is not large. With a series of government preferential policies on im-

port/export duties, production tax, discounted land and infrastructure lease, etc., the direct contribution to the national economy is not as high as expected. Indirect benefits such as the importation of post-modern technology and management capacity and the informationalization of socio-economic activities are arguably more important.

- Even though the capacity of the industry and its market is potentially large, the majority, 80% of the population, who live in rural areas are still outside of the wired world. The ICT industry in Vietnam needs to make great efforts to close the gaps with respect to telephone and Internet access and usage. Such deficits also offer the industry development opportunities which could prove to be very profitable.

5.5.7 Internet in Vietnam

According to Economist Intelligence Unit (T. Kelly & Minges, 2002), Vietnam was a late adopter of the Internet mainly because of government concern about the Internet's negative effects. Starting from November 19, 1997 when the Internet was officially introduced in Vietnam with a 64K connection in a pilot research project run by VARNET (Netnam then) in a limited research community, the number of Internet subscribers and users increased spectacularly from 2000. Until the end of the last century, the number of Internet subscribers was only 200,000 (Ministry of Information and Communications, 2007). Vietnam's Internet penetration level increased from 2.4% in 2003 to 17.7% by December 2006, slightly above the international average level (15.7%) (Vietnam Internet Center, 2007a). The total number of web sites with "vn" in the address rose from 2,907 in 2003 to 11,032 in 2005 and 39,477 in April 2007, including 502 sites with "gov.vn". The number of Vietnamese domain names by April 2007 is given in Table 5.6.

With an increase of 86% in the last two years, by May 2006 there were thirteen million Internet users or 16% of the population (83 million). At June 2007, about sixteen million people were users of the Internet, while another 70 million people conduct their lives without the use of computers (Vietnam Internet Center, 2007a). The growth of Internet usage in Vietnam over the 5 year period to 2005 is shown in Table 5.5. The leap from 7.5 million users in 2005 to 16 million in 2007 underlines a trend of rapid growth. At that point Vietnam had attained the world average of Internet usage per 100 habitants. In the same period, broadband Internet with ADSL increased quickly at 300% per year, reaching 227,000 subscribers.
However, Vietnam still lags far behind in the digital society. The Economist Intelligence Unit (EIU) ranked the country number 66th among 68 countries on "e-readiness" which measured six important indicators of each country's ability to promote business in a digital environment, namely connectivity and infrastructure, business environment, consumer and business adoption, legal policy environment, social and cultural environment and supporting e-services (Economist Intelligence Unit, 2007, p. 5). Similarly, IDC & World Times put the country at the bottom (53/53) of the information society index (Hochiminh Computer Association, 2006, p. 16). Vietnam was given a very low e-readiness rank because of its still poor technology infrastructure, low number of DSL connections, and a poor competitive environment that is "largely influenced by its own state-regulated telecoms industry" (Economist Intelligence Unit, 2007, pp. 7,15,16).

Table 5.6 Statistics of Vietnamese domain name by April 2007. Source: Vietnam Internet Center (2007c)

Type of Domain name	Quantity	%
Total	39477	100
com.vn	21833	55.30
gov.vn	502	1.27
int.vn	9	0.02
edu.vn	934	2.36
ac.vn	58	0.14
net.vn	762	1.93
org.vn	796	2.01
name.vn	48	0.12
pro.vn	54	0.13
biz.vn	264	0.66
Health.vn	19	0.04
info.vn	169	0.42
Secondary domain 2.vn	14029	35.53

Before 2000, solely played the monopoly of the Vietnam Posts and Telecom Corporation (VNPT) – a state owned enterprise (SOE) in the Internet and telecom markets. The Decree No. 55 in 2001 and the Ordinance of Posts and Telecommunication in 2002 opened up the markets to competition so that other companies now have the chance to provide telecommunications and Internet services, such as STP from Hochiminh city, Viettel from the Ministry of Defense, EVN – the Vietnam Electricity, and FPT, a joint-stock enterprise. Bandwidth connection to the world, mainly though Hongkong, Singapore and China, jumped from 64Kb in 1997 to 7Gb/s in 2006. Domestic connection in 2006 was 12Gb/s, mainly from VNIX of VNPT to other providers and the two cities Hanoi and Hochiminh, as shown in Figure 5.1 and Tables 5.7 and 5.8.

Table 5.7 Vietnam Internet connection capacity to the world in December 2006. Source: The author synthesizes from data of Vietnam Internet Center (2007a)

Unit: Mb/s

Country/ Vietnam Carrier	**Total**	**USA** (Fusion, VSAT)	**Korea**	**China**	**Hong Kong** (Hutchinson, Teleglobal Reach)	**Japan** (NTT)	**Singapore** (Singtel, Tsystem)
Total	7076	386	155	1085	3300	810	1340
VNPT	3875	310	155	620	930	620	1240
FPT	1705				1705		
STP	135					35	100
Viettel	1161	76		465	465	155	
EVN	200				200		

According to the online newspaper VietNamNet, in June 2006 VNPT and Viettel

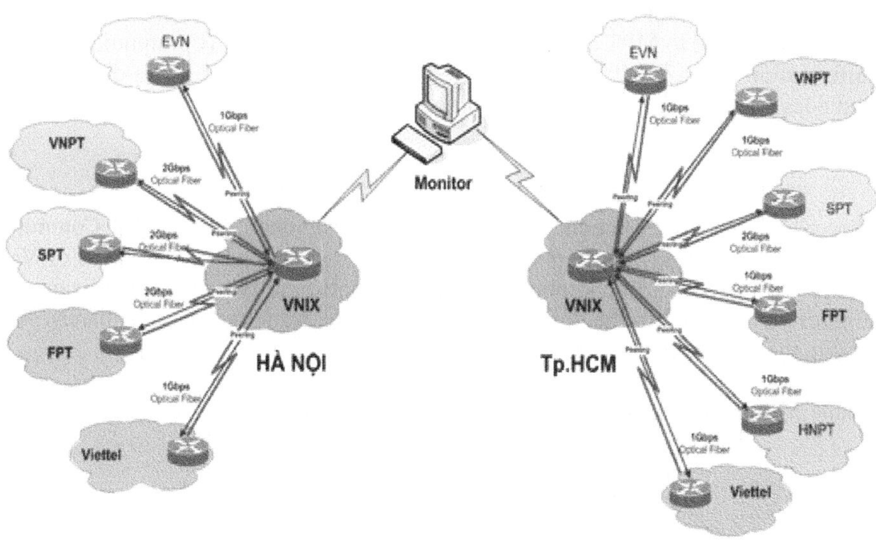

Figure 5.1 Domestic Vietnam national Internet exchange (VNIX) diagram. Source: Vietnam Internet Center (2006)

signed an agreement with Aiti (Brunei), CAT Telecom (Thái Lan), PLDT (Philippines), REACH (Hong Kong), StarHub (Singapore) and Telecom Malaysia (Malaysia) to start a USD780 million project to build a 20,000km submarine cable which will connect Malaysia to the US through Hongkong, Philippines, Guam and Hawaii main stations and four substations in Singapore, Thailand, Brunei and Vietnam. When the project is completed (scheduled for March 2008), Vietnam will have a 1.28Tb/s connection based on dense wave length division multiplexing (DWDM) technology. It will then be able to be a new regional hub for other countries, such as Lao, Cambodia, Thailand, Myanmar, Bangladesh (Hung, 2006).

5.5.8 Deregulation in the telecommunications sector

The deregulation of Vietnam's telecommunications sector has resulted in a fall in the costs of telecommunications and Internet access. In 10 years from 1995 to 2005, telecom and Internet tariffs have been reduced 9 times, and this trend has continued in recent years.

In 2001 the costs of international and domestic paid calls declined from between 10% and 15%. In 2004, the telephone tariff was reduced by 15% to 35%, which, together with other newly introduced services, resulted in a rapid increase in numbers of both telephone and Internet subscribers.

With the Prime Minister's Decision #217/2003/QĐ/TTg on Posts and Telecom Tariffs, the government recognized the right of business to determine and compete on prices. In 2004 telecom prices continued to go down as a result of competition between providers, Mobiphone, Vinaphone, Cityphone (Dinh, 2004). In 2005 at least two adjustments in telecom prices were made. The first on June 1 reduced 30-40% of the price of 1-2Mb line leasing and 40% of the price of IXPs and international telephone providers. The second adjustment on August 15 reduced telephone and dial-up costs for Internet connection from 4-20% (H. Hoang, 2005). In 2006 VNPT submitted to MPT a roadmap to reduce telecom tariffs (Communications and Information Magazine, 2006). The price of telecom services is going down rapidly but does not yet satisfy the needs of corporate users (see Chapter 11).

5.5.9 ICT Applications in businesses

According to a survey conducted by the Office of National Steering ICT Committee (K. M. Vu & Jones, 2006), investment in ICT accounts for less than 0.1% of Vietnam's total revenue. In addition, 70% of that small fund is spent on hardware, and the rest is allocated to the development of software, training and consultancy. As a result, the effectiveness of ICT use is not high.

Table 5.8 Market Share (%) of 6 Major Internet Providers in Vietnam. Source: Hochiminh Computer Association (2006, p. 21)

Rank in 2006	IXP	5/2003	5/2004	5/2005	5/2006
1	VNPT	66.29	58.83	46.72	43.13
2	FPT Tel	20.68	22.65	27.65	24.08
3	Viettel	1.09	2.53	9.68	18.61
4	EVNTel	N/A	N/A	N/A	5.65
5	SPT	3.49	7.15	7.06	4.46
6	NetNAM	6.32	7.04	6.67	3.08

Almost all enterprises have used simple office applications such as document composing, Excel, Powerpoint and e-mail communication. The most prevalent and broadly used software applications in enterprises are for financial management and accounting (about 88%). The other applications such as management have hardly been deployed (less than 8% of firms have such an application). In fact, not many enterprises have said that they are totally satisfied with their current software programs since they are designed for specific requirements without the opportunity of further being upgraded or integrated into a larger system. In regulation, the governmental fund for ICT expenditure allocated much for hardware procurement and reserved only a limited proportion (not exceeding 30%) to buy software, training and consultancy. With this ceiling policy, it was very hard for governmental agencies to buy commercialized software and qualified personnel to run the bought hardware (ILO/JIL & National Institute for Labour Studies, 2004).

ICT applications for commercial transactions consist mainly of publicizing and advertising enterprises and products on Websites, using e-mail, and Internet-based information searching. From about 1,000 websites in 2001, enterprises created 21,833 websites in 2006 (Vietnam Internet Center, 2006). However, online interactive transactions were very simple in the absence of online ordering or payment due to infrastructural difficulties and insufficient security measures like fire-walls, virus and malware protection, or public key infrastructure (N. C. Tran, 2005).

There is no doubt that significant achievements have been made in Vietnam in the area of ICT applications progressively from 2000 to 2005, but there are still problems with respect to slow application speeds and low quality of software products.

5.5.10 ICT applications in the government sector

In 1997 work was begun to establish a Vietnamese government intranet (CPNET), the first country wide network, connecting the head offices of provincial people's committees, ministries and the central government with the aim of providing and exchanging information. The first stages of the project have resulted in the establishment of IT departments in 102 ministries and provinces and 115 government bodies. At each governmental office (115 offices) there was a centre for integrating data with 5 servers supplying services such as e-mail, fire wall and 9 common software applications. This project has placed a high emphasis on hardware procurement and computer networks for all government agencies (K. M. Vu & Jones, 2006).

ICTs were introduced in the public sector to make improvements to internal management through the application and use of back-end public administration systems and databases. Computers were to be used in operational planning, the managerial tasks of governmental agencies in the areas of documentation, personnel management, budgeting, accounting, data storage, inventories and archives.

From 2000 to 2005, the Vietnamese Government has undertaken the Computer Computerization in State Administrative Management project, the SAMCom and locally often referred to as the "Project 112" , to build first foundation of e-government in Vietnam. The project was initially budgeted USD250 million. The World Bank agreed to a USD 94 million loan and the rest of the amount came from Asian Development Bank and the Government. Hardware procurement and the training of staff are major components of the project (K. M. Vu & Jones, 2006, p. 19).

Computer networks are a common application in public administration for exchanging information among administrative agencies. Most of the provinces have used software recommended by the Project 112, such as the MISA package for financial and accounting management. However, some provinces have complained that the software has not been adequately explained and that it has in many cases been found to be inappropriate for a given task. In such situations provinces may continue to use their own software or even develop new software rather than integrating and adopting the national standards developed under the Project (H. Nguyen, 2007).

A number of online applications have been initiated through over 40 pilot projects (K. M. Vu & Jones, 2006) that are to be fully implemented across government departments and agencies. Almost all of the staff at national government agencies now have e-mail, and there have been significant developments in the Vietnamese government's online presence, from 306 websites in 2005 to 502 in April 2007 (Vietnam Internet Center, 2007c).

The Vietnam Economic Review reported that different IT government groups hoped to launch their sites as the official government portal (such as www.gov.vn or www.egov.gov.vn by the Project 112, and www.vietnam.gov.vn or www.chinhphu.vn in the Office of Government). In the end, since 2006 www.chinhphu.vn has became the official portal of the Government which links to government and large business websites in the country. All 30 national and 64 provincial government websites have been developing their online presences to varying degrees of sophistication (H. Nguyen, 2007).

However, according to Vu and Jones, evaluation of these by end-users suggests that some of the website designs are unattractive and unfriendly, and that difficulties have been experienced in searching for needed information.

> E-Government initiatives have been far too focused on modernization and the purchase of hardware for government agencies. (...) All initiatives (Project 112 and PAR initiatives) are supply side focused with little thought to the actual needs of users, business and citizens, and how these needs can best be met... (K. M. Vu & Jones, 2006).

Information access is the other great issue of concern to managers. About 60% of government websites could be breached and controlled by foreign hackers (Government of Vietnam, 2004a). When experts at the Vietnam Security Network Centre (VSEC) studied security networks, they discovered that there are about 306 government websites that have been hacked, or have signs of being hacked. Regard-less of the exact numbers, the recent problems experienced by these government agencies raise a degree of concern about IT security in Vietnam. New policies on ICT procurement in the government sector and on ICT use in the government sector aimed at accelerating the diffusion of information in the public sector and provide more service to citizens and business (Government of Vietnam, 2007d).

Similar to NPIT2000, Project 112 was not regarded as cost effective. More seriously, it was banned in 2007 with the arrest of the project team leader - a vice minister of the Government Office, together with seven project team members due to fraudulence in project procurement (Anh Thư, 2007).

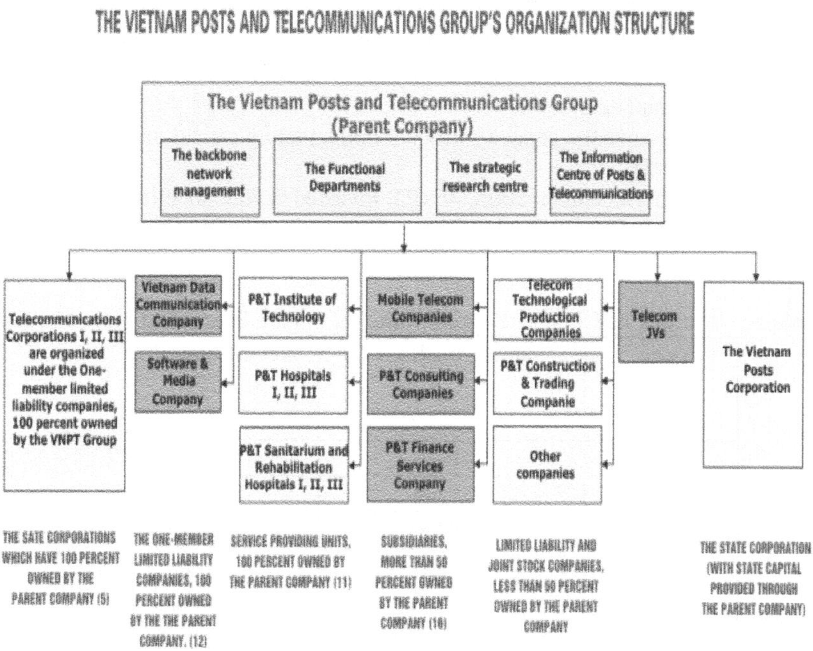

Figure 5.2 Organizational structure of VNPT at national level & its PTC. Source: VNPT (n.d.)

ICTs in rural and remote areas- the presence of Communal Postal and Cultural Points

"Điểm bưu điện văn hoá xã" - Communal Posts and Cultural Points (CPCP) are places intended to provide postal and communications services and limited library services to residents in communes where there are no Bưu cục- post offices. This hybrid economic and cultural model was introduced by the VNPT in 1998 with a subsidy from the government via the Vietnam Public Telephone Service Fund (Mai, Vu, Pham, Pham, & Nguyen, 2005) to serve both social welfare and business purposes.

Before 1998, the post and telecommunications (P&T) system in Vietnam was composed of three levels: P&T companies level one or provincial level (PTC1), P&T companies level two or district level (PTC2) and P&T companies level three or communal level (PTC3). Currently there are 3,000 PTCs for 9,019 communes or one PTC for 26,500 residents in an area of 110 km^2. The majority of PTCs were located in the centre of cities and towns whereas rural, mountainous and island areas were not covered with telecommunication services.

To narrow the digital gap between the rural and urban areas, VNPT initiated the CPCP model in order to:

- provide basic postal and communication services to residents in rural areas, especially in remote, island and mountainous areas, ethnic minorities and poor communes;
- satisfy the information needs of local government and people in the areas of economic development and the security needs of rural lands and borderlands;
- Stimulate the information needs to build a commodity based economy in rural areas where 80% people still live with the traditional self-reliant economy;
- Enhance the capacity of local human and physical resources (Mai et al., 2005).

From 1998 to December 2005, VNPT invested VND600 billion to build 8,075 CPCPs of which 7,495 CPCPs have been put into operation. The number of CPCPs built by year is given in Table 5.9 below.

Table 5.9 Number of CPCPs. Source: Mai, Vu, Pham, Pham, and Nguyen (2005)

Year	Newly built CPCP	Accumulated CPCP
1998	1643	1643
1999	1734	3377
2000	1200	4577
2001	692	5269
2002	705	5974
2003	619	6593
2004	718	7311
2005	764	8075

In a study of the technical skills needed for employees at CPCP in 2005, Mai et al. observed that, CPCPs operate as special dealers of VNPT under the Law of Trade and the Civil Code. The special condition is that unlike other P&T dealers, operators of CPCPs are given a concrete house and P&T equipment from VNPT. Each CPCP is operated by an employee who is usually a local resident and who works under contract with district post offices. Most (92.4%) of the CPCP employ-ees had graduated from secondary or high school. Half of employees received four weeks of training from VNPT (Mai et al., 2005).

By contract, employees are committed to manage the given facilities and equipment to provide postal, communication and cultural services, such as library services. They work in part without payment, for example when providing library services, but they do receive commission from their postal and communication services and compensation for maintaining the CPCP facility.

IT Human resource development and education

In 2004, the government adopted Decision 331/QD-TTg to approve a program on IT human resource development to 2010 as part of a series of measures to increase the number and quality of IT personnel. IT training in universities was encouraged to be delivered in the English language. The Ministry of Posts and Telematics (now called the Ministry of Information and Communications) was assigned to be in charge of setting up plans on IT human development.

A new law on education made it possible to set up private, non-state-owned schools. These policies are prerequisites to the achievement of change in human resource development in IT in 2006-2010. The Vietnam ICT Outlook 2006 warned however that: "these changes are actually slow and do not meet the demand. Filling the gap between human demand and the capacity of the education and training system is and will be imperative".

Two major sources of training in the country are formal training in universities and informal training in ICT companies. The accredited IT training and education system consists of 80 universities which enrol 10,000 new students who are able to graduate with bachelor's degree after 4 to 5 years. Besides formal education, there is an expanding network of "non-accredited" training of equivalent IT diplomas in cooperation with international training institutions and private universities. Private universities and companies have joined with foreign training institutions to provide informal IT training up to international diploma standard, for instance, FPT Aptech, a join venture between Vietnamese FPT Software Co. and Indian Aptech Worldwide, KOVIT between Korean Woosong University and Economics University in Hochiminh city .

Table 5.10 Number of IT training institutions in 2002-2006. Source: Hochiminh Computer Association (2006)

Year	University	College	Diploma (by foreign institutions)
2002	55	69	35
2003	57	72	40
2004	62	74	45
2005	70	85	53
2006	80	103	60

Some new universities and colleges specializing in IT were established recently, including the Hochiminh ICT University (upgraded from an IT college), IT University (upgraded from the IT Promotion Center of the National University of Hochiminh city), the private FPT University and the Vietnam-Korea IT College.

Vocational colleges also offer Cao-dang or practical engineer, a three-year training program in IT. Therefore, each year, about 20,000 new IT graduates have been produced from domestic universities and schools.

The growth in the number of IT training institutions can be seen in Table 5.10 above.

5.6 Main insights from Chapter 5

Insight 5A: There is a significant background and momentum in policy development (notably Vietnam Agenda 21 – VA21, and Politburo Directive No. 58/CT – TW") both for sustainable economic growth and the uptake of ICTs. However the planning horizon for the latter has been relatively short term, and the two areas of policy are in need of closer integration.

Insight 5B: Such policies recognize the severe threats posed to Vietnam by water shortage and pollution; environmental toxicity affecting lands and forests; climate change and similar problems.

Insight 5C: There is recognition that the meeting of human needs, and the development of human capacities to deal appropriately with problems or opportunities, are the prime concerns for both environmental and ICT policy. However current efforts at education, training and on-going support for human capacity development may be insufficient, both in scope and scale, to match the urgency of achieving sustainable economic development.

Chapter 6

Vietnam and the knowledge economy

6.1 Scope of the chapter

Drawing on literature, contributions at the 2000's Conference on the Knowledge Economy and the author's experience as a Vietnamese government officer in the ICT field, the chapter:

- Presents an overview of the role of knowledge in Vietnam's development since the Independence Day and through the 20th century.
- Gives an account of the Vietnam-Canada Information Technology Project (VCIT) and its influence on Vietnam's policy development in regard to ICTs and the knowledge economy.
- Analyses contributions by leading academics, professional experts, managers, political figures at the "2000's Conference on the Knowledge Economy", a milestone event in the development of Vietnam's ICT and Knowledge Economy policies and strategies.
- Canvasses a spectrum of viewpoints expressed at the 2000's Conference, and elsewhere ranging from those supportive of, to those skeptical about, the concept and potential value of Knowledge Economy strategy for Vietnam.

6.2 Vietnam's Knowledge Economy (KE) journey

This chapter summarizes a story about the journey of the concept of knowledge economy (KE) to Vietnam, from the first day of the nation's foundation 60 years ago, when Ho Chi Minh mobilized intellectuals by a call for patriotism, to the first years of the new millennium, when patriotism must operate synergistically with the free market in ruling the economy as well as the society.

The purpose of the chapter is an account of the journey of the Vietnamese state and the author personally in seeking the best ways to formulate knowledge and ICT policies, from import and adaptation of foreign ICT and KE policies in the mid 1990s to endogenous construction of such policies based on the needs and demands of the individuals and businesses in Vietnam; from technocrat belief in the power of new technology; and from critical research about the nature of development and the role of knowledge and ICTs. The chapter reviews how a foreign concept like KE was interpreted, analysed, enhanced, extended and localized within Vietnam, a developing country, so that it could be harnessed for the prosperity of the nation.

This chapter includes two main sections. The first section presents the position of knowledge in Vietnam with a focus on the author's own experience in the Vietnam–Canada Information Technology Project (VCIT) and ICT formulation in Vietnam. The second and main section summarizes the controversial debate on the theme knowledge economy and Vietnam in an important conference held by the government of Vietnam in 2000 to shape the Directive 58 on the use and development of IT to support the country's industrialization and modernization goals, the root of all ICT policies and legal documents as mentioned in chapter 5.

6.3 Vietnam and the knowledge economy

6.3.1 The role of knowledge in Vietnam before 2000

Sixty years ago Ho Chi Minh, the founder of Vietnam, paid special attention to the role of science, technology and the intelligentsia for the construction of the new independent country. In 1945 when the country won its national identity after 80 years of French domination, he appealed to Vietnamese scholars both in Vietnam and overseas to rebuild the country. Ho Chi Minh argued:

> To safeguard the independence, to build a wealthy people, a strong nation, all Vietnamese persons must have new knowledge in order to build the state (V. D. Nguyen, 2000, p. 142).

In response to his call, a wave of scholars from many parts of the world returned homeland to join the Northernmost Vietnam base and became known as the first generation of intellectuals and knowledge promoters of Vietnam, such as Pham Quang Le, Đặng Văn Ngữ and Lương Định Của.

Pham Quang Le - a military engineer working in France became Prof Trần Đại Nghĩa, the first director of the Military Arms Agency of Vietnam who was responsible for the creation of many important inventions under conditions of extreme hardship in the North-West forests which led to the victory of the Vietnamese Army. His inventions included the tripod antitank suicide bomb, torpedoes and the antitank guns DKZ and SKZ. Prof Đặng Văn Ngữ came back from Japan to build the Anti-Malaria Institute and nurture the next generation of Vietnamese parapsychologists; also Prof Lương Định Của from Japan who later was famous among farmers with his development of rice, jujube and pineapple species.

Former scholars from French colonial regimes were assigned important positions, such as the lawyer Phan Anh, who became the first Minister of Defense and Doctors Hồ Đắc Di and Tôn Thất Tùng who took the position of director of the Hanoi University of Medicine.

Intellectuals such as these, with their far-reaching vision and knowledge constructed a system of education which nurtured consequent generations of intelligentsia for Vietnam. Prof Vu Trieu An, an intellectual of that day, in a live interview on Hanoi Television HTV on September 2, 2005 to commemorate the 60th National Day said that in the times of war their "outstanding ideology fought against foreign aggression for the nation's independence".

Now that Vietnam has entered the 21st century, an important goal of country is to become a socialist oriented market economy.

What will be the role of science and technology, or in broader terms, knowledge for the development of the country? Le Kha Phieu, a Party General Secretary claimed:

> The modern science and technology revolution, the boom of hi-technologies, such as information technology and bio-technology, is creating an unprecedented qualitative change in production force, moving the human beings into a transition to a new civilization, the intellectual civilization.... In the information revolution, we have no choice but have to quickly access to the latest knowledge and technology to modernize the economy, step by step transfer the economy structure toward a knowledge economy with a higher competitive capacity and added value (K. P. Le, 2000, p. 25).

According to Tran Viet Phuong, a former PM adviser, the notion of a knowledge economy is not well understood in Vietnam, with little research taking place on the subject, and very limited availability of KE-related materials (V. P. Tran, 2000, p. 81).

6.3.2 The VCIT project and ICT formulation in Vietnam

After the reunification of the country, the Government Council (now the Government) set up a General Department of Electronics and Informatics in 1976, but it was soon dismissed due to internal conflict. In 1993, the Government issued the resolution No. 49/CP on information technology development in Vietnam in the 1990s with the goal of "leapfrogging" Vietnam's use of IT from very little to the level of a global contemporary standard by the year 2020.

Together with the establishment of the Steering Committee of National Program on Information Technology (SC-NPIT), a Secretariat referred to as the Office of the Steering Committee (OSC) gathered together a team of enthusiastic young staff. In 1996, the Vietnam-Canada Information Technology Project (VCIT) funded by the Canadian International Development Agency (CIDA) began to support the implementation of NPIT. The Global Alliance for Infostructure Advancement (GAIA), a consortium made up of NGL Nordicity, Branham, Hickling and Second Step Solutions, was contracted to be the Canadian Executing Agency (CEA) of the project and the Vietnamese Executing Agency (VEA), headed by the OSC Director was set up as its counterpart (UniCorn, 2000).

At the time, the author was working on the National Support Health Project at the Ministry of Health which was funded by the World Bank and was invited by SC-NPIT to be a Vice Director of the VEA and Vice Director of International Relations of OSC. Within GAIA were Canadian leading IT experts such as Art Caston, co-author of the best seller book "Paradigm shift: the new promise of information technology" (Tapscott & Caston, 1993) or Garth Graham, a former minister of Alberta, Canada.

To accelerate the introduction of ICT in Vietnam, the VEA requested CIDA and GAIA to change the initial purpose of the project from being "to strengthen the institutional capacity of the Government of Vietnam to plan, co-ordinate and manage the implementation of the NPIT" to "to strengthen the capacity of the Government of Vietnam to develop, manage and implement the National Program on IT". The basis for this request was that it was felt that the project focus was not to build a system of ICT management which had already been established (i.e. SC-NPIT, OSC), but rather to help Vietnam to implement the NPIT. The assistance required was not only advice and consultancy but rather additionally equipment and other practical forms of support to implement ICT pilot subprojects. The change was accepted by CIDA in order to:

> ... support Vietnam's reform process by strengthening its public sector capacity to develop and implement sound, equitable and environmentally sustainable economic and social policies (UniCorn, 2000, p. 11).

With a new logical framework analysis, the project was restructured into five strategic clusters: IT training, IT industry development, IT application pilot subprojects, IT policy development, IT management capacity strengthening. Gender consideration was usually taken into consideration in all VCIT activities, e.g. training, industry as a cross-cutting theme.

The main focus of the project was policy. The final report of VCIT remarked: "VCIT was one of the first IT Policy projects of this size and magnitude anywhere in the world" (GAIA Corporation, 2001, p. 5). Strategic policies and plans were drafted with VCIT support, including issues such as the development of IT standards and the localization of international IT standards, program design inputs to stimulate IT industry development, plans for IT applications and development, and specific programs and strategies to help increase the participation of women in the IT sector.

With ICTs in Vietnam at an undeveloped stage, and the great confidence of Canadians and international participants in the value of IT as a vital enabler for development, the Vietnamese program and project participants were swept along with enthusiasm, and possible negative related impacts were not considered in the early stages of the programs.

After the disbandment of SC-NPIT and OSC in 1999, though ongoing subcomponents of NPIT were still conducted by separate ministries and provinces, the coordination and management at national level (formally by Ministry of Science, Technology and Environment) was disrupted.

The VCIT project of which the author was an active participant was the only program left to maintain the momentum of NPIT at the national and strategic level. In 1999 and 2000, the VCIT project actively focused on a series of important issues to overcome, and even improve the situation. These efforts later had broad impacts on the ICT legal environment of Vietnam. For instance, the VCIT project:

- with the Government Committee on Organization and Personnel, studied the role of the IT professions in the reform of public administration and a new management mechanism integrating the information sector with the communication sector. This study later resulted in the establishment of the Ministry of Posts and Telematics in 2002 and the Ministry of Information and Communications in 2007;
- with the Department General for Posts and Telecommunications, studied and developed the National Information Infrastructure Proposal which later led to the 2005 Plan of Telecommunication and Internet Development in Vietnam to 2010;
- with the Ministry of Education and Training, studied IT in the Education Master Plan which would be the draft version of the 2004 Prime Minister's Decision 331/QD-TTg on the program of IT human resource development to 2010;

- with the Ministry of Trade, studied and proposed an E-Commerce Framework which finally became the 2005 Law on Electronic Transaction;
- with the Ministry of Finance, studied the policy of IT procurement in the government sector which finally became the 2006 Regulation of IT Procurement by government budgets in agencies and organizations;
- with the Ministry of Information and Culture, studied copyright protection regulation which later become the 2007 Prime Minister Directive on strengthening the protection of software copyright;
- with the Ministry of Science, Technology and Environment and related agencies, drafted a new IT master plan which became the Information Technology Master Plan, 2001-2005 and the Government's Decree on Development of the Software Industry, 2000-2005.

One of the most important activities of the VCIT project was the support to the Party Commission on Science and Education to develop the IT 2020 Vision for Vietnam. The work started in March 1998 when VCIT reviewed the draft IT 2020 Vision prepared by the Commission's task-force and facilitated a number of meetings and discussions with members of that secretariat. The VCIT project provided members of the secretariat, SC-NPIT and OSC/VEA with comments and recommendations on the IT 2020 Vision and on developing and implementing a strategy (GAIA Corporation, 2001, p. 70). The author was assigned to draft two sections: the outline of the IT master plan 2001-2005 and a conception proposal on IT law for Vietnam.

The height of the VCIT project, as well as a turning point in the history of ICTs in Vietnam was the sponsoring of a Conference on Knowledge Economy and its Issues to Vietnam held in June 2000 in Hanoi.

6.3.3 Formulation of ICT policies in Vietnam: from importation to self initiation

From time and experience with the VCIT project, NPIT 2000 and recent ICT activities, ICT policy as an input to development of Vietnam, and the author's experience in ICT policy development, have evolved significantly. Policy development has shifted from adaptation to self initiation; and from ICTs as an ends in themselves to ICTs to serve strategic national purposes. For the author, policy involvement has "zoomed in" from macro conceptualizations to more specific key domains: for example, projects on IT law and IT master planning.

At the beginning of the VCIT project in 1996, when planning for visits to Hanoi of Canadian policy experts, the Vietnamese partners in VCIT often advised:

> They must just tell us their experience and we'll adapt it to Vietnam. The experience comes from outside. The initiative comes from inside (GAIA Corporation, 2001, p. 3).

Most of the ICT policies and regulations, such as those on intellectual property, software copyright or IT law were imported into Vietnam, then adapted. The 2006 Law of Information Technology was based on the cyber law of Malaysia and the IT law of South Korea. The author collected documents on this legislation in 1999, then proposed similar legislation to the Government during the 2000-2005 period.

The VCIT Project supported the Government of Vietnam's capacity to understand what happens to the society and economy when IT is deliberately used to change things. However, ways of thinking about the future of Vietnam were changing. The goal of becoming industrialized and modernized was being qualified by the idea of becoming a "knowledge society". The Government of Vietnam needed to make difficult choices about how to spend its money wisely. It needed to know what are good and bad public policies in using ICT for development. But what guidelines and goals were now needed by the Government for the shaping of ICT policy?

At the end of the VCIT Project, some key persons of the project management team realized that the importing and adapting of foreign ICT policies was no longer in the interests of Vietnam. The author of this book asked Garth Graham, the Project director, whether the establishment of a "knowledge economy" could secure a developing country's necessary level of sustainability, or whether such a goal was rather a new form of exploitation by developed nations. Graham answered that with knowledge, people are enabled to create things to substitute conventional materials, thus, reducing their dependence on the limited availability of natural resources. This debate emerged in a research forum on the uses of ICTs for development which raised the questions about the development of a knowledge society for Vietnam as below:

> In what sense is a knowledge society different from other forms of society?
>
> What might slow down or block Vietnam's transition to a knowledge society?
>
> As the context of ICT use for development changes both inside and outside Vietnam, just what will the capacity to create and apply information technology policy actually require?
>
> How does Vietnam in turn assist the learning of external agencies that are facilitating local action to shape the environment in which technologies are applied? (GAIA Corporation, 2001, p. 8)

These questions, formulated in late 2000, are key issues in this book.

6.4 The "2000's Conference on Knowledge Economy and its Issues to Vietnam": consideration of the concept of a "knowledge economy" by Vietnamese scholars

6.4.1 Background: the procedures for formulating legal documents

This section explains the process of formulation of ICT policies in Vietnam and the position of the 2000's conference in relation to the Directive 58.

According to the 1996 Law on Promulgation of Legal Documents (National Assembly of Vietnam, 1996) legal documents in Vietnam include the Constitution, laws and resolutions promulgated by the National Assembly, orders and decisions of the country's President, resolutions and decrees of the Government, decisions and directives of the Prime Minister, and other legal documents issued by ministries. Texts issued by the Communist Party of Vietnam are not legal documents. Yet according to Article 4 of Constitute of Vietnam in 1992 (National Assembly of Vietnam, 1992), as the Party is the leader of the Government and the country, these documents, for example Directive 58 provide basic principles on the basis of which the Government develops its own strategies and plans.

To issue a legal document, based on the legal development agenda of the National Assembly and the requirement of management of the related issue, the correspondent agency (e.g. the Government) sets up a drafting taskforce; the taskforce will review existing legal documents of Vietnam and the experience of other countries and drafts a proposal. The proposal is sent to related agencies, organizations and individuals whose representatives can be gathered in workshops or conference to provide direct comments and discussion. The revision is submitted to the signatory agency (Prime Minister and the Government, the Politburo members, for instance) to have a final discussion and vote. If no revision is needed, the document is officially issued (National Assembly of Vietnam, 1996).

6.4.2 Key Conference themes debated in reports and discussions

The "Conference on Knowledge Economy and its Issues to Vietnam" in 2000 was held to gain inputs from managers, policy makers and intellectuals of relevant fields on the subject of the Conference which laid the foundation for the issuance of Directive 58 by the Political Bureau of the Communist Party. Organized by the Party's Commission on Science and Education, Ministry of Science, Technology and Environment and Ministry of Foreign Affairs on 21-2 June 2000, this conference, with the participation of 153 members who were Party, National Assembly and Government leaders and leading scientists of Vietnam, focused on the following themes, as summarized by Le Xuan Tung, Chairman of the Party's Central Commission of Ideology and Culture:

> The concept, content, and trend of KE.
>
> Possible changes in thinking and awareness to catch up with the trend: to adapt and create a KE for Vietnam.
>
> The national vision, strategy in relation to KE.
>
> Difficulties, opportunities and challenges of KE to Vietnam.
>
> Will science, technology, hi-technology play an important role in the country development?
>
> Will the investment on human resource development, in which education and good treatment of talented play a key role for KE?
>
> Where would be the roles of the government? (X. T. Le, 2000)

The rest of this chapter focuses on these key themes of the Conference and the debates and discussions related to them. Much of the discussion below has been taken from the Proceedings of the Conference compiled by Tran Minh Tien and Ho Ngoc Luat (M. T. Tran & Ho, 2000b). The discussion, except where specified as quotation (translated by the author), is summarized here by the book author.

6.4.3 Vietnamese perception of a knowledge economy

Professor Phan Dinh Dieu, former standing Vice Chairman of SC-NPIT claimed that knowledge can be defined by its scope and the kind of human understanding required to understand it. Similar to scientific and technological knowledge, human's knowledge of society, organization and management becomes richer and more diversified and plays a more important role, sometimes a decisive role in the wealth creation of an economy. He said,

> Thus knowledge of knowing 'what' is information, knowing 'why' is scientific knowledge, knowing 'how' is technology, knowing 'who' and 'with whom' is social knowledge, and knowing where' and 'when' is the know-how of business and commerce. (Phan, 2000)

Several Vietnamese scholars adduced that Marx had predicted a knowledge economy in his scheme of historical evolution. Tran Dinh Thien (D. T. Tran, 2000), for example, claimed in Marx's "Economic Manuscripts 1857-1861" (in a Vietnamese version of K. Marx and F. Engels, "Collected Works", volume 46, Part I, the Political Publishing, Hanoi, 1998, pp. 166-7) that 3 forms of economies have evolved in history, as shown in Table 6.1.

Table 6.1 Scheme of Three Economic Forms. Source: D.T. Tran (2000)

	1st Form	**2nd Form**	**3rd Form**
Position in the scheme	Natural economy, equivalent to agrarian economy and traditional peasant society	Market mode, equivalent to market economy of mechanic industrial age and industrial society	Communism mode, equivalent to a high developed economy (KE can be an expression) and a real freedom society
Mode of existence	Self-reliant, self-contained in separated communities	Division and exchange is limited by local and national borders	Division and exchange mainly by a borderless global network
Major advantages	Natural resources	Technical tools = materialized labor	Human knowledge (science – technology), skilled labor
Basis of power	Direct dependent on an individual or land ownership	Dependent on materials via capital (materialized labor)	Individual freedom upon their limitless and comprehensive development

Similar to Tran, Prof Dang Huu, Chairman of the Party Commission on Science and Education, classified three modes of economies by their production forces. They are 1) agrarian economy, or labor economy which is typified by manual labor, low productivity and land as main capital; 2) industrial economy or resource economy which is characterized by machine and natural resources, and 3) knowledge economy that mainly based on knowledge and information (H. Dang, 2000).

Originating from Marx's writings on the role of science as a direct production factor, Dang argued that an economy becomes a knowledge economy when science becomes the direct and most important production force. He defined knowledge economy as an economy in which

> the production, dissemination and use of knowledge play the most decisive role in economic development, wealth creation and quality of life. The major composition of KE is knowledge-based sectors which rely on the latest achievements of science and technology (H. Dang, 2000).

Dang Huu went further to infer that a KE may have various synonyms: "information economy", "network economy" and "digital economy" to focus more on the critical role of information technology for economic development; "learning economy" to focus on everybody's whole-life learning as the main driving force of the economy; "knowledge based economy", "knowledge driven economy", or "knowledge economy" to highlight the decisive role of knowledge and science in economic development, or sometimes simply "new economy".

6.4.4 Characteristics of a knowledge economy

For Dang Huu, a KE is a transition to a fast growing, quick changing and high risk economy. Mass production and standardization result from the flexible production of goods and services based on high technologies. Production through the use of technology is the most important, advanced and typical form of production. Businesses such as Microsoft, Yahoo and Dell become "knowledge business" and their workers became "knowledge workers". Due to investment in new technology as the top success factor, technology parks such as Silicon Valley are set up. Dang argued that as hi-tech production consumes less energy and power, and discharges less wastage; a KE is a clean and non-pollutant economy, a sustainable economy which would be expected to avoid the risk of regional and global economic crises that regularly overshadow international relationships. He said:

> A KE is an abundant not a scarce economy. Intellectual property rights are more important than the property rights over funding, natural resources and land. Those who possess more intellectual property would win (H. Dang, 2000).

A KE is a global economy. A product or a market is globalized: a product can be produced and sold anywhere, yet few products are produced in a single place. In this economy, enterprises facilitate development by both competing and cooperating with each other. The invention of a new technology creates new companies. On the one hand giant companies become global monopolists. On the other hand, big companies have to divide into smaller transnational companies. The change of corporate structure and relationship also spreads to the society in the form of democratization. Hierarchical and command management structures become non-hierarchical and decentralized, and network models that are more flexible, democratic, and adaptable can mobilize the creativity of all stakeholders.

Dang explained that society becomes a "learning society" in which human capital is the most critical element in the creating of knowledge because:

> Regular creativity and innovation is the most important force in development. There are many paradoxes in a KE, e.g. i) wealth is made from the unknown, the known is valueless; ii) The culture to find the unknown is the information network; iii) The discovery of an unknown negates the already known; iv) the higher the use value is, the cheaper the thing is, sometimes 'for free'; v) In the previous economic mode, the more scarce the thing was, the higher was its

> value. In a KE, products that are used more have a higher value win (H. Dang, 2000).

Prof Vu Dinh Cu, a Vice Chairman of the National Assembly, predicted that in contrast to dry technology, such as micro mecha-electronics, "wet technology" based on organic molecular could fill the "gap between humans and nature". For Vu, "wet technology" would be the basis for creating extreme-efficiency-and-pollution-free intelligent machines. Vu pointed out the differences of a knowledge product from other products as the dilemma of knowledge or intellectual products is that the cost of producing the knowledge is much higher than the cost of the container. Attempts to prevent hyper profitability due to monopolies and pirated copies are greater. In the case of conventional products, such as cars, the buyer has exclusive ownership, including possession, use and control. The buyer of a knowledge product has only the ownership rights to use the knowledge content as specified on the container (D. C. Vu, 2000). He said:

> The higher the level of knowledge and technology in a product, the higher the value-added rate. For example E. Sanchez and D. Myric's production of the movie 'Blair Witch Project' cost $500,000 but made a net profit of $M 140, which means 280 times of production cost (D. C. Vu, 2000).

The author would make a note here about Vu's example that often the value is more to do with its appeal to people than actual knowledge – many movies, songs etc. have been created quickly with little knowledge, but if they are popular and well distributed, the creators can make millions.

A number of Vietnamese scholars view hi-technologies in the areas of information, biology, materials and new energy as the major basis for the shift from the industrial age to the information age (T. B. Le, 2000; Ngo, 2000; Phan, 2000; M. T. Tran & Ho, 2000b; D. C. Vu, 2000). Drs Tran Minh Tien and Ho Ngoc Luat stated that information technology (IT) is a key element of a KE because of its extensive impacts on production, culture and social activities:

> The IT revolution is a critical factor to form a KE by increasing productivity, automation and exporting capacity, and reducing costs and stock of production. (…) The revolution is expected to open a new chance to backward countries to fill the gap with others. IT is a crucial foundation of the integration and globalization process. With networking capacity, the Internet is making the world smaller and bringing economic activities beyond nation/state borders. The geographical distance becomes shorter than ever. IT help to create new knowledge and 'added value' to conventional sectors. The information economy becomes the most dynamic economic sector (M. T. Tran & Ho, 2000a).

Tran Dinh Thien, however, did not agree that a KE comprises only high technology. He argued firstly that in reality, there has been no linear and consequential development but that there always exists an interference of 2 to 3 waves of production in a certain country and in the world in a certain history as Toffler pointed out (Toffler, 1980); secondly, a KE does not only contain the creation of knowledge but also the absorption, assimilation, utilization and dissemination of knowledge. Not every country has the same capacity to cope with those five latter tasks; thirdly, a KE does not include only hi-technology or the use of hi-skills or knowledge labor but also a process in which knowledge penetrates into and rules over all economic activities. The capacity to absorb, assimilate, use and disseminate knowledge is the most critical success factor in a KE. He thus argued that "the process of concretization of knowledge is also the major production process that is decisive to the development" (D. T. Tran, 2000).

It was also claimed in the conference that although globalization and the transition to a knowledge economy enlarges the divide between the rich countries and the poor as the former have more advantages in science and technology, funding and power, this process is not totally negative to developing countries. A KE opens new opportunities for developing countries to "take a short cut to wait in front" as new industrial countries have done in recent decades (D. H. Nguyen, 2000).

6.4.5 The role of intellectual property protection and ICTs

According to Dr Pham Dinh Chuong, Director of the National Office of Intellectual Property of Vietnam, intellectual property is defined as the product of intellectual creation, that can be stored or represented in a container, and that is transmissible and copy-able with the making of a copy not erasing the original source (D. C. Pham, 2000).

Dr Pham believed that IP becomes important in the context of a trading relationship, and that IP protection contrasts with the protectionism of primary products. On the one hand, free trade with a removal of protectionism; on the other hand, new protections of IP are strengthened with new sanctions. Dr Pham Dinh Chuong stated that IP protection is manipulated by developed countries and multinational corporations in many ways, i) using IPR to control creation and development in other countries; ii) putting pressure on less-developed countries and blocking their capacity to serve people's welfare. The threat of trading penalties in response to IP disputes reinforces the conclusion that IPR protection contains many risks and traps for developing countries (D. C. Pham, 2000).

The head of the IP authority of Vietnam accused developed countries focus more on the production and trading aspects of IP, increasing the intellectual content in products and services. The scope of IP protection is expanded and extended to new areas and possibly to the protection of non-traditional objects. For example, some decades ago, industrial protection covered only the patent and trademark systems, know-how and business secrets. Recently protection is extended to untraditional areas such as biotechnology (gene, micro organism) and to ICTs, such as IC design or software. Dr Pham argued that developed countries are modifying IP protection principles flexibly and extensively. To counteract, in developing countries there is a higher rate of illegal copying, imitating and counterfeiting technologies and products (D. C. Pham, 2000).

In an independent report to the National Assembly in 2001, Prof Nguyen Ngoc Tran, Vice Chairman of the Science, Technology and Environment Commission of the Assembly, argued that IPR are inappropriate; that unlike physical goods which are worn out by use, knowledge products, such as a law or an idea do not deteriorate, but are enriched by being used. In his view, the creation of the knowledge economy by capitalist countries on the one hand reduces their dependence on natural resources from developing countries and on the other hand, supplies them with profits from the knowledge goods and services that they produce. He suggested that through the creation of the knowledge economy, world capitalism has created a new form of exploitation, domination and colonization which was foreseen by Marx and Engels (N. T. Nguyen, 2001).

6.4.6 Proposed strategies to develop a KE in Vietnam

Principles to follow in developing a KE for Vietnam were proposed by Prof Nguyen Quang Thai, Vice Director of the Development Strategy Institute as below:

> Change ways of thinking. Although Vietnam issued numerable policies on science and technology promotion, such as the Law on Science and Technology, many policy makers and bureaucrats keep thinking that "second class" technologies transferred from developed countries are more relevant than high-technologies.
>
> Have relevant long and short-term plans. On the one hand, vision and long term planning is necessary to define targets and goals; on the other hand, short term plans are also very important to implement the visions and strategies.
>
> Select break-though areas, including i) Human resource development and skills training, ii) Technology infrastructure, with high bandwidth and low price Internet, iii) Market research to avoid some products becoming "surplus" after even just entering the market (such as basa fish, coffee) (Q. T. Nguyen, 2000).

Dang Huu recommended other principles as follows:

> Reform management operations to bring into play as much creativity as possible in the people through a healthy legal environment which is able to liberate production forces. The Government needs to govern by law and should not interfere with business activities. The governmental administration should be reformed and a new enabling legal framework.
>
> Develop human resources, increase the intellectual standards of the people, reform the education system, and reward talent.
>
> Improve the capacity of national science and technology (H. Dang, 2000).

In relation to building a learning society, Tran argued that the investment in human resource development through education and training is one of the essential strategies:

> We must change our view, from regarding education and training to serve social welfare as its major function to viewing it as an important investment, even most important investment (in the development of a KE – added by the author) (D. T. Tran, 2000, p. 111).

Similar to Tran, Prof Phan Dinh Dieu argued that human beings are the crucial capital for a KE, therefore, it is essential to promote an advanced and healthy system of education and training, a learning society where whoever wants to learn can learn. He said:

> Improvements to the living standards and hence the professional capacities of lecturers. At the moment their salaries are too low to live on.
>
> Reform of the examination system and of selection and recruitment to overcome problems such as learning only for the sake of obtaining degrees, not for real knowledge, i.e. fake degrees.
>
> Review and restructure of the curricula and training materials to be more appropriate to each level of education.

> Acceleration of the use of the Internet and IT in education and training.
>
> To do the four above tasks, the most critical thing is to mobilize funds and investments from all sources in the society for education and training (Phan, 2000).

6.4.7 Skeptical views: the knowledge economy as a new mode of domination

Some of the Vietnamese speakers at the Conference were mistrustful and skeptical about the KE concept. It was felt that in the context of KEs some countries have a superior position to the rest of the world and might try to manipulate the world economy. Dr Nguyen Xuan Thang, Vice Director of the World Economy Institute, for example, argued that

> The US, taking unfair advantage of being the leading producer in IT, wanted to attain an 'Agreement on Information Technology' in March 1997 at Geneva to remove all customer duties by 2000 of 200 IT products, ranging from software, hardware, telephone, semi-conductors and promote the free-trade in e-commerce (X. T. Nguyen, 2000).

There exists a pessimistic view about the utopia of a global homogenous community where all nations have the same level of development (Q. A. Nguyen, 2000). Prof Dang Ngoc Dinh noted that the gap between the rich and the poor is also a gap in knowledge and the capacity to create knowledge (N. D. Dang, 2000). Prof Nguyen Quang Thai appeared skeptical about the view of a borderless world with virtual governments and the end of territorial sovereignty. He argued that the role of the state is always important to guide and organize the society and protect the country from new exploitation by KE (Q. T. Nguyen, 2000).

Prof Dang Huu suggested that ICT, especially the Internet could be new challenges to national and local cultures. He said:

> cultural exchange enabled by the Internet, on the one hand expands access to dominant and diverse cultures, but on the other hand may adversely affect national cultures (H. Dang, 2000).

Dr Vo Tri Thanh, in his final comments at the Conference, suggested that Vietnam needs to study KE very thoroughly from the viewpoint of economics because in terms of the science of economics, there is no adequate and relevant theory of KE. He argued

> At the moment, we cannot measure knowledge, but have to use other indirect indicators, such as telephone per capita, Internet per capita. Secondly, knowledge which is both a private and public good is not well studied in economics (M. T. Tran & Ho, 2000b, pp. 237-240).

Vo Tri Thanh believed that development and wealth can only be achieved on the basis of scarcity and value allocated to resources:

> Some speakers had claimed that a KE is abundant and 'give for free' but from an economist's eye, no economy in the world can exist and develop if it takes a scarce goods (knowledge, material) for free. In reality, it is very costly to produce knowledge (M. T. Tran & Ho, 2000b, p. 240).

In a knowledge economy, the exchange is asymmetric, thus the government needs to take action requiring talented leadership as one of the most important success factors.

Dr Nguyen Xuan Thang reminded the Conference that recommendations in relation to KE today were exactly what had been said 25 years ago - that education, the fostering of talent, and the development of science and technology were keys to success, yet Vietnam was still one of the poorest countries in the world (M. T. Tran & Ho, 2000b, p. 240) .

Prof Vo Dai Loi and Dr Nguyen Quang A were skeptical about the recommendations concerning IT, biotechnology and new materials in facilitating the development of a KE. They believed that technology by itself is not the answer – politics, economics and sociology are all very important as well (M. T. Tran & Ho, 2000b, p. 240).

Prof Hoang Tuy argued at the Conference that Vietnamese people are lacking in imagination. He said that contrary to the prevailing myth that Vietnamese people seem to be ready to build a KE, with their cleverness, studiousness, and qualifications based on hard work, conceptions about study and education in Vietnam have encouraged more emphasis on cramming for examinations than on creative thinking or imagination. He criticised the extent to which the Vietnamese copy ideas from outside the country rather than inventing their own, and believed that the Vietnamese people lacked characteristics such as a "broad business mind", "farsight and vision", "boldness", "aggressiveness" and "determination" to create the synergies required for participation in the global economy (T. Hoang, 2000).

In relation to the role of individuals in a KE, Dr Nguyen Quang A argued that for a long time in Vietnamese society the collective has been of greater importance than the individual; individuals have had to put the society as a whole ahead of their needs. The strong opposition to individualism had constrained the individual's ability to freely develop, think and create. He felt that the treatment of people was based too much on the attainment of degrees and relationships with the authorities (Q. A. Nguyen, 2000).

6.5 Main insights from Chapter 6

Insight 6A: There has been a continuity of effort to envision an appropriate framework for KE policy and strategy in Vietnam, starting with reliance on imported models in the 20th century then evolving to more endogenous models in the 21st. An understanding of this trajectory is illuminating for understanding current and emergent KE policy settings.

Insight 6B: At the milestone "2000's Conference on Knowledge Economy and its Issues to Vietnam" education and human capacity building was seen as an issue of vital importance to the welfare of individual citizens, and as a cornerstone of a KE for Vietnam.

Insight 6C: The diverse viewpoints of Vietnamese scholars and experts, both supportive and skeptical, about the potentials of a KE are of value in developing real-world KE policy. The critical voices of KE skeptics are especially valuable in risk assessment and risk management in the formulation and implementation of KE policy.

Chapter 7
Identifying key issues

7.1 Scope of the chapter

- Using data collected during fieldwork interviews, selectively supplemented by data from the literature, this chapter identifies and outlines key issues relating to sustainable economic development, ICTs and the knowledge economy that were "on the minds" of interviewees.
- From hundred of hours of interview transcripts, analysis revealed the major themes that are presented in the chapter:
    - Pressures on the domestic economy of Vietnam
    - Pressures of globalization
    - The capacity of government to cope with these pressures
    - Environmental depletion and exhaustion
    - Potential contributions of a knowledge economy, including examples of success.
- The issue of social human factors, including education and the development of human capacity, runs as a cross-cutting issue across all these themes.

7.2 Fieldwork for this chapter

Data for this chapter was derived mainly from Round 1 of the fieldwork, consisting of 70 interviews with senior personnel from the government, education and research sectors (coded by Sxx - see Appendix 2 for names of organisations and/or titles of the interviewees) on the subject of sustainable economic development in Vietnam and the possibility of leveraging knowledge and ICTs to improve sustainability. Some selected quotes of interviewees from the second round (coded by Ixx – see Appendix 4) have also been included.

Of the 70 interviewees of the first fieldwork round:

- Twenty nine (41.4%) interviewees can be categorised as "eminent scholars and researchers".
- Twenty three (32.9%) interviewees can be categorised as "policy-makers and managers".
- Eight (11.4%) can be categorised as representatives of non-government organizations.

See more details about the interviews in the three rounds in Chapter 1 and also Appendices 2 and 4.

In the interviews and from relevant literature the author gathered views about different areas of concern in relation to sustainability for Vietnam, ranging from Vietnam's internal capacity to external risks for the country. Areas mentioned by the respondents in the 70 interviews (respondents generally nominated more than one area of concern) have been grouped in themes as follows: 1) domestic market economy pressures (mentioned by 23 (31%) of the respondents) 2) globalization pressures, mentioned by 28 (40%) of the respondents 3) government capacity, mentioned by 32 (46%) of the respondents; 4) environmental depletion and exhaustion, mentioned by 37 (53%) of the respondents 5) the concept of a knowledge economy and its impacts, mentioned by 53 (76%) of the respondents; and 6) the cross-cutting theme of social and human factors, mentioned by 31 (44%) of the respondents. Themes 1-5 are discussed in turn below, with comments on theme 6 interwoven in these discussions.

7.3 Domestic market economy pressures

As indicated above, 31% of interviewees saw domestic market economy pressures as an issue of high concern. Views expressed are outlined here.

7.3.1 The shift to a market economy

The fast revival of the market economy in the Doimoi paradigm has both positive and negative effects for development. In the short run, it is the main driver for development but in the long run, the free and anarchic development of the market economy can threaten sustainability.

From 1954 to 1975, a centrally planned economy replaced the market economy in the governance of the North of Vietnam. Under this system the government controlled the factors of production and made all decisions about the use and distribution of income in five year and annual plans.

After the reunification of Vietnam in 1975, the market economy was also suspended in the South of Vietnam. Farms were collectivized, and a centrally planned approach was applied across all sectors of economic activity. This phase ended in 1986.

From 1986 the Doimoi paradigm mandated that a market economy should be applied throughout Vietnam. Doimoi has recognised private ownership of farms and companies. It has encouraged deregulation and foreign investment. The free market economy has proved its dynamism and effectiveness in the growth of the economy in the last 2 decades, as shown in the section on economic growth.

All interviewees applauded the shift to the market economy, which has totally changed the face of the country. An expert in the area of Economics remarked:

> Over the past decade the economy quickly recovered from the 1997 global crisis and achieved a high GDP growth rate over 7% per year. The economic structure became more oriented towards the industrial and service sectors. The contribution of the agricultural sector to the economy has declined by 20%, while contributions of the industrial and service sectors have grown by 40% each. State ownership has been widely replaced by other forms of ownership. Thus there has been a steady decline in the percentage of the economy under

> state ownership, and a strong increase in private ownership, including foreign ownership (S32, interview, Nov 23, 2004).

Another senior academic said,

> in general, the economy of Vietnam is developing more quickly and with more stability with a proper preparation for the future. The key driver of growth is Vietnamese business (S32, interview, Nov 23, 2004).

S01, a person with extensive experience in Government, argued that the current economy of Vietnam was in fact a mixture of the emerging market economy and the remnant of the old centrally controlled, self-reliance regime. He believed that state-owned enterprises which heavily depended on government subsidy were a source of corruption, waste, loss and low efficiency, and believed that it was necessary to move even further away from a command economy where government decides what to do and there is no room for new creativity (S01, interview, Oct 28, 2004).

Despite the progress under Doimoi, many interviewees still saw the levels of efficiency and effectiveness of the economy of Vietnam as being too low. They felt that part of the problem is that the economy is mainly based on the physical resource sectors - such as agriculture, aquaculture, mineral and oil exploitation - with low profit rates.

7.3.2 The application of new knowledge

Productive performance is also affected by the limited extent to which knowledge, science and technology are applied in production and management, which makes the economy less competitive. Several researchers, for example S20 (interview, Nov 15, 2004) and S32 (interview, Nov 23, 2004) believed that shortcomings in the application of new knowledge to production and management were major reasons why the efficiency of investment has deteriorated. The incremental capital output ratio increased from 3.0 in 1990 to 5.1 in 2004 (i.e. $5.1 investment is needed to gain $1 marginal increment in GDP), a much less cost-effective picture than in other fast growing economies which have a ratio of about 2.5 to 3.

7.3.3 Urbanisation, commercialization and industrialization

The survival of traditional professions such as agrarian and handicraft enterprises in the countryside, which have existed for thousands of years, suddenly became vulnerable because of the quick growth of urbanization, commercialization and industrialization. Before 1990, urbanization appeared slowly in Vietnam, with about 500 large and small urban centers nationwide where residents were housed in tiny "tube houses", a Vietnamese term to describe a long but narrow house like a tube in cities and towns (V. T. Le, 2004). These tube houses are tiny but extremely expensive (S05, interview, Oct 21, 2004). After 1990 the number of urban centers has increased rapidly and expanded in alignment with positive changes in socio-economic conditions with 649 in 2000 and 656 in 2003 (Q. A. Le et al., 2004).

Concrete buildings and industrial zones have been erected on fertile rice fields along roads (S04, interviews, Oct 28 and Nov 10, 2004). These new urban centers or industrial zones have discharged a great amount of unprocessed civil and industrial wastes to land and rice fields, as attested by I37, a farmer in Haiphong (interview, Mar 7, 2006). Farmers, after getting some compensation and losing their inherited land, became redundant (S52, interview, Nov 17, 2004). The Vietnam Farmers Association estimated that each year, about 200,000ha of land are lost due to urbanization, the equivalent to 1.5 million peasants becoming jobless (Vietnam Farmer Association, 2005). In addition, the expansion of urbanization and new industries, such as chemicals or services, take up large areas of fertile land formerly used for rice cultivation, as I36 indicated during an interview:

> The area of cultivation in our farm reduces from 400ha to only about 100ha now: 200ha were taken to build a golf course and 50ha to build a chemical factory. In other words, our water surface for fishponds is taken away to make room for other business (interview, Mar 7, 2006).

7.3.4 Pollution of tourist sites

The free and anarchic development of the market economy has had negative impacts also in new sectors. Tourism, the green economy which is often regarded as a clean economy, also troubles the environment with unorganized and short-sighted growth. S29, a senior manager the Vietnam National Administration of Tourism, revealed that the recently awarded World Heritage site Phongnha-Kebang was polluted to an alarming level. No good plan of exploitation and protection was prepared and implemented so that waste has been discarded everywhere with impunity by both tourist providers and their customers. He said that Halong, another World Heritage site, is badly polluted, not by waste, but by oil discharged from tourist ships. He added that:

> ...this is the common situation of many other tourist sites, such as Babe Lake in the North, the Dailanh tourist zone in the Center and Phuquoc Island in the South. Local tourist authorities may be aware that long-term protection is essential to continued economic development, but this understanding is not shared by many investors or even by local fishermen who are now working as service staff in the tourist sector (S29, interview, Nov 18, 2004).

7.3.5 Focus on short-term economic profit

Excessive focus of effort and resource allocation to short-term economic profit, and ignorance of the market-run economy about the benefits of protection of the future makes unbalanced growth and divisiveness in society. The Vietnam Agenda 21 stated:

> Due to limited development resources, there are not enough physical conditions for implementation of sustainable development. Most of the investment has poured into the works serving immediate benefits, little remained for regeneration of natural resources and environment protection (...) Due to too much emphasis put on economic development especially GDP growth, little care has been paid to the natural system (...) (Government of Vietnam, 2004d)

7.4 Globalization pressures

As already indicated 40% of interviewees saw globalization pressures as an issue of high concern. Integration into the global market, characterized by the membership of Vietnam in regional and international treaties, such as ASEAN or WTO, on the one hand opens new doors for Vietnamese business and people to join the world, on the other hand, imposes new risks to them.

7.4.1 The illusion of fair and free trade

Globalization was seen as an indispensable trend required by the demands of international commerce and enabled by scientific and technological achievements like ICTs. However, it was the perception of these interviewees that rich and powerful countries take advantage of this trend to dominate the world market. A senior government official said that real equal and free trade has never existed, but the rich become richer by impoverishing the poor:

> Big countries use knowledge and power to dominate and delimit the world market, control globalization, international institution, create market barriers and exploit natural resources from inferior countries (S59, interview, Jan 6, 2006).

Or, in the words of another senior government person:

> The rich victimize and push the burden to the poor while reserving the natural and environmental resources for themselves. For example, the US keeps Alaska oil reserve and import oil from the Gulf. Japan closes its nuclear plants and advises countries like Vietnam to build new plants (interview, Nov 3, 2004).

A leading economist went further:

> Fair and free trade is a great cheating. The greatest unfair liar of equal and free trade is the US. It forces others to open their market and provide access to their natural resources, buying those extremely cheap while subsidising its own farmers to create a dirt-cheap ceiling price on agricultural goods. Less developed countries have no way but to follow its rule (S18, interview, Dec 22, 2004).

Many interviewees agreed that poor countries, though aware of the sustainability issue, have to sell their natural resources to meet today's development needs.

> Eventually, resources will be exhausted within these nations before being exhausted on the global scale. At a business level, enterprises, such as PetroVietnam, have to sacrifice long-term to gain short-term benefit, such as accepting a 50/50 share with foreign joint investors, while being aware that if the resource could be kept in Vietnamese ownership, its value would be greater in the next several decades (S50, interview, Dec 14, 2004).

7.4.2 Poor countries competing with each other

In addition, poor countries compete with each other in attracting investment, contracts and market share from overseas. For example if two developing countries, China and India, both want to gain an outsourcing contract from the US, China can offer \$80 if India offers \$100. As a consequence, poor countries offer themselves to be exploited by rich countries, driving each other to ever-greater comparative disadvantage. Since labor in Vietnam is even cheaper than in China, should Vietnam systematically try to undercut China, India and other developing countries in the world labor market? Or should Vietnam – as a senior expert and academic in technology pondered – try to avoid the cheap labor spiral and focus instead on a high skill, high value labor policy? (S02, interview, Oct 27, 2007)

One problem in moving along that path is that, in the global free market, Vietnam is a newcomer. With just 15 years of market economy experience, it is hard for the country to compete with 200-year old capitalist economies. An expert in socio-economics information suggested that Vietnam

> must improve many of its weak internal mechanisms and organizations, social and personal thinking to fit to the global game. Otherwise it would be hard to maintain sustained development and compete with others, and the gap would become larger (S20, interview, Nov 15, 2004).

7.4.3 Mechanisms and organizations

The weakness of mechanisms and organizations could be seen in the negotiation process with the WTO when Vietnam business was not informed about opportunities, threats and conditions involved in joining this treaty and had no input to the government on such issues (S52, interview, Nov 17, 2004). In fact, documentation about commitments to WTO was regarded as confidential and not made available to the public until November 5, 2006 after Vietnam had already became a member of the WTO. An anonymous article in VnExpress.net claimed that the absence of a business voice in negotiation and the late publication of the documentation would impose disadvantage and loss for Vietnamese businesses in trade with foreign competitors (K.D., 2006).

7.4.4 The role of international institutions like the United Nations

In the view of interviewees, international institutions like the United Nations, with admirable goals like peace, friendship, stability and development, and admirable programs such as poverty reduction, in reality have been of very limited benefit to poor countries in the last 50 years. The assistance is not considerable. As the interviewees saw it, half of the world is dying and the divide is becoming larger than ever. The World Bank, IMF and WTO were blamed. The World Bank was seen as too much influenced by the US and other rich countries, as S01 argued

> ... debates in Seattle and Cancun are examples of the struggle between the poor who realized they were losing too much under the international treaty against the rich (interview, Oct 28, 2004).

7.5 Government capacity

Forty-six percent of the 70 interviewees saw the capacity of government to deal with problems of sustainability as an issue of concern.

7.5.1 Slow progress of reform

In the public sector, the Administration Reform Program which started since 1986 has progressed very slowly. A member of the National Assembly remarked that the system of government, though it has been reorganized many times after each Party Congress, is still very cumbersome. "Red tape and corrupt practices tend to block the growth of the economy and society" (S07, interview, Oct 26, 2004).

7.5.2 Low levels of cooperation

In addition, low levels of cooperation among line ministries and government agencies makes the government a compartmentalized system (in Australian parlance 'silos') with low effectiveness in dealing with all-of-government problems like environmental pollution or crises. "Each arm of government tries to exploit natural resources for its own interest and ignores others' interest or the common interest" (S59, interview, Jan 6, 2005). "As a result, reaction of the government tends to be slow and ineffective, and it is often not clear which agency is responsible" (S41, interview, Dec 2, 2004).

7.5.3 Awareness of modern administrative approaches

S01 said that many government personnel are slow in changing their awareness of modern administrative approaches that anticipate the needs of the public and deal with long-term perspectives, rather than solving incidents (S01, interview, Oct 28, 2004).

7.5.4 Lack of clarity in the legal system

Other interviewees expressed concern about the lack of clarity of the legal system, commenting that its poor enforcement could neutralize the effectiveness of good policy.

For example there could be a great burden of losses in construction projects due to unclear urban planning in Hanoi and other cities, especially if coupled with an unsatisfactory compensation policy (interview, Nov 4, 2004).

7.5.5 Recycling policies

Regarding environment protection, a member of the National Assembly said that Vietnam still did not have adequate policy on the recycling of resources. He argued that Singapore was importing water and also recycling it. Even Thailand, a country with more water reserves than Vietnam, recycled water used in production processes. He argued that "Water, if not well managed, will be seriously scarce, and Vietnam, with 100 million people would not be able to import water like Singapore". He also said that sometimes Vietnam's policies were controversial, or even contradictory, for instance the government wanted to protect forests but at the same time encouraged people to export more wood or wood-derived products. "How can a forester keep closed forests and wood producers can [still have] have wood? This is a cause of corruption and theft", he said (S07, interview, Oct 19, 2004).

7.5.6 Shortcomings in democratic practices

Shortcomings in the democracy of Vietnam were seen as part of the reason for sub-optimal government capacity. A scientist interviewed said that accountability is attenuated because members of the National Assembly and government cabinet are not directly elected by citizens. As a consequence,

> these persons do not have a particular responsibility to the people but rather to the mechanism that appoints them to their posts. This issue of accountability, together with the extent of the cash economy and a wage policy that does not encourage creativity, are among the main roots of corruption (S56, interview, Dec 21, 2004).

S01 argued:

> The most important needs are: a comprehensive institutional framework to ensure all abilities can develop, and a transparent legal system which forbids only things that are harmful to the society and lets people do whatever their abilities allow and the law doesn't prohibit. A move to a transparent and clear e-government where people can know, discuss and control the economy could help to liberate the enormous creativity of the Vietnamese (interview, Oct 28, 2004).

7.6 Environmental depletion and exhaustion

Fifty-three percent of the 70 interviewees saw it as a matter of high concern that Vietnam is facing serious problems of damage to the environment. Among this group, some had a broad view that global environment problems, such as oil exhaustion and climate change would affect many aspects of life in all countries, including Vietnam (S50, interview, Dec 14, 2004; S01, interview, Oct 28, 2004), while others were more specifically concerned about environmental problems inside Vietnam. Those who said they did not believe the problem of environment could affect their business were mostly people who work in the education, entertainment, finance or ICT sectors.

7.6.1 The consumption of energy and materials

Some interviewees were concerned that when a society undergoes economic development people tend to consume more energy and materials – thus the environment becomes more polluted, and competition for usable land intensifies. “As economic development has accelerated in Vietnam, land has become very scarce in cities and also in rural areas, and extremely expensive in comparison with incomes” (S04, interview, Oct 28, 2004). Other interviewees focused on the coming energy crisis when Vietnam’s and international oil resources would be depleted and an energy strategy based on new power, such as nuclear, would be necessary. Two experienced government ministers (S57, interview, Dec 23, 2004, and S03, interview, Oct 29, 2004), forecast that by 2017, nuclear power would be used in Vietnam, but all those who have knowledge about the technology in Vietnam would be retired at that stage, therefore training of human resources in nuclear technology must be started from now.

7.6.2 The treatment of waste

To many interviewees, pollution was a top issue. Treatment of sewage and waste water from industry, and wastes from power plants were increasingly expensive (S03, interview, Oct 29, 2004). There was also concern that Vietnamese people did not have a proper sense and understanding in resource saving or environment protection. One academic suggested that Vietnam must recycle and treat waste from all factories and cities or "would die in it" S58 (interview, Dec 22, 2004).

Many natural resources like land, water, and forests have been used up or become worse in quantity and quality, according to a socio-economics expert (S56, interview, Nov 15, 2004) who pointed to evidence that land is being overexploited and forests are being reduced. A scientist stated that 74.3% of Vietnam's land area (24.7 million ha) has problems of steep gradient, acidification, impoverishment or erosion; and the quality of forests is declining.

7.6.3 Appropriateness of agricultural planning

Another interviewee with experience in agriculture management said that agricultural planning has been criticized for lacking appropriate directions: for instance, some plants grown in large areas like gum trees, cassava, and maize extremely impoverish the land; or aquaculture in the Mekong river delta that entailed destruction of mangrove forests to build shrimp ponds, could actually de-oxygenate and acidify water and soil (S53, interview, Dec 7, 2004).

7.6.4 Decline or extinction of many plant and animal species

Besides natural resources, natural ecological systems such as endangered marine environments face decline or extinction of many plant and animal species. A senior government official in aquaculture sector revealed that aquaculture output growth was largely due to extensive cultivation on brackish water and near to shore, not by intensive cultivation or productivity increases. The extension of aquaculture constantly narrowed habitats and ecosystems like tidal mud and mangrove forests that are the nourishing source of bio-diversity and the home of both near-shore and 90% of off-shore creatures. Aquaculture of shrimp, a major export of Vietnam, mainly based on natural habitats, was cheap and profitable in the first cycle of five years but would soon degrade unless investment in environment protection is taken (S59, interview, Jan 5, 2005).

7.7 The concept of a knowledge economy and its impacts

The majority of interviewees, 76%, had various concerns about the role of knowledge, a knowledge economy, and sustainability.

7.7.1 Knowledge and technology-based exploitation

To some, the mode of domination in the world is changing from physical exploitation to more sophisticated knowledge and technology-based exploitation. Many interviewees, like S01, S41 and S52, shared the view with Nobel laureate Joseph

Stiglitz (2002) that raw materials are becoming cheaper while high tech commodities are becoming more diversified and expensive.

S02 argued that knowledge economies are selling high technology products at exorbitant prices and buying raw materials dirt-cheap. However, not all of those interviewed agreed that the protection of intellectual property is a new form of exploitation. S02, for instance, stated that knowledge is "a commodity that can be sold, and none can lay blame for exploitation on those who sell a commodity" (interview, Oct 27, 2004). He said that knowledge used to be a common property, but has now become a special high-class commodity that links to private ownership. From an economic perspective, one person has the right to sell, and the other person has the right to buy, and this is not exploitation.

But he did admit though that,

> ... from a philosophical and political point of view, there are unfair tricks that become normalized. At the beginning you claim that you sell me a product of half of its price. When I get used to it, you sell it for full price, and no one knows whether it is a fair price or not (S02, interview, Oct 27, 2004).

7.7.2 The depletion of resources

The outstanding difference between a knowledge economy and a physically based economy, to many interviewees, was the extent of dependence on nature. When the depletion of resources directly affects material-based sectors, e.g. manufacture or agriculture, the knowledge-based economy does not seem to be directly affected, and the resources seem to be inexhaustible.

Some others trusted that technology would be the solution for the depletion of the nature. "Natural fuels, if they become exhausted, need to find substitution, based on knowledge and high-tech" (S02, interview, Oct 27, 2004).

7.7.3 Level of technology

In that context, the level of technology application in the economy of Vietnam is very low in economic sectors such as industry (S03, interview, Oct 19, 2004; S54, interview, Dec 21, 2004), fishery (S59, interview, Jan 5, 2005), agriculture (S36, interview, Nov 30, 2004; S39, interview, Dec 1, 2004) and tourism (S29, interview, Nov 18, 2004).

Machines used in Vietnamese factories were often two or three generations out of date. The utilization of raw materials and energy in production was not highly cost-effective. When natural resources are being depleted, such indicators should be improved to reduce the cost of production and environmental pollution. According to these interviewees, the World Economic Forum gave a lower competitive environment ranking to Vietnam from 62 in 2003 to 77 in 2004 owing to its un-improved institutional and technological environment.

7.7.4 Research and development

One expert (S07) pointed out that the application of science and technology did not match the socio-economic demand in the last 20 years, and had not become a direct production force. Current policies did not fully mobilize the endogenous capacity of

Vietnamese people to create and adopt science and technology into production. He said that at the government agency in charge, there was no technology transfer center or database about technology transfer existed to facilitate this process (S07, interview, Oct 19, 2004). He explained the reasons in the following way. Firstly, major research projects were often initiated by researchers not based on actual production demands. Thus the results of research were rarely applied in reality. Secondly, it was said that Vietnamese love the latest fashions. While this might be good enough to engage quickly with the latest technology, it might not be good enough thoroughly to master a technology and bring it to production. Fashionable research and development projects were spreading limited resources and efforts. Few people knew if the outcomes of those projects were actually feasible, beneficial and applicable, as he said,

> Researchers and science and technology managers in Vietnam jumped from IT to hi-tech projects even though not many understood what were hi-tech. When they became bored with hi-tech, they moved to nano-technology. If others had software, we would talk about software, though we didn't know if we were really good at software. Can they feasibly become spear-heading sectors in Vietnam? Can Vietnam master those technologies? (S07, interview Oct 19, 2004)

S27 suggested that hi-tech in the Vietnamese context should not be in "outer space" but the best possible technology that can be applied in economic sectors to enhance productivity and quality, e.g. the know-how of breeding, or ways to cultivate rice better (S27, interview, Nov 18, 2004). S54 argued that hi-tech should be technology that is friendlier to the environment. He considered that although Vietnam was eager for new knowledge and technology, it had no in-depth direction or strategy regarding any particular science and technology (S54, interview, Dec 21, 2004).

7.7.5 Examples of successful business initiatives

Several provinces have been successful in attracting investment through front-runner policies to transform the local economy from agriculture to service or industry sectors. Examples given were Vinh Phu province in the North and Binh Duong province in the South, where large industrial zones have been developed. These initiatives took geographical advantage of being convenient for transportation and near big economic centers. The growth of these initiatives was encouraging and they improved GDP, though this meant suffering trade-offs in environmental pollution (S01, interview, Oct 28, 2004). Many people like S01, S43 or S22 believed that with a higher GDP, the provinces would have more possibility of recovering the environmental losses, but others like S57 and S18 argued that the environmental cost would be much more expensive, or even unrecoverable.

An impressive example of the dynamic utilization of knowledge, technology and management in private business was the application of new methods of production, business and trade by the NamHoa Trading & Production Corp., which is managed by young, capable people. The company in Hochiminh city specializes in wooden toy production imports machines from Taiwan, and wood as raw material from Italy, enabling Namhoa to produce high quality products which could compete with the cheap but low quality toys from China in conservative markets like Germany and Japan (S63, interview, Jan 25, 2005).

In summary, during recent years, the socio-economic development of Vietnam has relied much on the exploitation of natural resources, low labour productivity, and production technology and consumption patterns that have used extensive energy resources and other raw materials, and also released large amounts of emissions.

7.8 Main insights from Chapter 7

Insight 7A: Pressures in the domestic economy and from globalization are seen as interlinked, driving the nation along courses of action that seek high returns in the short term, but rapidly degrade the environment and deplete non-renewable resources.

Insight 7B: Government is still not fully geared to the conditions of a market economy, and thus is only partially capable of optimising the opportunities of the free market while managing the risks. Selective but widespread reform is still required of government structures and processes, especially measures that improve capacity for co-ordinated, whole-of-government action.

Insight 7C: Environmental degradation is widespread. Significant human and technological inputs, underpinned by education and research, are needed to reverse the current decline and achieve sustainability and improving living standards. Examples of success need to be further recognized and publicized. The appropriate deployment of ICTs and knowledge strategies is an essential part of this scenario.

Chapter 8
Rural Vietnam

8.1 Scope of the chapter

- Uses fieldwork interviews selectively supplemented by literature to explore issues of sustainable economic development, knowledge strategies and ICTs from the viewpoint of people from, or closely connected to, rural Vietnam.
- Focuses on the interplay between knowledge transfer and adaptation in rural production, in a setting characterized by poverty, and under- employment or unemployment – especially among rural youth.
- Addresses key challenges in the area of adaptation to spreading urbanization, and the quest for improved value-adding in agriculture and other rural enterprises while also arresting and reversing environmental degradation.

8.2 Fieldwork for this chapter

This chapter presents an account of the rural and agricultural sector which used to be the largest economic sector of Vietnam. This sector is now facing rapid development and significant risks due to that development. The chapter seeks to convey a sense of what actually happening in the rural and agricultural sector of Vietnam concerning sustainability, knowledge and ICTs. The interviewees for this chapter include government officials, researchers, business operators and representatives of organizations. Their responses supplement the account of sustainability issues derived from literature and presented in Chapter 5.

This chapter reports on interviews with 7 senior government officials, 14 researchers, 3 business managers, 2 representatives of organizations and 5 villagers. Direct quotes from 26 interviewees are included.

8.3 Rural life and sustainable economic development

Agricultural and other rural production plays a crucial role in sustainable development in Vietnam. In the view of interviewees rural production is facing serious challenges to sustainability.

These challenges include population growth, deterioration of the environment, disadvantage in the pricing of products, poverty, low levels of productivity, and problems of government agencies in supporting farmers. These issues are explored in more detail below.

Interviewees were unanimous that sustainable agriculture and rural life are fundamentally important as a cradle of national endeavor.

The key factors in the generation of a livelihood for tens of millions of Vietnamese through rural production are labor, land and water.

S53, a former senior researcher in the area of agricultural planning, regarded this sector as:

> an important element of the environment which also in many cases affects the development or deterioration of the environment. Therefore strategies concerning the use of agricultural resources must be a key component of strategies for sustaining agriculture and rural development (interview, Dec 7, 2004).

8.3.1 Reports and statistics

According to the "Final report of the transition of the agricultural structure and rural economy in 20 years of Doimoi" prepared by Ministry of Agriculture and Rural Development (MARD), the proportional contribution of the agricultural sector in the national economy declined quickly, but still accounted for approximately 20% of GDP (Ministry of Agriculture and Rural Development, 2004). Villages and hamlets are home to 54 ethnic groups and about 74.1% of the whole population (Khong, 2002).

More than 20 years of Doimoi, including deregulation and privatization in the agricultural sector has changed the face of rural communes of Vietnam. The changes include the application of technology, new management processes, and proactive engagement and cooperation with international stakeholders.

The MARD report showed that the sector has achieved a steady and high growth rate of 5.5% per year for 20 years since 1986. That result secured domestic food supply and changed the country from being a chronic importer to the world's second largest rice exporter. Together with rice, other agricultural products also achieved leading positions in the world market, for example pepper (14.3% share of world market), Robusta coffee (40%) and cashew nuts (9.5%). The growth created new jobs, improved farmers' living standards and, spectacularly reduced the poverty rate from 58% to 29% in a decade from 1992.

In statistical terms, the environment has been improved, with forest coverage increasing from 33% in 1999 to 36% in 2003 (Khong, 2002), By 2004, the yield of seafood fishery and cultivation increased by a massive 213%, as compared with the 1995 yield of 2.4 million tons of seafood. Of the total current seafood production, 60% derives from cultivation and 250,000 tons are exported. From being unknown, Vietnam became the second largest exporter of shrimp to the US market (S. C. Tran, 2003).

However, the robust growth of the overall economy in a short period entailed a high demand for food production from the agricultural sector. This in turn has caused or exacerbated a range of environmental problems.

Population growth is a source of stress to the Vietnamese countryside. Gourou (1940, p. 466) argued that population density is the best indicator for measuring the intensity of the exploitation of the environment. The enormous growth of the population in Vietnam reduced the amount of land per capita dramatically. In 1943 the population had been just 22 million (X. Hoang, 2004, p. 72). By 2005 it rose to 83 million, and was predicted to be one million more each year (General Statistics Office of Vietnam, 2006a).

It was argued that the population of Vietnam has increased 5.3 times in the last 85 years, compared with an average rate of 3.6 times increase in the world population (Vietnam News, 2005). S04, a senior researcher in the area of socio-economic trends (interview, Oct 28, 2004) pointed out that the population density in Vietnam is twice higher than that of other developing countries in the South East Asia (excluding the Philippines) and of China, the most populous country; and tenfold higher than that of the developed countries (see also Table 5.1 in Chapter 5).

By 2005, 62 million people, accounting for 73.4% of the national population, were living in the countryside (United Nations Population Division, 2006). According to the World Bank (1998) farming in a 3 ha rice paddy is 5-6 times more profitable than farming a paddy smaller than 1 ha. However, one third of Vietnamese peasant households cultivate on small pieces of land less than 0.06 hectares and also, farm land is very scattered, and is not suitable for mechanization or large-scale production (World Bank, 1998).

8.3.2 Coping with population and environmental pressures

The burden of population growth may seriously jeopardize or at least slow down improvements to people's improved living standards and other advances in the social and economic development of the country. When population density increases, individual farm holdings become ever smaller and are thus not suitable for mechanization and economies of scale in production.

Agricultural production is strongly dependant on external and natural factors and profoundly impacts the quality of environment. On the one hand, global warming and climate change with respect to rain, sunlight, cyclone, and drought already appear to have had a strong impact on harvests. In 2003-2004, 80,000 hectares out of 260,000 hectares of coffee were lost due to drought in the Central Highland (Ministry of Agriculture and Rural Development, 2004).

Water (from the Vam Co Dong river) for irrigation has become polluted and scarce due to new industrial constructions, such as dams and harbors. To supply water, many deep wells were being drilled to suck underground water (I51, interview, Apr 7, 2006).

Natural resources and ecological systems have been severely degraded or destroyed. On the ground, 74.3% of the land area or 24.7 million ha in Vietnam have problems such as sloping, acidification, impoverishment and erosion (Ministry of Agriculture and Rural Development, 2004).

Intensive farming requires the extensive use of pesticides and chemical fertilizers. In an interview an environmental expert, S53, estimated that at two litres of pesticide per hectare or each year four million hectares of land in Vietnam are affected by 8 million litres of poisons. These toxicants severely contaminate land, water and the living environment. He also said that the water-born environment is being degraded by the uncontrolled economic activities. Mangrove forests in the Mekong River delta are widely destroyed in order to build shrimp or fish ponds which in turn oxygenate and acidify water and soil (S53, interview, Dec 7, 2004).

The MARD report pointed out that water, especially underground water is overused for irrigating and aquaculture, as in the case of coffee planting in Central highland (Ministry of Agriculture and Rural Development, 2004, p. 36; C. H. Nguyen, 2002).

It has been well documented that many plants and wild animals are in danger of becoming extinct, from big mammals like tigers and elephants to small creatures like frogs and snakes.

Sea fishery has hit its unsustainabity threshold. S59, an expert in aquaculture and fisheries planning, estimated that the yield limit of the East Sea (South China Sea) was around 1.3 million tons while the exploitation output in 1998 was already 1.13 million tons. With even more investment in fishing boats, the stock of seabed fish was reported to have reduced by 30% from 1984 to 1994, and the output and size of hundreds of sea foods such as sardine, lobster, abalone and squid have decreased (S59, interview, Jan 6, 200, C. H. Nguyen, 2002).

8.3.3 Pricing issues and intangible costs

S53 talked about pricing issues. In the market, the current pricing of agricultural products is extremely disadvantageous to farmers whose products such as rice or fruits are perishable. In the value chain, farmers work the hardest but earn the least:

> For instance, a hectare can harvest 10 tons of cabbage or about $260 at the 2004 market price, while a refrigerated truck to transport the cabbage to market would cost $1,700 (S53, interview, Dec 7, 2004).

Intangible costs, for example water for irrigation, are not counted in production costs, though huge amounts are used in agriculture, for example, 54,600 m3 per hectare for a sandy shrimp crop, or 6,000 m3 water for a hectare of coffee and 12,000 m3 for rice (L. T. P. Nguyen, 2003).

If fully counted, the cost is uncompetitive with agricultural products of developed countries where agricultural products are heavily subsidized by industrial profits. In the international market, the price of agricultural goods is unstable. For example, in 1994 when the price of a ton of coffee dropped tragically from $3,000 to $300 in the 2001-2004 harvests, tens of thousands of hectares of coffee plants were destroyed by Vietnamese farmers (S53, interview, Dec 7, 2004).

8.3.4 Rural youth, surplus labor and poverty

In group discussion in Longan province, 50 km south of Hochiminh City with three farmers coded I51, it was explained that their children could not keep their profession as farmers due to the shortage of farm land and the low income from cultivation.

Rural youth thus cannot in many cases continue in the farming tradition. They have to find some another job at site or move to cities to look for a better fortune. About one million youth join the redundant rural labor force each year (General Statistics Office of Vietnam, 2006a).

Poverty is a risk of sustainable development. According to Trong Vu Khai, Vietnamese farmers have very low incomes owing to low rates of productivity and profit. Revenue from a hectare in Vietnam is about USD 1,500 in comparison with USD 5,000 to 10,000 in other Asian countries. On average, a farmer only earns about VND 113,000 (~USD 8) a month (Trong, 2004). In an emerging market economy the market price of daily consumption becomes more expensive. These rising costs especially affect the poor, and the gap between rich and poor is widening. Though poor people do not consume as much as rich people they must exploit the environment harder to make ends meet. As S22, a senior government official concerned with trade and industry, said in interview:

> If farmers are poor, they will cut down the forests to grow rice or hunt wild animals even though they know these actions are illegal, but they have no other means of subsistence (S22, interview, Nov 16, 2004).

8.3.5 Women and ethnic minorities

Other problems for the sustainability of rural development include:

> Ninety percent of the poor live in rural area and 60% of the farmers are women who have limited education in technology and understanding of gender equity" (S53, interview, Dec 12, 2004).

> Forty-nine ethnic minorities in mountains and highlands covering three quarters of the country area are even poorer. They suffer inadequacies in health, education, and with language barriers, though the government has many programs to ameliorate the situation (V. N. Dang & D. D. Le, 2004).

S55, a senior academic, argued that:

> Poverty is not only in terms of money but also in the form of deprivation in health, education, information and entertainment services. Farmers are the principal and direct players in [the countryside's] development and the final persons who decide whether to keep or destroy the forests, the land and the riches of the sea. If they are poor, they cannot think about long-term sustainability (interview, Dec 21, 2004).

8.3.6 Agricultural technology

S53 (interview, Dec 7, 2004) explained that the application of mechanization and other technologies in the agricultural sector is very slow and at a low level. Almost all agricultural production is done manually. Industry, especially post-harvest processing industry, is not sufficiently supportive of innovation among "front line" producers in agriculture, forestry and aquaculture. Due to the trend of replacing traditional breeds with new or imported ones, the diversity of genes of local plants and domestic animal breeds has declined severely. Consequently, the control of insect pests is more difficult and native biodiversity is threatened.

8.3.7 Human livelihoods and the ecosystem

S55 (interview, Dec 21, 2004) argued that rural Vietnam consists not only of economic elements but also an interaction between humans and the ecosystem. This system can be trapped in a negative spiral of development: population growth, environment degeneration, poverty and socio-economic and cultural divide. The relationship between population density, environmental deterioration and poverty is complex: the interaction of the three elements creates a spiral of overpopulation, resource exhaustion and destitution. To a certain threshold, that incremental spiral whirls robustly with its entire elements i.e. economic, socio-cultural, and environmental but at some point a critical value of at least one element falls below a critical value, causing the system to collapse.

To maintain sustainability, it is commonly claimed that the needs of development today must be held in parallel with the needs of environmental protection for tomorrow. In the case of rural development in Vietnam, some researchers like S59 or S52, did not accept this approach. S52, a senior scientist, said in interview:

> Sustainable development, in the view of developed westerners, is a harmony of economic, social and environmental factors. WWF and IUNC request Vietnam to protect wild elephant and prohibit farmers to enter forest. So how about the destiny of persons who are killed by elephants and have no food to live on? .... Why do the developed countries force the poor to protect the environment, when they, the US, Canada and Australia, do not sign the Kyoto Treaty which is more important to protect the global atmosphere? Protecting the environment is good, but we cannot protect by sacrificing the poor people. Why should this country be concerned with protection when others are not? This is a conflict of plutocracies and nations' interest. Shrimp exporting makes millions of people well-off though the environment, worse-off. When a hectare with rice earns only VND 10m (about $700), and they receive VND 100 m (about $7,000) with shrimp, can you tell them to stop destroying the rice field or mangrove forest? (S52, interview, Nov 17, 2004)

8.3.8 Government's capacity to improve sustainability

In the view of S53, awareness about rural and agricultural sustainability is often vague among provincial leaders. Few leaders go to sites to meet farmers; thus they cannot understand farmers' problems or the solutions to problems.

A government report noted, for example, that in the early 2000s government management over-estimated the market's demand for sugar cane and coffee, causing an over-production crisis resulting in the need for farmers to destroy thousands of acres of sugar cane and coffee crops (Ministry of Agriculture and Rural Development, 2004, pp. 36-37).

S59 (interview, Jan 6, 2005), a leading researcher and planner of the Aquaculture Planning Institute, Ministry of Fisheries said in the interview that environmental resources and aquatic systems have broken down due to a lack of coordination between sectors and line-management agencies. There is a conflict of interest in relation to natural resources; each sector tries to exploit natural resources for its own interest and does not take into account the interests of others or the common good.

8.4 Knowledge and sustainability

In the previous section, primary challenges to rural development have been given as over-population, urbanization, the negative aspects of rural technology, poverty, shortcomings in governance, and finally the degradation of environment. This section will examine the contribution of knowledge to the sustainability of rural development by presenting an overview of the following areas: awareness of family planning and population growth, learning to adapt to urbanization, knowledge of how agriculture might move from low to high value; and human capacity building.

8.4.1 Awareness of population issues

Being aware of the dangers of population growth is one of the first necessities for dealing with the problem. Since 1960, the Government has consistently been pursuing population programs which, together with many other activities, have focused on (N. Q. Vu & Nguyen, 2004):

i. increasing management capacity among family planning staff
ii. communication and education of people on birth control
iii. improving the quality of population data, and
iv. research on population characteristics.

The majority of Vietnamese people now understand that family planning would help them to maintain a higher standard of living. Moreover women and family would be healthier and have more chance to join political and socio-economic activities. Thanks to increasing awareness and knowledge among people, especially women, on the importance of birth control, family planning patterns are changing. For example, the proportion of couples who applied contraceptive methods increased from 53.7% in 1993 to 75.4 in 1997. As a result, the fertility rate fell from 3.8 babies born per woman in 1989 down to 2.3 in 1999, and the population growth rate dropped to 1.7% during 1989-1999, 0.5% lower than that of 1979-1989 period (Government of Vietnam, 2000).

Despite this progress, one million babies are still born each year. The awareness of problems and possible birth control methods has clearly helped Vietnamese rural people and the nation as a whole to diminish the threat of excessive population growth.

8.4.2 Learning to adapt to urbanization

Traditional housing space in Vietnam was almost an organic component of nature, using natural materials that were mostly integral to the local environment. The housing "lived with" nature. Modern architecture, in contrast, tends to "conquer nature", spawning an ever-expanding landscape of concrete structures that obliterates farming land. Much architecture tends to ignore amenities that nature grants. Fresh air is replaced by fan-forced or conditioned air. Nature is subordinated to technology. It is ironic that, at a time when societies supported by ICTs (the Internet age) are developing as open systems, buildings increasingly become artificial, closed containers. In other words, nature tends to be neglected in urban planning, and together with it, the farmer. Prof Hoang Dao Kinh (2004), former Director of the Department of Conservation thus, argued that urban planning must respect and harmonize with, rather than conflict with, nature. And to do that it is necessary to learn from nature, and from traditional indigenous architecture.

Prof Pham Ngoc Dang (N. P. Dang, 2004), director of Center for Envirionmental Engineering of Towns and Industrial Areas, argued that urbanization in Vietnam was mainly done by concretizing rich farmland along roads to the industrial or urban zones, for example Nomura industrial zone in Haiphong, North Thanglong or Noibai in Hanoi, or nearby residential areas like the new 1200MW Quangninh thermo power plant at Hakhanh village in Halong city. S04 (interview, Oct 28, 2004) added that such planning is easier for urban planners and costs less for governmental and other stakeholders in the short run. However he was worried that once land is concreted over, it would be difficult to return it to cultivation – for example if the food shortages of the past ever recur. Clearly, short term profit overshadows long term benefit but long run cost could surpass short run profit. S04 therefore suggested that planners should build infrastructure in poor lands – not prime agricultural land – to encourage developers to build new towns or factories. To do that, planners and management must seriously know the value of land resources, and determine where to develop the industrial and post industrial economy, and where to reserve land for agriculture.

Another aspect of the urbanization of rural communes is that planners have too often taken the easier path and focused on erecting new buildings in rural areas, while neglecting existing building stock. The result is that older neighborhoods are often left without modern facilities, such as water supply, sewage, power or telecoms. Consequently these are full of shabby cottages as in the cases of Kimlien, Vinhtuy or Dichvong villages (N. P. Dang, 2004).

For agricultural villagers who lose their land it is very difficult to find a new job. Only 2% succeed in moving across to the industrial sector, and 33% to the service sector, while 50% have no realistic option but to remain agrarian work-seekers, totalling 7 million redundant laborers. The labor jam in rural areas indicates that urbanization and industrialization have not taken much account of rural needs, particularly the plight of land-lost villagers who lack both social and technical skills needed for alternative employment, and information about how and where new jobs may be found. The Ministry of Agriculture and Rural Development recognizes, and proposes that, vocational education needs to be delivered to 1 million villagers each year (Ministry of Agriculture and Rural Development, 2004, p. 37).

8.5 The quest for more knowledge-intensive, value adding, rural production

Indigenous knowledge accumulated in Bat Trang, Dong Ho (see Chapter 9) and in 1,450 other craft villages in Vietnam is an intellectual base of diversified professions, such as silk reeling, textile dyeing, processing of primary products derived from agriculture (e.g. leather) or extracted from nature (e.g. minerals), waste recycling, handicrafts, production of building materials, porcelain and ceramics. Le Quy An et al. believed that those craft villages are efficient workshops for rural economic development and for surplus rural labors of about 10 million. Many people attempt to cross over (part time or full time) from primary agricultural production of commodities (high mass content) to value-adding forms of secondary or tertiary production of goods or services (higher knowledge content). In rural areas 27% of farmer households earn from both farming and other occupations, and upwards of 13% of rural households are now engaged in careers other than farming (Q. A. Le et al., 2004).

In addition to building up old knowledge, Vietnamese people should also learn how to get more value with less input suggested interviewee S01 (interview, Oct 28, 2004).

8.5.1 Coffee production

As an example of how better information can increase profits and reduce risk, S01 (interview Oct 28, 2004) pointed to coffee production.

The Central Highland of Vietnam suddenly became one of the biggest coffee-growing areas in the world. In 2004 this area stormed the world market with around 900,000 tons of coffee beans. However, due to the sudden increase of supply, the price of coffee in the world market plummeted from $3,000 to $300 per ton, a bitter disappointment for Vietnam. In addition to the problems of revenue, it was found that coffee cultivation severely depletes the ground water in the region. In the droughts of 2000 and 2004, around 80,000 hectares were short of water. S53 and S01 in separate interviews projected that if the production area of 260,000 hectares was reduced by half, and the proportion of processed coffee sold (compared to sale of raw coffee beans) was increased from 0.5% now to 25%, farmers would earn double and ground water as well as rain forest in the Central Highland would be better preserved. To do that requires, of course, not simply technological knowledge but economic and strategic knowledge from the producers, and the capacity to provide appropriate planning and co-ordinational support and infrastructure (including transport of goods and personnel) from the Government.

8.5.2 Paper production

S01 (interview Oct 28, 2004) also illustrated the case for improved knowledge transfer through an example from paper production.

Phong Khe was a craft village in Bac Ninh province that had possessed a long tradition of paper production. In the favor of market economy policy, family producers changed into enterprises that applied new technology in the most optimal manner. Only stages that were most critical to the quality of the products were automated, otherwise, labor was still used to cut down cost. The investment of Phong Khe was only VND 200 billion, a quarter of the largest state own paper factory, Bai Bang. Yet Phong Khe hired top national and international experts with competitive salaries: 15 out of the 20 best experts in paper production in Vietnam now worked in the village. As a result, Phong Khe family enterprises overtook Bai Bang both in terms of production quantity and quality and became competitive to Singapore or French producers. This example illustrates, within the framework of market competition, technological and managerial knowledge transfer is a driver for rural enterprise to move away from poverty to prosperity.

8.6 Knowledge Oases

8.6.1 The metaphor of a knowledge oasis

The author applies the metaphor of "knowledge oasis" to describe the knowledge locked in Vietnamese villages and farmers' minds. S67, for instance, was an ex-farmer who after retirement opened an eco-tourist service on an island in the Mekong River (S67, interview on Jan 27, 2005). The owner of the family run enterprise was very successful and knowledgeable in the new business and in maintaining the environment, but his knowledge was trapped in his island, and not shared. There was no electronic connection from his island to the rest of the world. Farmers do not share their knowledge, partly because they are afraid of losing know-how, and thus losing the power to improve wealth and self-esteem. Farmers have no trust, no time, and no incentive; and there is no means to support knowledge sharing. More discussions of S67 and the knowledge oasis are given in later chapters, especially Chapters 16.

Individual human beings are the most important factors in development, and their welfare should be its ultimate target. In the real world the growth of globalization and the increasing knowledge intensity of economies on the one hand brings great opportunity for developing countries but on the other hand widens the divide between urban and rural areas and the gap between people who are information-poor and information-rich. In that process, those who can quickly adapt, and who are able to learn and find out new ways to deal with the new situation are more likely to succeed, as exemplified by the cases S67 (interview, Jan 27, 2005) and S62 (interview, Jan 25, 2005) recounted to the author during his interviews with them.

8.6.2 Knowledge oasis on island

War veteran S67 was a former farmer in Kien Giang province. One day he visited his daughter who lived on an island in the Mekong River in Long An province. The farmer realized that the place could become attractive to tourists thanks to the beautiful landscape. Emptying all of his pension money he invested in a small eco-tourist restaurant on the island. In 2000 when the My Thuan bridge connecting Hochiminh City with the South East delta was completed with an Australian Overseas Development Aid (ODA) grant, tourists stormed Long An to explore the wild southern gardens on water. From 300 to 400 customers visited his restaurant each day. Bit by bit, he learned how to satisfy different groups of customers, for example, visitors from within Vietnam preferred to fish by themselves, while international visitors mainly wanted to have clean bathrooms. S67 revealed that with tourists, he could charge for fish at VND 60,000 or even 85,000 per kilogram when the normal market price was only a half. His business, though small, was very successful and profitable in comparison with the rice farming of his fellow villagers. Being a successful case of self-learning for sustainable development S67 was ready to share his experience with other villagers throughout Vietnam but had no channel to do that. In other words, his knowledge is locked in his brain as a coconut palm grows alone in an oasis.

8.6.3 Knowledge oasis in a mindset

S62 was another example of learning capacity for success. He was assigned by his brother, a rich investor, to build an 18 hole golf course in the Longthanh district of Dongnai province, 20 km north east of Hochiminh City. At the beginning, he hired a group of architects from the U.S., the Philippines and Singapore to design and construct the first 18 holes of golf course on scenic hills. However, he was not satisfied with the foreigners' work because they leveled the hills to build the course, and thus did not take the advantage of the natural terrain to make the course different to others. So S62 spent two years in visiting golf courses overseas and seeking documents about golf-course design. Finally, he mastered the art. He dismissed the foreign experts and tried to design and manage the construction by himself.

When the author interviewed him in January 2005, the second 18 holes were in service and a project to expand to a further 18 holes was in progress. Although not following any conventional design methods, his course on the one hand satisfied technical requirements but on the other hand was unique and very challenging to most golf players. In addition to the natural terrain, other advantages included a mixture of native and exotic vegetation and a system of resort estates inside the area. To run the business, S62 hired a U.K. professional golf manager and built a computer network to manage all the golfing facilities, playgrounds, client services and amenities and maintenance. The club website was set up to advertise the business and handle transactions with customers.

The club's url is http://www.longthanhgolfresort.com.vn/. At present, the golf course of the family owned business Longthanh Golf Club and Residential Estate offers:

> (...) 27 holes open for play. Each hole has its own characteristic shape and unique view that offers a wide variety and separate challenge to novice and professional alike. Fairways are lined with beautiful oil palm trees and other similar species, the emphasis being on evergreen in order to keep the course in pristine condition throughout the year (S62, interview, Jan 25, 2005).

This business turned an impoverished land into a new gold mine not only for the investor but a source of income for thousand caddies and other workers in the district. The design of the golf course was so eminent that other investors wanted S62 to work for them on similar projects. S62 had to refuse the offers because his brother and principal wanted to keep his ability as a business secret.

8.6.4 Lessons from the oases

In different ways, their self-learning capacity helped S62 and S67 to move from lower-value physical (mass) based production to much higher value enterprises based both on greater knowledge (mind) content and leveraging of environmental assets for greater profitability as well as sustainability. With brain power and creativity, those people learned for themselves the way to rise and and achieve desired results in the knowledge economy. Despite criticism that formal education in Vietnam still under-emphasizes practical knowledge (Chapter 6), these cases suggest that other Vietnamese people could learn and succeed similarly. S43 (interview, Dec 7, 2004) remarked that a learning society culture was gaining adoption widely in Vietnam with many people exerting themselves to study both within and outside school walls. S44 (interview, Dec 7, 2004) claimed that without technology learnt from Japan, Vietnamese shrimp farmers could not so quickly have achieved their current impressive levels of production and export. Those interviewed agreed that such self-learning should be encouraged.

In the view of knowledge promotion in the national interest, however, S62 and S67 are examples of knowledge oases where accumulated knowledge was locked in the brain of individuals and could not be shared, either because of no media channels (S67) or because the knowledge became a business secret (S62). Locked knowledge could be classified as knowledge oasis in a Coconut metatype (further discussed in Chapter 16) where knowledge holders cannot share their accumulated wisdom effectively with others whether through conventional or ICT channels. Without networking, operating at a sufficient scale, the society can not make full use of the power to multiply individual knowledge.

8.7 Main insights from Chapter 8

Insight 8A: The rural life and economy of Vietnam are under great pressures including population size, spreading urbanization, environmental degradation, slow technological change, and limited capacity for change-management in all levels of government. Youth, women and ethnic minorities are particularly affected by the problem of surplus labor in rural areas.

Insight 8B: There is increasing recognition that educational and knowledge strategies are crucial to addressing problems of sustainable economic development in rural Vietnam, while at the same time preserving core national values and traditions.

Insight 8C: Dramatic knowledge-based advances in areas such as coffee and paper production encourage a view that the nation can "think its way through" rural problems. So do successes in areas such as tourism. However there is a problem of "knowledge oases" where pioneering small businesses remain isolated cases, and small-business know-how spreads only slowly across the rural economy.

Chapter 9
Indigenous Knowledge

9.1 Scope of the chapter

- Examines three cases of indigenous, traditional art and craft, namely Dong Ho Printing, Bat Trang Ceramics, and Quan Ho Folk Music.
- Focuses on the problem that forms of artisanship of deep cultural significance to Vietnamese identity may be unsustainable in an open market economy, and with the loss of environmental resources that have nourished them for centuries.
- Explores how such indigenous creative industries are faring with the onset of the knowledge economy, and whether ICTs are of present or potential assistance for their survival.

9.2 Sustaining indigenous knowledge

Fundamental to the very identity of Vietnam are traditional folk professions which nourish cultural heritage. Vietnam's economic recovery since the war with the US is widely recognized as remarkable. Assuming the economic progress of the country continues and becomes a text-book case of success in terms of the free market and the WTO global framework:

- What would be the national cost-benefit implications if, in the free-market process, heritage crafts and livelihoods were lost, together with the environmental conditions that have nourished them for generations?
- Can ICT-supported measures to maintain the knowledge base of this folk industry assist its continuity and market position?

9.3 Dong Ho printing village

In the case of the Dong Ho printing village there is a distinct danger to sustainability caused by changes to customer tastes due to modernization, difficulties in maintaining an adequate skill base among folk artisans, and the reduction in supplies of raw materials.

The cultural asset at stake in such a case is referred to as "intangible" – in other words the social meaning and value of the activity cannot readily be measured in objective commercial terms. UNESCO identifies intangible cultural heritage as a mainspring of cultural diversity and a foundation for sustainable development. Intangible cultural heritage means:

> the practices, representations, expressions, knowledge, skills – as well as the instruments, objects, artefacts and cultural spaces associated therewith – that communities, groups and, in some cases, individuals recognize as part of their cultural heritage (UNESCO, 2003).

Passed on from generation to generation, this cultural property is constantly recreated by communities and groups in interaction with their environment, nature and time. Cultural heritage nurtures human creativity and fosters shared meaning for individuals, communities and the nation as a whole.

In the transformation from a closed agricultural economy to a globally networked economy and driven by the imperative of economic growth, this cultural heritage that has accumulated in the multi-millennia history of Vietnam is in danger of extinction.

This issue was studied on February 20, 2006 in the Dong Ho folk-printing village in Bacninh province, 35 Km North of Hanoi where the author and Ms Tran Kim Anh, his assistant conducted two in-depth interviews at the village , one with I21, and the other with I23.

I21, one of the only two old masters of the art, said that Dong Ho wood carved printing is at least 500 years old, and has been practiced by some 20 generations in the village. The Dong Ho painting school, which passes on the important craft, flourished before 1944, when all 17 clans of the village were engaged in printing

Figure 9.1 Spring festival in Dong Ho folk printing. Photograph copied by the author from a brochure provided by I21

production. In contrast to modern printing methods that use toxic chemicals, Dong Ho printing is in harmony with nature, having no environmentally harmful effects. The five Dong Ho colors are made from natural materials, e.g. white color from shell powder and black from bamboo ash, and artisans carve the wooden printing blocks by hand. A poem of Hoang Cam, illustrated by a woodprint by I21 (Figure

9.1), conveys how the craft of Dong Ho printing embodies Vietnam's artistic values, social philosophies and humanitarian desires, and is part of the renewal of life and culture with the coming of spring.

According to I23 who managed I21's showroom in the village, learning the professional skills was very hard, but it was even harder to market the products with the tastes of customers changing so fast in the new economy. Along with the introduction of imported technologies of graphic production, Dong Ho paintings became less popular, especially among young people who preferred modern printing and would not buy Dong Ho pictures to decorate their houses in spring as before.

Painters struggled to find markets and consequently many valuable Dong Ho woodprint blocks were used as firewood while the prints piled up in the shops. Between 1970 and 1985, a small number of Dong Ho paintings were exported to Eastern Europe and the Dong Ho printing school narrowly survived, but many of the painters took on other crafts. By the time of the interviews, there were only two old craftsmen remaining, namely Nguyen Huu Sam and Nguyen Dang Che, who tried to preserve the painting craft. Apart from the decline in demand, suppliers of raw materials for Dong Ho, such as the special poonah paper and traditional colors, have decreased. Artisan masters like I21 who hold the indigenous knowledge are mostly over 70 years old or deceased, while the upcoming generation does not want to follow their profession. The precious local knowledge was in danger of passing away with them unless measures were taken to preserve it.

To maintain the art, I21 devoted four decades to collecting and preserving about 200 old and new woodprints, of which 150 are especially valuable. He taught the art to his children, children-in-law and grand-children, and explained at interview that he would be willing to teach anyone who wanted to learn the art. To sell the paintings, I21 established an art gallery in downtown Hanoi, managed by one of his daughters-in-law, and a showroom in the village managed by his son, I23.

I21 acknowledged that the government's open Doimoi policy gave him more chance to restore the traditional profession. From 1991 to 1995 he busied himself in reproducing woodprint blocks and publicizing his products. With the help of an art critic, a TV program was broadcast annually on VTV4 at the time of the new lunar year to introduce Dong Ho prints to Vietnamese people within the country and internationally. Government agencies such as the ministries of Culture and Information, Agriculture and Rural Development, and Trade as well as the Viet Nam Tourism and local administration made a concerted effort to help the Dong Ho villagers revitalize the age-old trade, including taking measures to boost production, to promote trade and publicity at home and abroad, and to honor craftsmen who have rendered outstanding services to the preservation and expansion of the craft. These efforts have had positive results in attracting tourists and customers back to the art. During the interview with I21, the author saw many visitors coming to I21's house to visit and buy prints.

I21 hopes that his profession has been saved, and that it will now exist perpetually. He trusts that this precious indigenous knowledge will bring good income to his clan and provide value to customers.

In fact the outcome of this story is still not yet clear. The growing commercialization of fine arts and the widespread use of cutting-edge technology have eclipsed many of Vietnam's traditional arts and crafts. Dong Ho printing is just one of modernization's potential victims, and despite the great efforts made by individuals and organizations to restore and develop the Dong Ho painting craft, the tradition is in continuing danger of extinction.

9.4 Bat Trang ceramic village

The village of Bat Trang, 10 km east of Hanoi has long depended on an ancient form of production: pottery. This fine craft village has achieved a revival in recent times. It has often been cited as an example of successfully sustaining a traditional profession, preserving indigenous knowledge, as well as utilizing ICTs to enhance its trade. The ceramics of Bat Trang are well-known because of their harmonious and unique designs, patterns and glaze, which are strongly imbued with national identity. Bowls, dishes, pots, cups, wine jars, vases and lamps from the village are seen throughout Vietnam and are exported to many countries. Annual exports to Europe, US, Japan, South Korea, Australia are valued at around USD13 million (N. C. Tran, 2005).

The expansion of Bat Trang ceramic production solved the problem of unemployment in the surrounding regions. All the villagers of Bat Trang, from teens to elderly, and 3,000 - 5,000 other laborers from neighboring communes could find some kind of job at a nearby pottery kiln. Income from economic development has laid the foundation for a process of urbanization from within the village with new cement roads overtaking small brick lanes, and rain water runoff filling up ponds and wells. The beautiful ceramics and the unique processes of craft production attract about 6,000 foreign and many local tourists per day. They come to observe and buy wholesale or retail. The old northern village which was the original Bat Trang is gradually disappearing.

The trade-off for these socio-economic achievements was that the expanding scale of production severely damaged the environment. With a high population density of 2,500 inhabitants/km2, and 1,000 kilns using coal and gas, each year the crafters were consuming about 70,000 tons of coal and 100,000 tons of clay and other materials for ceramic production. That production discharged 130 tones of dust, 225 tons of unused clay and 6,800 tons of cinder and an uncounted amount of smoke into the surroundings. As a consequence, the village was polluted by heat, dust and wastage and a high prevalence of air-born diseases. 70% population contracted respiratory health problems and 80% suffered from trachoma (GTZ & MPI, 2004).

To counter the problem, crafters have applied new technology to improve the quality of their production. The laborious work of grinding and mixing clay has been replaced by machines, thus gaining higher productivity. Coal kilns have been replaced by gas kilns which produce less ash and smoke, reducing dust pollution and the wastage of unused clay. The rate of successfully finished products rose to 90% compared with 20-40% using the old coal or charcoal kilns because the temperature is higher and more stable, and the operation of the kilns is easier. Gas kiln technology has improved the quality and reduced the cost of Bat Trang ceramics thanks to the saving of labor. More importantly, the introduction of new technology (gas kiln and associated mechanisms) has significantly reduced the pollution of the village. It has been argued (GTZ & MPI, 2004) that application of new technology and knowledge in Bat Trang could be an example for the other 1,450 handicraft villages in Vietnam to follow. The Bat Trang model could help in resolving the conflict between socio-economic development objectives and the protection of environment, and therefore such knowledge should be multiplied to other communes.

Together with the change in production technology, Bat Trang is also well-known for the use of ICTs for the production of traditional ceramics, and for the export of ceramic products. The aim of attaining new customers and markets drew many ceramic craftsmen to Internet-based marketing. Since 2000, e-mail and the Internet have been used by the villagers as a cheap and fast method of connecting to clients in non-Asian markets, particularly in Europe and North America. About 20 enterprises had a website as a place to advertise their products and contact points to foreign markets. This effort has been a catalyst for the entire community. Enterprises and individuals increasingly began to use ICTs. By 2005, 15% households in Bat Trang had computers, and 100% residents had access to telephones (N. C. Tran, 2005, S51, interview, Dec 14, 2004).

To gain more data about the case, the author and Ms Cu Thi Ha, his assistant conducted two in-depth interviews at the village (coded S37 and S38) with three Bat Trang ceramic industry leaders on Nov 30, 2004. The discussion focused on sustainable development and the role of knowledge and ICTs in the village.

The majority of ceramic workshops in the village belonged to families. The fieldwork data reconfirmed that in the short term, the move from coal kiln to gas technology improved the quality of products and saved the cost of production. However, in the long run and in a broader context, the interviewees were worried about the escalating price of gas that Vietnam imported from the global market. While gas doubled its price, the makers could not increase the output price sufficiently to keep pace. To overcome a price crisis, interviewees revealed that they were assessing other technology, such as coal gasification from China, with the assistance of Ministry of Science and Technology of Vietnam.

The impact of the rising global price of gas on economic and environmental sustainability at Bat Trang demonstrates the point that the chance of sustainability in a small tempo-spatial context can be affected by sustainability in a broader tempo-spatial context.

In addition to production, a series of sustainability related issues were noted from the fieldwork: less severe but nevertheless continuing degradation of the environment, erosion of traditional village community life and culture, and the loss of traditional knowledge. Firstly, though gas kilns discharged less smoke, the dust, waste and heat from the continuously expanding production made the whole village feel rather like a dirty oven. Secondly, when villagers became more affluent, they replaced the old-fashioned thatched cottages with multi-storey concrete buildings and small brick lanes with wide asphalt roads. The ancient typical village that tourists wish to see was being displaced by emerging urbanization. This was of great concern to S38:

> How to develop but also preserve a traditional fine craft village, not an industrial village? A village with five hundreds years' history must not appear with a modern factory or high building which can be seen elsewhere. People complain that Bat Trang is too modern, not a village with traditional hand crafts. Bat Trang should be developed with sophisticated and high value arts and crafts so that tourists can feel and experience the unique tradition of thousand-year-old ceramic production and learn that our products contain not only labor, coal or clay, but have been crystallized from a thousand years of history and culture, so that customers are willing to pay USD 10 a ceramic pot, even

> though an [apparently] similar [but mass produced] Chinese-made one only costs one tenth of that! (S38, interview, Nov 30, 2004)

The interviewees agreed that technology and knowledge learnt from outside could help the local people to lessen environmental devastation by either using more efficiently materials, polluting less the environment or even by creating new substitution for traditional material or fuel. With new value enhanced by knowledge (local or foreign) the producer receives higher revenue with reduced environmental cost. For example, to earn USD10, a Bat Trang craftsman with her historic pot would use ten times less amount of clay and fossil fuels and thus ten times less pollution than a Chinese counterpart. S38 argued that

> the uniqueness of hand made ceramic product was more valuable than industrial mass-production ones. A small tea pot sold in Taiwan market at $10 while a big vase priced only $3 with much more raw materials and fuel. That trade philosophy should be understood by every family so that they would not compete to each other, otherwise they would reduce the value of each other (interview, 31 Nov, 2004).

In the struggle for survival and development, the villagers of Bat Trang found ICTs to be a new, cheap and quick means of communication to connect and trade with the world. ICT applications in use ranged from basic word processing or email in S37's workshop to home-made websites where craftsmen could advertise their products, communicate directly with customers and saved the cost of transaction. The unprompted movement spread out quickly to many producers but soon challenged them with series of difficulties: money to hire a host servers, shortage of English and IT skills, qualified teams to maintain the sites, and the unavailability of broadband when dial-up connectivity was more expensive and slow. Furthermore, it was difficult for Internet users to locate the home-made websites which were not easily found through major search engines like Google or Yahoo because the website addresses were oases of information that nobody except themselves knew about.

To overcome the problem, Bat Trang Porcelain & Ceramics Association, with support from Mekong Private Sector Development Facility (MDF) and International Finance Corporation of the World Bank, set up Bat Trang Ceramics Export Promotion Center (BTEP) to facilitate e-business activities of the local entrepreneurs. BTEP's website, http://Bat Trang-ceramics.org, became a portal for the village where 28 pottery members could present their name, products and services to the world. Through it, customers could find Bat Trang's goods and prices and obtain an order form (containing appropriate contractual conditions) online. S38 claimed that genuinely interactive online sales could not be performed due to the unavailability of an adequate legal framework for e-commerce.

In an interview with S51, a senior researcher in science and technology policy, the author learnt that S51 had conducted a study in 2005 of nine traditional villages (including Bat Trang) on how e-commerce could be used in such business settings. The study confirmed earlier findings that the main difficulties for villagers to successfully implement e-commerce included: unreliable technological infrastructure, unavailable legal infrastructure, blocking of Internet access due to security concerns and the high cost of Internet connectivity (N. C. Tran, 2005).

At Bat Trang village this author heard first-hand of dilemmas faced by potters who were perpetuating precious family designs, or innovating from their special Bat Trang heritage. One potter explained that she could not post pictures of her ceramics into the web, as she feared that her designs – her family's secret and treasure – would be stolen. People in Bat Trang had some vague familiarity with the idea of intellectual property, but did not actually know how to protect themselves against imitators. There had been no action on formally registering unique industrial designs or patenting innovative production technology. S38 explained that the registration of a trade mark or design was often too cumbersome and expensive for a pottery producer. To protect their secrets, Quang Minh Pottery Kiln, for example, had a workshop where artists created new designs, then ceramics were produced to sell direct to customers. However the problem of imitation continues for many family enterprises.

9.5 Quan Ho folk music

The Dong Ho folk printing and the Bat Trang ceramics examples focus mainly on indigenous knowledge embodied in cultural goods combined with cultural tourism. The Quan Ho folk music performed at the Quan Ho Folk Music and Cuisine in Bac Ninh province combines the notion of cultural services with cultural tourism.

Figure 9.2 Adding value to traditional Quan Ho folk music. Source: the author

Born about the 13th century in the Bac Ninh province, the Quan Ho folk songs were to celebrate spring. Boys and girls came to adore Buddha and after that, gathered together in front of or inside the pagoda or in the field to sing Quan Ho. Originally, Quan Ho were exchange songs between two mandarins' families then they gradually spread out and became popular among the northern people. Groups were formed just for singing, and many marriages were cemented at these get-togethers. After centuries, Quan Ho became one of the most significant Vietnamese folk-song types and the most famous place to perform Quan Ho is in Lim festival, often in spring. Like Dong Ho printing, Quan Ho faced problems of sustainability because its rationale for existence was cultural value rather than profit.

With the coming of the new economy, Quan Ho Folk Music was commercialized so that customers could receive various services ranging from folk music by request, Quan Ho local food and drink, and information about the history of Quan Ho, as in Figure 9.2. The new business model opened up a new income source for Quan Ho singers and their managers.

There were perceived difficulties in the deployment of ICTs for indigenous knowledge promotion in the case of Quan Ho folk music. It was felt that the intangible traditional skills and know-how Quan Ho were very difficult to capture or transfer by modern technology. In practice, they were primarily handed down by word of mouth inside clans or from generation to generation (I21, interview, 20 Feb 2006). I30 explained the difficulty in capturing Quan Ho songs:

> Besides handing down orally, old artisans may also write down on paper. However, the obstacle is that it is impossible to set Quan Ho melody to modern music. A Quan Ho song composes of many repetitions of "i-a", pauses or castanets and rhythms that are too complex to recorded in modern stave. Therefore, the major method to preserve Quan Ho is learning through word of mouth from grandparents to grandchildren (interview, Feb 25, 2006).

She added "recording could be possible", but was doubtful that the totality of live Quan Ho performance or participation could be experienced or learned from recorded media.

Though ICTs may have high capacity to store and transmit indigenous knowledge and facilitate the creation and diffusion of indigenous knowledge such as Quan Ho folksongs, the fieldwork revealed that local experts in the heritage arts faced many difficulties in applying the new technologies.

It was noticed that none of the folk music singers of Quan Ho intended to use computers though they thought the Internet might help them to learn about the hobbies of the youngsters. I30 (interview, Feb 25, 2006) explained why she did not use the Internet: "Although I know that the Internet is a source of rich, accurate and quick information, being an art practitioner, my spare time for other activities is very limited."

Last but not least, respondents were confronted with awareness and psychological problems: many interviewees were too timid to go to a Net cafe for the first time in their life. Some others misunderstood the required prerequisite to use the Internet (I29, interview, Feb 25, 2006). I31 (interview, Feb 25, 2006) admitted: "Money is not a problem, but I don't have an electronics background so I never think of using computer." If they could overcome that timid feeling, for instance, with the help of an internet-literate peer, to try it for the first time, they could use it in the future.

9.6 Tradition, technology and the knowledge economy

9.6.1 Preserving heritage forms

Traditional art forms, such as folk music or printing, often need to be conserved in their age-old native styles and content. They are thus difficult to "update". Yet one of the most typical characteristics of the knowledge economy is innovation.

Despite the deep reservations among Quan Ho artists about the relevance of new technology, and innovation generally, the author noted individual attempts by young artists in the Quan Ho Folk Music and Cuisine Center who eagerly went to villages to collect folklore from old artisans (I29, interview, Feb 25, 2006, I30, interview, Feb 25, 2006).

I21, a member of the Vietnam Federation of UNESCO Clubs (interview, 20 Feb 2006), explained the importance of conservation of the original of Dong Ho engraving to the author as follows:

> If the paper (of the printing) is not made from the "zo" tree (poonah) which is now only hand-made by Phongkhe village in Yenphong district, the colors are not purely made from natural materials, such as the black from the ash of bamboo leaf, the yellow from the flower of sophora japonica or the shiny white from shellfish from Haiphong - that printing will not be Dong Ho.

In addition, English, IT skills and business skills were other problems for artisan masters to use IT (I21, interview, 20 Feb 2006).

Also noted were parallel governmental and international efforts to preserve local value. A workshop on experience and practice to preserve intangible culture and heritage was held in Hanoi from 15 to 17, Feb 2006 by the Agency of Cultural Heritage of Vietnam and UNESCO Vietnam to launch the UNESCO's Convention for the Safeguarding of the Intangible Cultural Heritage (VietnamNet, 2006a). Such activities could be regarded as some initial steps toward a content industry for Vietnam that was only at the infant stage (PTT, 2006).

9.6.2 "Top down" vs "Down top" approaches

In relation to their use of ICTs there is a marked contrast between the patterns noted in Dong Ho and Bat Trang, and an apparently comparable case in Gyandoot, India critiqued by Heeks (2005). Like May (2002) or Henwood (2004), Heeks has concerns about how far ICTs can really contribute to development. Discussing the ICT4D (ICTs for development) trend Heeks argued that:

> we are often blinded from this reality by the blizzard of e-development pilots, prototypes, plans and possibilities where "would" and "could" replace "does" and "has".

Heeks cited Gyandoot, a "top-down" initiative of computer kiosks in rural India, in support of this argument. Winning awards from the Stockholm Challenge and the Computer Society of India in 2000 with much publicity, later studies in 2002 nevertheless showed the kiosks of Gyandoot abandoned or closed owing to unreasonably low usage and few signs of developmental benefits (Heeks, 2005).

In contrast the ICT applications in Dong Ho and Bat Trang have grown not "top down" but "down top". They exemplify Meadow's contention that few people think about the long term:

> Although the perspectives of the world's people vary in space and in time, (...) the majority of the world's people are concerned with matters that affect only family or friends over a short period of time. Others look farther ahead in time or over a larger area – a city or a nation. Only a very few people have a global perspective that extends far into the future (Meadows et al., 1972, p. 19).

The Dong Ho and Bat Trang examples, compared to the Gyandoot example, suggest that incremental "grass roots", pragmatic action on ICTs in a development can be a better basis for long-term benefit than set-piece projects sponsored and guided from "the top".

ICT initiated from the grass-roots in Dong Ho and Bat Trang craft villages or in cyber cafes (see Chapter 13) seem to be full of vitality compared to top-down telecenter or imported projects that often die prematurely like the Gyandoot project. The Dong Ho and Bat Trang ICT applications are crucial bridges to help local businesses and villages to sell their knowledge-intensive products overseas.

However it was noted that at a certain level of development, the grass-roots ICT efforts of Dong Ho and Bat Trang faced a series of problems of funding, technology, management, or English competencies that require support from outside government and associations. In interviews there was debate about the professionalism and efficiency of ICTs applications developed by the crafters. I54 a leading expert in the ICT industry, argued that a do-all approach like Bat Trang was sub-optimal because in the value chain people with best skills in each task would work better than a potter who does it all, from making to selling his product in self-reliance mode:

> Regarding knowledge based resources like Dong Ho painting or Bat Trang ceramics it must be the responsibility of the national cultural reservation department not the local people to preserve the heritage. Similarly, regarding IT applications such as the use of website to promote Dong Ho printing, the assessment must come from the printing seller, not from the producer. When the producer is also the seller, the division of labor is not clear. To sell the product as his main objective, a professional seller would know how to find a proper market and the way to trade it, such as by advertisement or packaging. ICTs are only a part of this, not the whole solution. ICTs can be used in the marketing and promotion of traditional goods. Yet, that activity is not highly paid attention in Vietnam (I54, interview, Apr 8, 2006).

For these kinds of reasons it would seem that "bottom up" is not enough. Governments and NGO's need to link up with producers and their associations to facilitate optimal outcomes in ICTs for sustainable development.

9.7 Main insights from Chapter 9

Insight 9A: Indigenous, traditional art and craft occupations are endangered by competition and substitution in the open, globalized market economy, and by damage to environmental resources that have nourished their industries for generations. However the new conditions if well managed also open up opportunities, particularly through cultural tourism and Internet marketing of cultural products.

Insight 9B: Practitioners of traditional arts and industries are anxious to see their culturally unique knowledge survive and prosper in the context of new production and information technologies, and are cautiously open to change as long as this does not jeopardize the essentials of their traditional artisanship.

Insight 9C: Government policy frameworks which aim to assist the indigenous artisan sector to develop as part of the knowledge economy need to be subtle and holistic. In this sector particularly introduction of ICTs and other new technologies will succeed only if adopted "bottom up" and not imposed from above.

Chapter 10
ICT Nonusers

10.1 Scope of the chapter

This chapter on ICT nonusers, chapter 11 on ICT users and chapter 12 on virtual worlds are together relevant to the broad topic of this book project in considering, in the context of Vietnam, the extent to which ICTs are capable of linking individuals or groups together in a collaborative effort to achieve economic sustainability. This chapter

- introduces the concept of the digital divide, and notes the results of a 2003 study which identified actual and potential channels for information transfer in rural areas of Vietnam.
- reports on 16 interviews with nonusers of ICTs undertaken in early 2006.
- summarizes characteristics of the nonusers interviewed such as geographical distribution, age, occupation, income and level of education.
- answers questions about nonusers of ICTs relating to their understanding about possible uses of ICTs; their main channels/sources of information; and the factors perceived by them as obstacles to accessing ICT-based information, particularly the Internet.

10.2 The digital divide

This chapter is an account of ICT nonusers and how they manage to obtain the knowledge they need to function in the digital age. They are the people who are on the "wrong side" of the digital divide, who are supposedly disadvantaged compared to their digitally enabled fellow citizens.

The digital divide that occurs in many countries across the world may well be a significant factor in economic non-sustainability. It was claimed that those who for whatever reasons cannot use ICTs and the Internet may be relatively unable to contribute to, or benefit from transformations towards sustainable economic development.

Few doubt the reality of the digital divide concept, noting the disadvantages of those who are information poor in relation to social equality, job opportunities, or educational advancement.

There is a debate on the causes of such social stratification in the digital world. Two major viewpoints are:

- The optimistic theory of normalization which believes that at least in affluent nations, the networked community grows step by step over time until it includes the society as a whole (Resnick, 1988); and
- The pessimistic theory of diffusion which predicts an ongoing social stratification in the application of new technologies (E. M. Rogers, 1995, pp. 435-442).

If people are asked why they do not use computers or the Internet, their answers may represent a mixture of have-nots and want-nots (van-Dijk, 2005, p. 28). A survey done in Germany in 1999 showed the leading six reasons why 501 German nonusers did not buy a PC (ranked from top down): 1) Don't need a PC for job or private use, 2) Don't know what I would use it for, 3) No time or liking, 4) PCs too expensive, 5) PC means less social contact, and 6) Don't think I can handle a PC (ARD/ZDF & Multimedia, 1999). Similarly, a survey by the National Telecommunications and Information Administration (NTIA) in 2002 revealed reasons for not being online among the 42% of American identified as ICT nonusers – see Table 10.1.

In Vietnam, together with other factors, the digital divide is perceived as a critical obstacle to the capacity for growth. It is believed that, especially in rural areas, the digital divide contributes to a lack of citizen participation in decision making, villagers' inability to apply new farming techniques, and low levels of education; and also that being unskilled and "hungry" for information are important determinants of persistent poverty (World-Bank, 2003, pp. 23-24).

For Vietnamese ICT nonusers, fortunately, there remain other alternatives for knowledge transfer. Smith, Toulmin and Qiang (2003) identified poster boards, loudspeakers, radio, TV, newspapers, books, telephony and human intermediaries as real-world alternative channels of information in rural Vietnam as shown in Figure 10.1. These authors found that loudspeakers were the medium with the widest coverage as media of communication to rural people. As asserted by them in a World Bank report "The only medium which reaches all these key groups well is the loudspeaker system; radio and TV are second in effectiveness" (World-Bank, 2002, p. 46).

Table 10.1 Reasons for not Using the Internet, Percentages of Nonusers (United States, 2002). Source: Lenhart et al. (2003, p. 10)

	Major reason	Minor reason	Not a reason
I don't want it	52	16	26
I don't need it	52	19	24
I'm worried about online pornography, credit card theft	43	14	37
It's too expensive	30	18	42
I don't have time	29	17	49
The Internet is too complicated	27	9	43
I don't have a computer	11	n/a	n/a

Similarly, UNDP and MARD ranked loudspeakers as the top source of information for both availability and effectiveness, followed by the village head, local TV programs, radio and national TV. Neighbors or peers only held a relatively low position (rank #6) whether for availability or usefulness in that report (UNDP & Ministry of Agriculture and Rural Development, 2003, p. 11).

10.3 Fieldwork data about ICT nonusers

Fieldwork data about ICT nonusers was collected in the period January to April 2006 through in-depth interviews with 16 people who had never used a computer or the Internet before. The sample of nonusers comprised three groups of nonusers: rural people, urban people and folk artisans. Consistent with the purposive sampling approach of the book, in each province the author hired a knowledgeable local guide to help him in the tasks of identifying persons to interview, seeking their agreement to be interviewed, and arranging where the interview would be held, whether at the interviewees' workplace or home.

Agriculture still accounts for 20% of GDP (despite its recent decline as a proportion of the national economy). Rural villages and hamlets house 74.1% of the whole population (Khong, 2002). Therefore care was taken to include rural, farm based interviewees who were interviewed at their homes.

The following three open-ended issues (among others) were explored with the sample of ICT nonusers, not necessarily using these words but in a colloquial, conversational style that would be comfortable for interviewees:

- What kinds and degrees of understanding are there about ICTs and online sources of information and knowledge for tackling problems associated with economic and environmental sustainability?
- What are the major channels of information used in acquiring new information and knowledge?
- What are the obstacles that prevent nonusers from enjoying the benefits of ICTs? If they have tried to use ICTs, what are major problems they have encountered?

Most of the interviews were with individuals; but one group discussion was held between the researcher and three farmers in Longan province. Details of the interview procedure are given in Chapter 1 and Appendix 3.3.

10.4 Characteristics of the sample of nonusers

The sample of nonusers was comprised as follows:

10.4.1 Gender

Half of respondents were female (50%).

10.4.2 Geographical distribution

As distributed by region of Vietnam:

- 11 lived in the North
- 1 came from the Center, and
- 4 lived in the South.

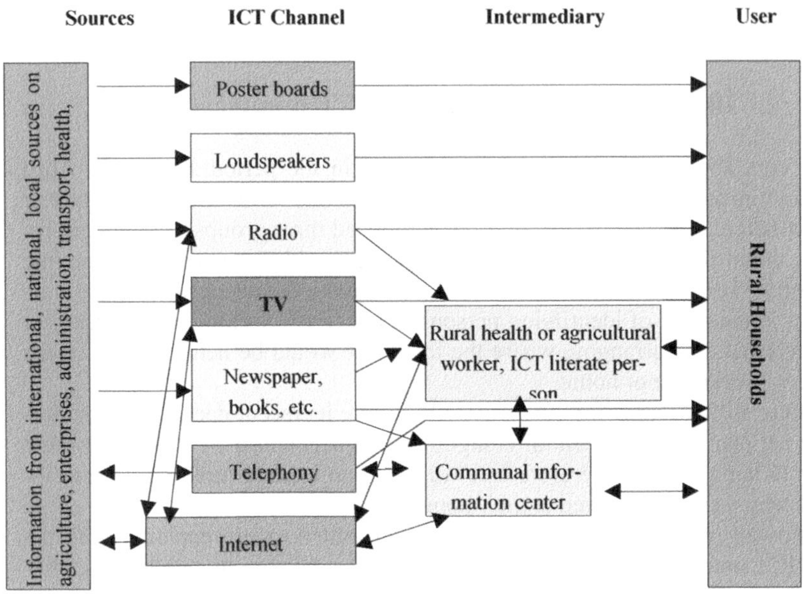

Figure 10.1 Information channels in rural areas identified by a previous study. Source: Adapted from Smith, Toulmin & Qiang (2003, p. 35)

As distributed by rural and urban location:

- 4 lived in agricultural villages
- 5 lived in suburban areas or towns
- 4 lived in villages that were being urbanized on the outskirts of a town, and
- 3 lived in cities. All three urban nonusers were persons who came from the countryside to the city to work or study.

10.4.3 Age

Table10.2 Age of ICT Nonusers in Sample

Age	#	%
20-29	3	18.8
30-39	3	18.8
40-49	3	18.8
50-59	4	25.0
60+	3	18.8
Total	16	100.0

Ages ranged from 25 to 71 and interviewees were distributed almost evenly in ten-year groups as shown in Table 10.2. No interviewee below 20 years of age was available – thus the author would recommend further study to explore the features of young nonusers.

10.4.4 Occupation

Among nonusers, farmers were the biggest group consisting of three in the North and three in the South. Their main jobs were cultivating rice, corn, or vegetables and aquaculture. Other nonusers were three folk-artists in Bacninh province, one owner of a large seafood restaurant in Hochiminh City, one director of a Quanho culture business center in Bacninh province and one student in Hochiminh city who came from the poor province of Quangtri in the Center. Seven groups of jobs were identified in the interviews as shown in Table 10.3.

Table 10.3 Occupations of ICT Nonusers

Job	#	%
Farmer	6	37.5
Fisherman	1	6.3
Folk Artist	3	18.8
Worker	2	12.5
Retired	1	6.3
Student	1	6.3
Business	2	12.5
Total	16	100

10.4.5 Income

The monthly income of nonusers in the sample varied widely from VND 100,000 to 4,000,000. The average income of a northern/central farmer and a student was much lower than that of other jobs (VND 400,000 and 1,237,500, respectively). Other nonusers had an income comparable to that of a user. It would be necessary to remark that 1) farmers were more economically self-reliant, and their incomes were often in the form of crops such as rice or maize rather than money; 2) in 2006, USD 1 was VND 15,983 at the official exchange rate, yet was about VND 2,826 in purchasing power parity (Central Intelligence Agency, 2007). Thus, a farmer's actual income, if exchanged into money, could be relatively high. For the reader's reference, 1kg of rice costs about VND 5,000 and a student's lunch at a minimum costs about VND 5,000.

10.4.6 Education

Categories commonly used to describe formal educational levels in Vietnam are primary, secondary, high school, vocational training, higher education, and 'other' – and these categories were used in the study. Among the nonuser sample, farmers had lower education (from primary to secondary levels) than the other occupations (from vocational training to higher education). Several of the nonusers had attained high school graduation or bachelor degrees, or were university students. Chapter 11 which reports on interviews with intensive ICT users includes a diagram which compares the educational levels in the nonuser sample with those of the ICT users interviewed.

10.5 Understanding about ICTs and online sources of information

Almost all nonuser interviewees were aware of the important role played by information and knowledge in competing within the demanding contexts of work and life. However few people knew much about the actual use of ICTs, or of the kinds of information content available online. Instead, they were using other channels to get information, though not all of them were seen as effective and useful.

10.6 Major channels of information used by ICT nonusers

As mentioned above, in a study five years before, the loudspeaker was identified as the major channel of information for farmers (P. Smith et al., 2003; World Bank, 2002). However this study suggests that the loudspeaker might be losing its leading position. As depicted in Figure 10.2, the major channels were:

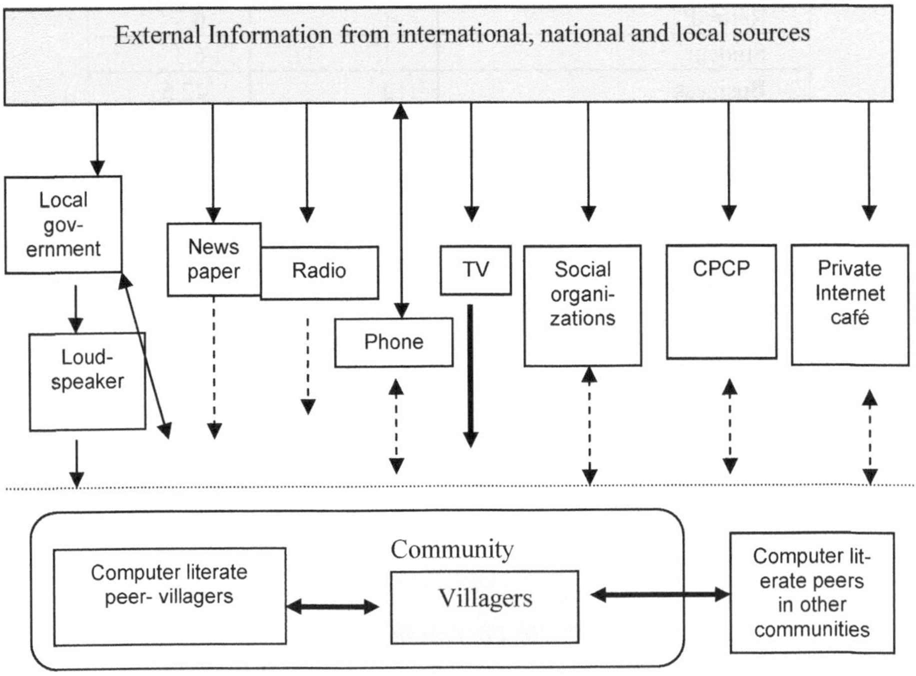

Figure 10.2 Main sources of external information for ICT/Internet nonusers

- **Television and peers** (people who lived, worked or studied nearby). Nine of the 16 interviewees regarded TV and peers as the main source of information. While TV provided various kinds of general information, peers supplied specific knowledge, such as job availability, working know-how or learning-related information. TV and peers overtook loudspeakers as the top sources of information.

- **Loudspeaker systems** were a channel through which the local authority announced local guidelines or government regulations to the people e.g. hygienic cleaning, farming schedules, health or vaccination timetables. Three interviewees recognized the existence of loudspeakers in communes but revealed that they were not important to them. Information from loudspeakers was often deemed to be less attractive and very narrow (I28, Feb 25, 2006). Furthermore, the transmission time was very limited and often conflicted with the availability of villagers who were usually too busy, often working at a second job (I37, Mar 7, 2006);
- **Newspapers and books** were not read by rural nonusers in contrast to nonuser residents in urban areas (I27, Feb 25, 2006; I26, Feb 24, 2006; I53 Apr 7, 2006);
- **Telephone**: Thanks to the expansion of the public telecommunications infrastructure recently, more households possessed telephone sets, mainly fixed phones. Seven interviewees reported that they could use the telephone to exchange information about the market or jobs with their peers;
- **Radio**: Four interviewees claimed that the radio was a source of information but two of them (I27, Feb 25, 2006 and I37, Mar 7, 2006) added that they spent very little time listening to the radio;
- **Communal government and social organizations** (For example, Farmer's Associations, Women's Associations or Veterans' Associations): Social organizations sometimes organized meetings to provide villagers with information, such as know-how to raise cows or goats, or to obtain loans. Only three out of the 16 interviewees viewed communal government or social organization meetings as sources of information, but said they nevertheless did not attend because either they did not have time (I37, Mar 7, 2006 and I51, Apr 7, 2006) or that the information provided was not relevant or beneficial (I51, Apr 7, 2006);
- **Computer and the Internet**: To all nonusers those technologies were something far away. One interviewee did not have any idea of what benefit the new technologies might offer (I37, Mar 7, 2006).
- **Communal Post and Cultural Points** (CPCPs): According to Vietnam Posts (Vietnam Communist Party Poliburo, 2000), as a joint initiative between the Ministry of Posts and Telematics and the Ministry of Culture and Information, about 8,000 CPCPs with Internet and telephone connection were set up by Vietnam Posts and Telecommunication Group in a total of 9,069 communes. That initiative was an effort to enable villagers in poor and remote areas to have access to ICTs (see Chapters 5 and 13). CPCPs were not identified as an information source in the interviews reported here, and this may cast doubt on its cost-benefit as a knowledge transfer channel.

An important insight gained in the nonuser interviews was that computer literate peers, or neighbors played an important role for nonusers in providing new knowledge, with geographic proximity and relevance of content of the information sources as key factors. With peers, nonusers can seek the most necessary information conveniently and informally; and they can check if the knowledge they receive is usable or not. It could be argued that some years ago, general information from TV or loudspeakers might have been adequate, but that in recent times villagers need more specific, practical knowledge for working or living, and traditional one-way media such as loudspeaker systems or even broadcast television cannot keep pace.

10.7 Obstacles for ICT nonusers in accessing the Internet

Interviewees confirmed that there were a range of obstacles facing rural residents in gaining access to the Internet, the major source of digital information. Even villagers who lived very close to cable networks for the Internet, TV and telephone in Bacninh or Haiphong did not access Internet services. Major obstacles for nonusers were not identical for everyone but varied by profession, residential locality, age and economic status, as outlined below.

10.7.1 Financial cost of ICTs

The cost of ICTs was a forbidding barrier to farmers. In the North, the income from farming was too low, about VND 100,000 a month (approx. USD 6) in relation to an Internet installation fee of from VND 80,000 to 470,000 (I27 and I28, Feb 25, 2006, I35, Mar 7, 2006) or a cable TV subscription fee at VND 55,000 a month (I37, Mar 7, 2006). In the South, I51 (Apr 7, 2006) claimed that "only a rich family can afford to buy a computer at 3.5-5 million VND... Many households are very poor or live in slums".

10.7.2 Widening rural-urban gap in infrastructure availability

The gap in infrastructure availability between urban and rural areas has become larger. The telecommunication infrastructure, including Internet, telephone, cable TV, developed quickly and became more available in urban areas. However in communes like Dongho in Bacninh province, or Nhanbinh in Longan province, Internet service was not available. The nearest cyber café to Nhanbinh, for example, was 7km away. Students at secondary or high schools were not taught IT because there was no computer lab there (I51, Apr 7, 2006).

10.7.3 Subjective difficulties

In regard to subjective difficulties, time constraints were often mentioned. Farmers claimed they were very busy doing a second job to get more income (I26, Feb 24, 2006, I35, Mar 7, 2006). They also stated that their educational limitations, in areas such as English and business skills, precluded them from using the Internet (I24, Feb 24, 2006). Consequently, villagers were reluctant to try using the Internet because they believed they lacked necessary skills and knowledge (I51, Apr 7, 2006). Also, farmers did not see potential benefits from the new technology, as said by I35:

> Work at communes is often simple, do not require technology. The Internet is something far away. We do not know what benefits ICTs can bring about for us. (Mar 7, 2006)

10.7.4 The elderly, the young, and helpful friends

While most older nonusers would say no to ICTs, younger persons appeared more enthusiastic in approaching the Internet. Many children were observed surfing the web in Internet cafes within communes. Though being more enthusiastic towards the new technology than their elders, some juniors nevertheless experienced the barrier of psychological reluctance. Shyness and timidity stopped many interviewees from their first trial of a new technology like the Internet, as I26 revealed:

> The feeling of a person who never uses the Internet is very shy... Although I already know the Internet is a new channel rich of diversified information and there is an Internet cafe next door, I dare not try (I26, Feb 24, 2006).

If asked to try the Internet by a peer, people might overcome that adverse feeling, as I53 recalled: "I used a computer for the first time thanks to some friends who often went to Internet cafes [and who] asked me to go with them" (I53, Apr 6, 2006).

While almost all farming interviewees had no intention to use the Internet or a computer, some revealed that they might encourage their children to use ICTs in the expectation that their children could learn the emerging technology. Along these lines, I27 (Feb 25, 2006) bought a computer set for his son who studied in Hanoi. Similarly, the old artisan I21 asked his children to set up a website to market his Dong Ho prints via the Internet.

10.7.5 Schooling seen as a bridge over the divide, but difficult to attain

In both the North and the South of Vietnam, villagers wanted to send their children to school in the hope of them having a better future, yet low incomes prohibited them from achieving this dream. In the South in particular, three farmers in a group discussion listed various reasons: schools were far (6 to 10 km), transportation was not good and separated by rivers or canals, thus students had to cross these by boat, parents could not afford to buy bicycles, clothes and books for their children to study further at a high school (grade 10 to 12), only one such high school being available for a district (I51, interview, Apr 7, 2006). In consequence, many children in rural areas quitted school after grade 8 or 9 to come back to the traditional rice pad, as explained by I37 (interview, Mar 7, 2006).

10.8 Main insights from Chapter 10

Insight 10A: Almost all nonuser interviewees were aware of the importance of information and knowledge in the demanding contexts of life and work, but computers and the Internet were not seen as directly relevant to this group. Most of the older nonusers were not interested in using ICTs, but some said that they might encourage their children to use ICTs. Younger people appeared to be more enthusiastic about use of the Internet, although some reported initial fear of using the technology for the first time, and needed the encouragement of peers to assist them in making a start.

Insight 10B: Television and peers were identified by the sample as their main source of information, TV providing various kinds of general information, peers providing specific information such as job availability and knowledge needed for work and learning. Nearly half of those interviewed reported that they could use the telephone for exchanging information with their peers about the market or jobs. There was a small awareness in the study group of radio and communal government and social organizations as possible information sources. Newspapers and books, and CPCPs were not identified by the group of ICT nonusers as channels of information.

Insight 10C: The kinds of factors that were viewed by the nonusers as obstacles to accessing information on the Internet varied with the different occupations, residential localities, ages and economic status of the individuals. The financial costs of ICTs; the lack in rural areas of availability of telecommunication infrastructure and Internet services; time constraints; educational limitations; perceived lack of skills and knowledge; perceptions of the Internet as being non-relevant to work; and a lack of awareness of potential benefits were obstacles identified during the interviews.

Insight 10D: The villagers in the nonuser group viewed schooling beyond a basic level as desirable for a better future, but reported that it was difficult to attain because of factors such as the distance of schools, transportation and terrain difficulties, and the costs of bicycles, clothes and books.

Chapter 11
Intensive ICT users

11.1 Scope of the chapter

Explores the characteristics of users of the Internet in Vietnam, what they use the Internet for, and the kinds of issues confronting them in obtaining the information and knowledge that they require. More specifically, the chapter

- outlines the characteristics of the user interviewees in the study: 15 ICT users and seven game players (all of whom were under the age of 35)
- describes the experiences of the computer and Internet users interviewed, including changes in life and work styles relating to their Internet use, and advantages and problems they have encountered.
- provides background information on the issues of computer gaming, the emergence of online "virtual" communities, and IT literacy. Gaming and virtual communities are explored in greater detail in Chapter 12.

11.2 Assessing the situation of Vietnamese ICT users

Chapter 10 sought to provide some insight into the situation of about 70 million Vietnamese people who are living and working without computers and the Internet. This chapter seeks to do the same for the fast growing minority who are ICT users.

During the last decade, the Government of Vietnam has been putting in place a series of policies and practical measures to accelerate ICT deployment, with the expectation that these will assist citizens take advantage of the opportunities that the new digital technology can offer (Vietnam Communist Party Poliburo, 2000). As a result by April 2007 the Vietnam Internet Center, the agency in charge of Internet management in Vietnam, estimated that about 15.7 million people were using the technology, 2.8 million higher than the figure of April 2006 (Vietnam Internet Center, 2007b).

However knowledge is sparse in the setting of Vietnam, a developing country, about such questions as:

- For what purposes do people use computers and the Internet? and
- What are the changes, benefits and difficulties that users of the technologies are experiencing, and how do they deal with these?

Often people and governments can confuse aspirations for human wellbeing and the actual contributions that technology can make. Despite their powerful capabilities ICTs do not deliver a technologically-driven utopia. Van Dijk struck a note of realism when he stated: "despite its image of being interactive, most Internet usage, apart from emailing, is relatively passive and consuming" (van Dijk, 2005).

In response to such concerns, fieldwork was undertaken to explore the situation of ICT users in Vietnam. New users are constantly joining their ranks, and with the growth of Internet participation and traffic, day and night, they build ever more densely interweaving networks of human communication in Vietnam's virtual space.

This chapter presents the results of the author's fieldwork investigation to find out who the Internet users are, what these people use the Internet for, and what issues confront them in obtaining information and knowledge that they require.

11.3 Fieldwork data about intensive ICT users

The main source of information for this chapter is data collected from January to April 2006 during in-depth interviews with 22 intensive users of ICT in Vietnam.

Two types of ICT users were included in the sample: one category whom the author designated just as "users" and another designated variously as "players", "gamers" or "game players".

A user was defined as a person who had at the time been using a computer or the Internet for any purpose, while

A game player was a person who played online or offline computer games as their primary purpose for computer use.

Twenty two respondents were chosen in work places and Internet cafés from the North to the South of Vietnam. Selecting interviewees in these environments increased the chances of the sample being made up largely of intensive, rather than possibly occasional, users. About 10% of people refused to answer either because they were busy or did not want to be exposed. In Internet cafés, the author and his assistant walked in and selected one or two persons at each site. In addition, they also selected interviewees from a private shop, a university, a research unit and two governmental agencies. Interviewees were asked conversation-prompting open questions such as:

- For what purposes do they use ICTs?
- What are difficulties that they are experiencing in their usage, and how do they deal with these?
- Does the usage of ICTs help them in their work and life, and if so, how?

11.4 Characteristics of the sample of ICT users

The 22 respondents comprised 15 users and 7 players.

11.4.1 Gender

Within the group of computer and Internet users, the ratio of female to male users was 53.3% to 46.7%. No female could be described as a game player. All 7 such players were male. Several female teenagers were playing games in Internet cafés visited by the author and his assistant, but when questioned, they were found to be undertaking other online activities as well as gaming.

11.4.2 Geographical distribution

In contrast to the nonuser sample, the majority of whom lived in rural areas (Chapter 10), the majority of users sampled were currently living in cities or towns. Four of them (18%) lived in rural areas near suburbs.

11.4.3 Age

Reflecting national statistics on age distributions of ICT nonusers and users, a higher proportion of younger interviewees were selected for the ICT user sample than for the nonuser sample. The age range of the combined user sample (i.e. users plus players) had an almost normal distribution (see Figure 11.1). No user was aged below 15 or above 35. Game players were even younger. More than half of them were teenagers. The youngest gamer was only 12. The oldest player was not over 21.

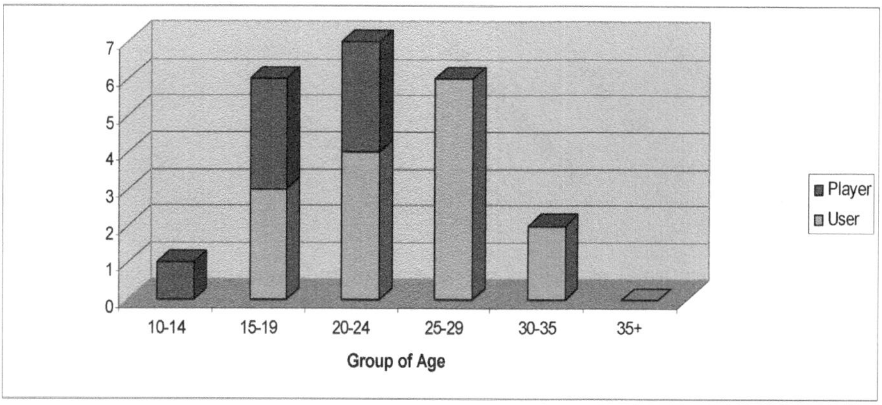

Figure 11.1 Age of general users and game players

11.4.4 Occupation

Users worked in various fields, as follows: 3 students, 2 government officers, 2 university staff, 1 researcher, 1 public relations officer, 1 medical doctor, 1 security guard, 1 hairdresser and 1 sales manager. The occupations of game players were much more homogeneous: they all were students from high school or university except one jobless player who had just graduated from high school.

11.4.5 Income

Users other than students had an income varying from VND 600,000 to 3 million per month with a mean of VND 1,216,000. Students belonging to the "user" category were given by their parents an average monthly allowance of about VND 544,000.

Players mostly did not have any actual revenue but an allowance from their parents varying from VND 90,000 to 1.5 million a month.

As a yardstick for interpreting these figures the market price of 10kg of rice, one of the most popular commodities in Vietnam, cost about VND 50,000 and a student's lunch was from VND 5 to 20,000.

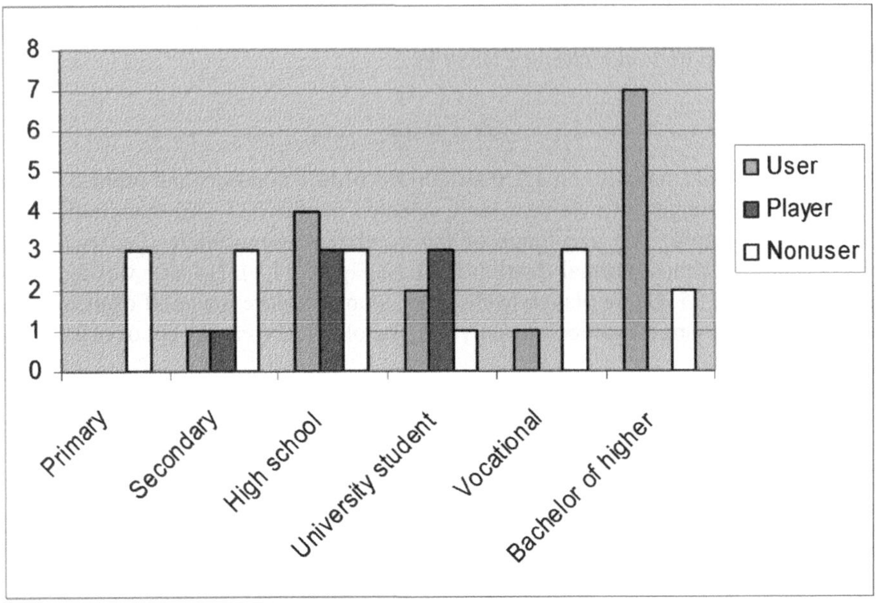

Figure 11.2 Educational profile of users, players and nonusers. Note: The ICT nonuser sample is described. One non-user's educational level is unknown.

Many students reported that they spent a considerable amount of their money on gaming. I19, a university student (interview, Feb 16, 2006), for instance, claimed he paid about VND 500,000, one third of his money, to play Volamtruyenky, one of the most popular games among Vietnamese youth in 2006. Similarly, I18 (interview, Feb, 16, 2006), a student of sociology and human culture, spent 5 hours and VND 15,000 each day, totalling VND 450,000 per month, for Gun Bound. Players in rural communes spent much less in total, but still a high amount in proportion of their stipend. For instance, I22 (interview, Feb 20, 2006) a high school student, paid VND 100,000 monthly, one third his allowance for the Internet: the same amount could buy his rice for a whole month.

11.4.6 Education

The educational background of users, players and nonusers (see Chapter 10 for an account of the nonuser sample) is shown in Figure 11.2. Users in general had a higher educational profile than nonusers. Many users had bachelor's degree or higher. Most players were students from high school or university. One gamer who participated in the sample was only in grade 7.

11.5 Experiences of interviewees as computer and Internet users

The next section examines a number of factors which may help provide a better understanding of Internet use patterns in Vietnam. Five main factors which emerged during interview conversations as contributing to patterns of online use were:

- time spent online
- length of experience of computer use
- access places
- educational level, and
- purpose of online activities.

Other interacting factors included gender, living location and hobbies.

11.5.1 Time spent online

There seemed to be no association between hours spent on the computer and income and experience, but there did appear to be a clear difference between rural and urban users. The average number of years of experience of the urban users of computers (8.9 years) and the Internet (6.0 years) was longer than that of the three rural users in the study (2.3 years and 2.3 years, respectively).

Also, in terms of online hours per week, the urban users interviewed spent approximately four times more hours than the rural users (23.6 hours vs. 5.2 hours).

Players spent much more time with the Internet than users, with a mean of 47.4 hours per week (about 7 hours per day). I08, a player in grade 12 even claimed that he sometimes played 20-22 hours a day (interview, Jan 24, 2006). Young players were exposed to computers and the Internet very early, for example, the 12 year old interviewee I33 (interview, March 6, 2006) disclosed that he had experienced computer use for 3 years. Some even touched a keyboard before going to primary school as revealed by I07, a gamer in Hanoi (interview, Jan 24, 2006). Data is shown in Table 11.1.

Table 11.1 ICT/Internet experience

	City	Rural	Player
Average years of using a computer	8.9	2.3	5.6
Average years of using the Internet	5.7	2.3	4.1
Average weekly hours of use	23.6	5.2	47.4
Maximum weekly hours of use	56	8	100

11.5.2 Access Places

As for where interviewees used a computer, the 5 main places identified (in descending order of use frequency) were: Internet café (81.8%), home (50%), work place (22.7%), school (13.6%) and friend's house or post and telecom companies (PTCs) (each 9.1%).

There was not much difference in frequency of use between a normal Internet café and a PTC's Internet café. For the purposes of this chapter Internet café and PTC are regarded as one type of access place.

Internet cafés (other names: Internet dealer or Internet service center) were the most popular places where users could come to play games, chat, surf the net or make an international call by VoIP at very low rates (VND 2-3,000 per hour). These places were usually full of students from nearby schools. In the sample interviewed more than 80% of interviewees cited net surfing as a purpose of use, and more than 50% of the gamers accessed the Internet from Internet cafés. Figure 11.3 is a picture of an Internet café in Hochiminh City.

Figure 11.3 A private Internet café in Hochiminh City. Photo: the author

In the context of Vietnam a perhaps surprising finding from the interviews was that many users and players had Internet access from home. Employees also accessed the Internet from their work places. Few persons said they accessed Internet from a friend's house. It might be of concern to government that schools did not play much of a role in providing computer and Internet services to students (only 7.7%).

11.5.3 Purposes of online activities

To answer the question "What are the primary purposes for which you use the Internet/ ICTs?" users and players named as their top 10 purposes the activities shown in Tables 11.2 and 11.3. In Table 11.2 the purposes are cross tabulated with educational level and also compared to U.S. data in Table 11.4. In Table 11.3 the purposes are cross tabulated with other factors.

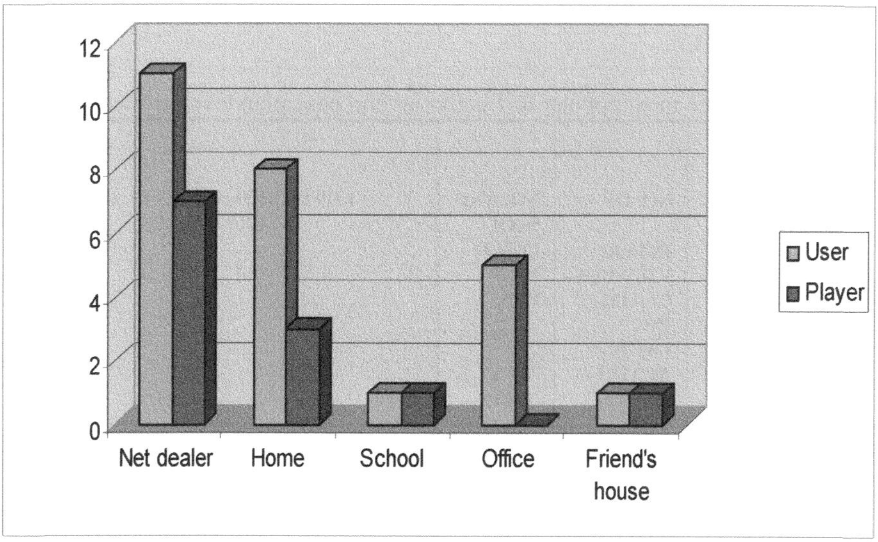

Figure 11.4 Access place (number of persons)

Of the sample of 22, 90% of users had at least an email account, all with Yahoo. However, users did not spent much of their time on email but on other activities, which are listed in descending order from most-mentioned: chatting, playing online multiplayer games, searching for news or information, downloading online music, downloading educational materials, playing solo games, work, other entertainment, creating online content, making friends and relaxing. While the majority of users did not chat with a stranger but only with old friends, several male users made friends with unknowns (I33, interview, Mar 6, 2006; I19, interview, Feb 16, 2006).

Usage purposes cross tabulated with education

There existed differences between usage patterns of those who did, and did not have higher education. Bachelor's degree holders spent less time on chat and computer games than others, and spent more time using computers and the Internet for work, study, news and information. University students were the most active online users, exploiting many functionalities of computers and the Internet, with relatively high usage rates for email, chat, and multiplayer games followed by study, online music and searching news and information. University students also created online content.

High school students used less diversified applications than university students, and spent more time on emailing and chatting, followed by multiplayer or solo games, news and information and other activities. Several male high school students played truant from school to play games at Internet cafés. Only one school student in lower grades was included in the sample: he (I33, interview, March 6, 2006) played multiplayer games, listened to music or studied mathematics on line when he had spare time – a relatively limited range of activities.

Usage patterns cross tabulated with other factors
Besides education, types of online activities also varied with other dimensions such as gender, experience of computer use, hobbies, and urban versus rural locations, as shown in Table 11.3.

Table 11.2 Top ten most popular ICT activities and education level in Vietnam

RANK	**ACTIVITY**	**NO. OF INTER-VIEWEES CLAIMING EACH ACTIVITY**	**NO. AND % OF INTER-VIEW-EES**	**EDUCATIONAL LEVEL OF INTERVIEWEES**			
	Multiple responses invited from each interviewee			**Secondary school**	**High school**	**University student**	**Bachelor or higher**
			N=22 (100%)	n=2 (9.1%)	n=7 (32%)	n=6 (27%)	n=7 (32%)
1	Email	20	91%	1	6	6	7
2	Chat	14	22%	1	6	5	2
2	News	14	64%	1	3	3	7
4	Online/ interactive games	11	50%	1	3	4	3
4	Music	11	50%	2	2	4	3
4	Study	11	50%	2	1	4	4
7	Offline/ solitary game	5	23%	0	3	0	2
7	Work	5	23%	0	3	0	2
9	Other entertainments	3	14%	0	0	0	3
10	Create online Content	2	9%	0	1	1	0

Table 11.3 Top ten most popular activities cross tabulated with other factors

ACTIVITY	NO. OF INTERVIEWEES CLAIMING EACH ACTIVITY (Multiple responses from each)	Male	Female	User	Player	Rural	Urban
Email	20	11	9	14	6	5	15
Chat	14	9	5	8	6	3	11
News	14	7	7	11	3	2	12
Online/ interactive games	11	10	1	4	7	0	11
Music	11	6	5	7	4	2	9
Study	11	6	5	7	4	3	8
Offline/ solitary game	5	2	3	5	0	2	3
Work	5	2	3	5	0	1	5
Other entertain-ments	3	0	3	3	0	0	3
Create online Content	2	2	0	2	0	1	1

Gender

Types of use made by female and male interviewees were somewhat different. Male users interviewed reported spending more time on chat and multiplayer games than females who reported spending a relatively higher proportion of online time searching for news and information, playing solitary games, entertainment, study and work. In searching news or information, male interviewees looked for politics, sport (I45, interview, Apr, 5, 2006) or lottery information (I15, interview, Feb 15, 2006), while female interviewees spent more time on social, business and IT information (I06, Jan 24, 2006) or on topics relating to women or children (I10, Jan 24, 2006). However there was no difference between male and female interviewees for emailing and listening to online music. These findings are reasonably similar to usage patterns of American online male and female users reported in 2001 (Howard, Rainie, & Hones, 2001).

Length of experience of computer use

This factor also appeared to be associated with the purposes for which interviewees used ICTs. The very experienced (>5 years) Internet interviewees in Vietnam spent the largest proportion of time dealing with email and instant massages, playing multiplayer games, undertaking specific searches for news and information, and performing professional work, while new users (< 1 year) spent most time emailing and chatting, followed by information surfing, offline or solo games and music.

Age

Among the interviewees people of different ages made different uses of the computer. Older users spent less time in chatting than younger users. Beside education, other co-factors associated with age, for example, marital status or family duties, may limit the time that people can spend online. In his interview I34, a medical doctor, explained why he seldom chatted online: "Before, I chatted a lot, but since I have got married, I never enter a chat room except when I sometimes chat with my sister overseas" (interview, Mar, 2006).

Living location

Although there were only two rural users in the sample, the purposes of use they reported were decidedly less diverse than the urban interviewees. Their online activity was restricted to email and chat. They also spent considerable on offline game playing owing to the low quality of connectivity available to them (I22, interview, Feb 22, 2006). In terms of creating online content, I23 (interview, Feb 22, 2006), a rural businessman, hired an IT company to create an website for his micro-business in contrast to an urban student I45 who himself as super-moderator created the site for an online forum of Vietnamese fans of the Italian AC Milan football Club.

15.5.4 Usage patterns in other countries

Europe

The spring 2000 Eurobarometer's study showed that email was seen as the most popular "killer app" of European online users, followed by searches for educational materials, downloading of software, searches for information or news, playing of computer games, listening to music, buying things online, and watching TV (Norris, 2001, pp. 222-225).

United States

In the U.S. "level of education" was one of the most important factors associated with patterns of behavior among ICT users. Computer and Internet usage time was greater among Americans with high education levels than those with low levels of education (Fallows, 2004). Howard, Rainie, & Jones (2001) also showed that Americans with a bachelor's degree or higher level of study used significantly more applications in the areas of information, email, work, education, and business, while Americans with lower educational levels used more entertainment and chat applications. Very experienced users (six years or more) spent more time on email, browsing, professional work, news, and schoolwork than new users (less than one year) who spent most of their time on email and browsing, followed by entertainment and professional work (Lebo et al., 2003).

While the Vietnamese online users spent most of their time using simple applications like email, chat and games, American users (UCLA, 2004 and Fallow, 2004) exploited the web with some quite complex activities such as shopping and buying online, searching for health or travel information, tracking credit cards, banking and paying bills.

Table 11.4 Top ten most popular Internet activities in the U.S. (%). Source: 1: (Lebo et al., 2002, p. 10); 2: (Lebo et al., 2004, p. 29)

US Internet Activities	**% In 2000**[1]	**% In 2003**[2]
	N=2096	N=2009
Web surfing or browsing	81.7	90.4
Email and Instant message	81.6	77.2
Finding hobby information	57.2	46.7
Reading news	56.6	52.0
Finding entertainment information	54.3	45.6
Buying online	51.7	44.2
Finding travel information	45.8	34.6
Using instant messenger	39.6	...
Finding medical information	36.6	36.1
Playing games	33.0	28.5

Lebo et al. (2004, p. 95) wrote:

> The impact of the Internet ....cannot be overestimated. A technology that practically did not exist in American homes only a few years ago is now a standard feature in nearly two-thirds of homes.

Popular Internet activities were noted by Lebo et al who undertook longitudinal studies of Internet use in the US. Examples of such activities are listed in the table below (Lebo et al 2002; Lebo et al, 2004).

In a 2002 UCLA study, "experience" was found be an important factor influencing usage patterns (Lebo et al., 2003). In that analysis new users were defined as persons with 2 years or less experience with the Internet, and very experienced users were those with 5 years or more. If usage trends in Vietnam follow those reported by Lebo et al in their 2004 US study, then activities such as online buying, searching for medical or travel information, and tracking of credit cards will become much more common in Vietnam in the next few years.

Korea

Studies about ICT usage in developed countries, such as South Korea, have highlighted factors which are likely to influence computer and Internet use. According to the Ministry of Information and Communication of Korea, after 10 years of the Internet in Korea, age, gender and region were factors affecting digital divides within the population. A survey of 17,347 residents across Korea undertaken in June 2004 showed that:

- 95% of the Korean population aged 6 to 29 were online users, compared with 86.4% of people in their thirties, and 27.6% of those in their forties,
- 74.4% of men compared with 62% of women used the Internet, and
- 70% of urban inhabitants used the internet, while the percentage of rural users was 64.2% (Internet-World-Stats, 2004).

The survey indicated that Koreans in general spent an average of 11.5 hours per week on the Web. The percentage of online time spent on email was found to be 33% and the percentage of time spent on information searching was 73%. Online gaming was a popular pastime.

11.5.5 Changes in living, working or learning style experienced by Vietnamese interviewees

Many of the online users interviewed said the changes they had experienced were remarkable, while others had little or no sense of a significant change arising from the spread of ICTs and the Internet. To some, the changes seemed positive, to others adverse. Interviewee I34, a medical doctor in Haiphong (interview, Mar 3, 2006) explained his perception of little significant change by saying: "Not much change – I still have time to meet my friends or play soccer".

Interviewees noted that in the urban centers there were changes in the ways that young people lived and studied, and because most of the ICT users in Vietnam are young, these changes are likely to be relevant to the future of Vietnamese society. Students have been changing the way they study as their usage of the Internet has advanced in accessibility, speed and functionality. Although those interviewed believed few or any schools or teachers in Vietnam were providing lessons or assignments online, many students stated that they could locate useful documents from the Internet for their studies. One student undertaking a Master's degree in Law said that she searched for Vietnamese legal documents which had been much harder to find before the Internet was available. She added:

> I search for online English learning programs or documents for my study. In the past, when I found a document, I had to take notes by hand. Now, I can download the whole document and use them in my essays or assignments (I06, interview, Jan 24, 2006).

At their workplaces, staff used intranets and the Internet to exchange documents. I48 (interview, Apr 5, 2006), a government official in Hochiminh City, described E-office – a locally created inter-agency system which allowed all departments of their ministry to link together:

> With the system we can transmit and receive messages from and to other agencies here in Hochiminh City and to the headquarters in Hanoi. We can search for legal documents or we can exchange email or instant messages.

Such changes have not only occurred in modern offices in cities. A rural seller, I23, claimed that he could use the Internet to advertise his Dong Ho printing products online to many countries in the world, and he even could check queries that clients lodged about his paintings from his home. Such changes are fundamental, and may have far-reaching impacts on the way Vietnamese villagers deal with the outside world in both business and personal life.

The Internet is replacing conventional telecommunication facilities like telephones or conventional postal services to become a powerful tool of connection. As noted by I12: "When I did not have the Internet, I used the telephone to talk with my friends. Now I can exchange mail or chat with them (interview, Jan 25, 2006)". With the new media, friends are able to communicate with each other at any time and from any place. I04, a rural student moving to Hanoi to study, said "I can meet and talk with remote friends". Urban clerk I06 felt happy because she could "meet by face less, but still can help each other. By chatting, I can talk with my friends via the Net" (Interview, Jan 24, 2006).

In addition, people exploited the Internet for information because of its affordability, as I04 (interview, Jan 24, 2006) observed, "I can search for information on the Net without buying a newspaper or book". The implication of such confidence from this rural young female interviewee is really crucial because it means that with the expanding provision of Internet services from Internet cafés, information now is not only more accessible, but also more affordable for rural people.

All respondents used computers and the Internet as a new source of entertainment. They might play a solo game or a multiplayer with invisible challengers. They could listen to online music or download a video clip for their mobile phone. Women searched for new fashions and designs while men sought sport or lotto news, or even surfed the Internet just as a habit. As already noted, chatting and emailing were in themselves favorite amusements. When visiting Internet cafés the author and his assistant noted that Vietnamese youngsters swarming around the monitors seemed attracted by the sheer charm of ICTs.

When a youth claimed "I do not feel any change at all" (I07, interview, Jan 24, 2006) it did not mean he believed that the Internet did not alter his life. He explained that what he meant was "because accessing the Internet has become my habit". In other words, for young Vietnamese like him who were born with the Internet and who have engaged with it for as long as they can remember, it has become indispensable. As I10 (interview, Jan 24, 2006), a public relations officer in an advertising company, asserted:

> Using the Internet has become an essential habit of not only me but many people. Without it, we think that we would be slower than others, slower than what is happening elsewhere. It is a tool for me to send a message without moving myself.

When taken to extremes, however, engagement with the Internet has become almost an addiction for some of the interviewees themselves, or for people they know. The female public relations officer I10 (interview, Jan 24, 2006), for instance, disclosed: "The Internet is my indulgence. I feel happy when I win a game." I05, an administrative clerk at a university (interview, Jan 24, 2006), admitted: "I spend more time on the Internet because I like it. So I finish my work sooner to surf the Internet." Similarly, I45 (interview, Apr 5, 2006), a university student said: "Before, when I had spare time, I went out. Now, I enter the Web." Such interviewees are spending ever more time in the virtual world than in the physical world. Why and how this happens is conveyed by the following comments from student interviewees:

- I04 already mentioned above said, "I can meet and talk with remote friends".
- I18 (interview, Feb 16, 2006), Humanities student, said: "I have less time to contact my friends".
- I08 (interview, Jan 24, 2006), a multiplayer gamer who sometimes spent 20 hours playing, exclaimed: "Yes, many changes. I spend much of the time to learn how to play and spend less time with my family. I feel exhausted, and my study suffers." (Vietnamnet, 19/07/2006)

ICTs have changed the lives of many Vietnamese youth in ways that can be described as revolutionary. It is appropriate to quote the comment of the sociologist I01 to conclude this section on the ICT users' experiences of change:

> The Internet has both positive and negative impacts on users. The ICTs with many channels of unhealthy information may strongly affect the personality and lifestyle of adolescents whose awareness is not yet developed and is sometimes even deviatory and immature. These persons, who do not know what they should or should not do, absorb information without any selection... The Internet is changing a whole young generation in good or bad ways – we don't know which (interview, Jan 19, 2006).

11.5.6 Technological advantages

Why has the Internet and other ICTs succeeded in challenging TV and other traditional media and attracted so many Vietnamese youth in a such a short time? Interviewees gave as reasons the multiple, integrated advantages of ICTs such as their power to process data and to provide information; the diversity of applications that were offered; and their widespread availability.

ICTs are mighty tools for storing and manipulating data. I01 (interview, Jan 19, 2006), remarked: "ICTs and the Internet are more advantageous than other media because they help me to process and store data very conveniently for my study. In the old days, I had huge bookcases for storing my files. Now I can keep all of my data in a small USB flask disk." Other respondents added, "I can save documents in a computer. Otherwise, you have to write it down" (I05, interview, Jan 24, 2006) and, "We can even print the document" (I07, interview, Jan 24, 2006).

Above all, the Internet was perceived as a new source of information. I06 (interview, Jan 24, 2006) argued "The greatest advantage is information. All you need is a computer with Internet connection". I03 (who had begun using the Internet just two months before), in comparing the technology with other media, remarked: "The Internet is more advantageous than other media because I can find various types of information from the Web" (interview, January 20, 2006). I45, a student who acted as a volunteer super moderator of the virtual forum AC Milan club fan, added "I also read newspapers and watch TV, but with the Internet, information is the most up to date (interview, Apr 5, 2006)."

Interviewees saw as a major advantage the ability of the Internet to overcome constraints of time and place, making content and software available at any time. I34 argued:

> With TV or radio, it is not always convenient to watch or listen to the programs we like. With the Internet, I can find information anywhere at anytime. We even can listen to music or download movies and free software, like anti virus (interview, March 3, 2006).

These technologies also reduced spatial distance and made things seamless. I12 (interview, Jan 25, 2006) explained further:

> The Internet is more advantageous in timeliness. I can get information as promptly as the provider launches it online, thus, I do not have to wait for many stages that information has to go through with other media such as arranging, printing and publishing. Information on the Internet is always available, while information via television depends on the broadcasting schedule.

I22 (interview, Feb 20, 2006) noted that "Communication by email or chat is much more convenient than postal mail". Moreover the new technology helps users become proactive in selecting information they need, and to become a source of information to others as well. I10 (interview, Jan 24, 2006) said "Other media, such as newspapers and TV, are one way and passive channels. The Internet enables us to be active and we feel closer".

For the above reasons, the Internet and ICTs have created a new environment with endless novelties. Many people see them as a new source of entertainment, not conventional but smart entertainment, because:

> In comparison with play-station-games like PS2, role-player-games (RPG) require you to be much smarter because you are not fighting against a machine but human beings (I45, interview, Apr 5, 2006).

Interviewees in rural communes also realized the superiority of the Internet to other media. I22 (interview, Feb 20, 2006), commented:

> I don't have a chance to use the telephone. There is the Good morning program on TV about the coming examination but I cannot see it as I have to go to school at that time. With the Internet, I can search for such information whenever I have spare time. I don't care much about the audio/visual advantage of the Internet over other media. But with it, I can contact university students who originated from my village, my commune or my school to help me with my studies or in getting information about the university.

11.5.7 Challenges in relation to ICT use

On the one hand, ICT users saw computers and the Internet as superior to and more multifunctional than all media before them, thus empowering people with more social advantages. On the other hand, ICT users also acknowledged challenges and threats that accompany engagement with computers and the Internet. Some difficulties mentioned by interviewees were common to ICT users internationally, while others were specific to the Vietnamese context. Issues raised most often in the interviews are outlined below.

Health problems

All interviewees expressed concern about health problems faced by ICT users. The protracted use of the Internet with millions of repetitive actions could cause severe physical trauma likes the repetitive strain injury (RSI) that some interviewees reported suffering. I01 (interview, Jan 19, 2006) experienced sore eyes, and hand strains from manipulating a mouse, when using the computer for long periods. I12 (interview, Jan 25, 2006), a government officer, felt pain in her back and neck when using computer equipment. I45 (interview, Apr 5, 2006), a university student, reported that he had just had an operation to help with his short-sightedness, and added: "Using the Internet for long periods is really harmful. We can get sleeplessness, appetite or weight loss, eye problems, or sickness." I09 (interview, Jan 24, 2006), a game player, admitted: "If I play throughout two whole nights, I am exhausted."

Of no less concern were mental problems related to computer use. When people grow increasingly dependent on the technology, they become addicted. This prompts feelings of anxiety about misallocation of time and failure to carry out responsibilities. I10 (interview, Jan 24, 2006) claimed: "Chat kills a lot of time." I06 (interview, Jan 24, 2006) disclosed "when I intend to log out, suddenly I meet a friend whom I have not seen for a while. I have to chat with her and cannot complete the work I intended to do."

It would appear from the experience reported in the interviews that health and safety issues relating to computer use need ongoing and systematic monitoring and attention in Vietnam.

Inappropriate content

Interviewees expressed their concerns about unwholesome information on the Internet. Both older and younger interviewees were worried by the abundance of pornography and other content that they described as evil. I07 (interview, January 24, 2006), a high school student, said unhappily that "pornography and sex abounds on the Internet with erotic stories, movies or games." I06 (interview, Jan 24, 2006) and I45 (interview, Apr 5, 2006) both said that "children should not have access to such information." I10 (interview, Jan 24, 2006) went further with her consideration:

> If there were no control of Internet content by a governmental agency, people might become dramatic or extremist, such as the crisis of suicides among school students in Japan or Korea who set up forums to join together.

Concerns were expressed that adolescent users:

> often have limited social knowledge and experience and become bewildered accessing the enormous amounts of information from the Internet; and may

> not be equipped in relation to bad information. That can have negative impacts on their personality development (I48, interview, Apr 5, 2006).

I01 (interview, Jan 19, 2006), a sociologist, remarked: "Unhealthy information may affect teenagers whose awareness is too premature to select which information is good, thus possibly having bad affects on their personalities and lifestyles."

In extreme cases, schools were so concerned about the adverse effects of the Internet that they even forbade their students from entering Internet cafés. I22, a student at a high school in Ho town (interview, Feb 20, 2006) said:

> Many students dare not go to net cafés either because they don't know about IT or their teachers forbade it. My school also prohibits students from using the Internet because usage of the Internet means playing games or accessing of bad websites. My school caught many students in the net cafés and expelled them from school.

Finding appropriate strategies for coping with the problems of harmful content, yet not restricting freedom of information is another issue that needs systematic attention in Vietnam.

Reliability and cost of information

Interviewees were concerned that information on the Internet is not always reliable and free. I05 (interview, Jan 24, 2006) argued that "information from the Internet is not always correct and complete, and sometimes we have to pay to get it". As with other mass media, the Internet has a powerful impact, both positive and negative, on the opinions of its audience. I47 (interview, Apr 5, 2006) remarked:

> From a social aspect, the Internet can become a powerful tool for some organizations and individuals to blacken the names of other organizations or individuals, and mislead the opinions of those who do not have the capacity to be cautious in accepting the information.

Vietnamese content

Despite the increasing availability of local e-newspapers and publications in Vietnam, interviewees still felt there was a lack in the quality and quantity of the Vietnamese content. I34 (interview, Mar 3, 2006) remarked: "The Internet does not change much of my working style because if I need a professional document, I have to search in foreign websites." In addition, I07 (interview, Jan 24, 2006) expressed concern that "Information in Vietnamese websites is not always most up to date."

Infrastructure and security

In relation to infrastructure, the quality of Internet connections was a source of complaint by almost all ICT users interviewed, except those who lived in Hanoi and Hochiminh City. Even in a big city like Haiphong, a major harbor 100 km east to Hanoi, users were discontented with the service: "The connection is often disrupted" (I33, interview, Mar 3, 2006). In rural areas, the infrastructure was even worse in both its coverage and quality. I22 (interview, Feb 20, 2006), in a rural Internet café in Bacninh province, 30 km north east to Hanoi, said: "The connection in rural Net cafés is very slow. I have often wanted to send a message, but had to wait for so long that I had to drop the message." In addition, many users expressed their concerns about Internet security and how they were aware of the dangers of being attacked by viruses or spam (I12, interview, Jan 25, 2006; I05, interview, Jan 24,

2006; I34, interview, Mar 3, 2006 and I06, interview, Jan 24, 2006) and also of being hacked (I48, interview, Apr 5, 2006).

Government control of Internet use

The role of the government in controlling use of the Internet was criticized by many respondents. Users and Internet café managers were asked about the effectiveness of government policy measures which require every user to show their identification card (in order to limit the side effects of the Internet, such as surfing porn sites or playing game too much). Many users responded that such measures were not realistic. I07 (interview, Jan 24, 2006), for example, said:

> The requirement of the government for security purposes, such as showing identification cards when entering an Internet café, is neither relevant nor manageable.

I32 (interview, Mar 6, 2006), an owner of an Internet café in Hai Phong made a similar point.

Vietnamnet, a popular Vietnamese online newspaper argued that no Internet dealer could avoid violating the Circulation No. 02/2005/TTLT-BCVT-HHTT-CA-KHDT on management of Internet dealers. Such regulation was not only cumbersome and unrealistic, but also excluded children under 14, the minimum age to have an ID from Internet cafés (Vietnamnet, 19/07/2006).

English language competency

The English language was a prohibitive barrier to the majority of Vietnamese users. Not only less educated people, but also intellectuals interviewed admitted that they had problems with English, the Internet's dominant language. I05, a university researcher in the Humanities and Social Sciences (interview, Jan 24, 2006) said, "I do not know much English, especially technical terms." Another social researcher, I04 admitted "English is my biggest difficulty" (interview, Jan 20, 2006). I34, a doctor working in public health (interview, Mar 3, 2006) explained why he only read Vietnamese websites: "The reason is that it is compulsory to know English to read professional documents on the net. Therefore I only enter Vietnamese websites.' Interviewee I07 (interview, Jan 24, 2006) expressed a dilemma mentioned by almost all interviewees: "Foreign websites keep updating instantly, but you need to be good at English."

IT literacy

One of the major necessary conditions for narrowing the digital divide is improving IT literacy. Official statistical reports showed a steady increase in the last decade in the number of IT training facilities in universities, colleges and informal education institutions, and the number of IT graduates. By June 2006 in Vietnam 80 universities, 103 colleges and 60 foreign training institutions had IT faculties with intakes of over 20,000 students per year (Hochiminh Computer Association, 2006, p. 23). The interview-based data from this study revealed that the education system nevertheless still plays a very modest role in preparing electronic readiness. In the finding of this study, only 13.6% of users accessed the Internet from school.

Several ICT users interviewed said they had difficulty in finding information, and often attributed this to a perceived shortfall in their Internet searching techniques.

However many users were confident about using ICTs because they had learned by themselves and/or learned partly from peer users and partly from Net café managers. Peer users could be people who sat nearby, or a virtual rival in a game. Self-learning clearly played a crucial role to equip users with IT literacy. That situation supports van Dijk's remark:

> Most digital skills are not the result of computer courses but of learning through practice in particular social user environments ... practice is more important than formal education (van Dijk, 2005, pp.77,90).

I02, an Internet café manager (interview, Jan 20, 2006) suggested setting up an association of Internet users to help users gain information skills and a better understanding of the ICT environment.

Rural users

Rural students had more difficulty in using the Internet than others. I22 (interview, Feb 20, 2006) revealed

> In the net café, many students only use the applications which have already been installed by the Net café manager like chat or email. They don't know how to access a website or search for information.

Farmers lacked IT literacy simply because there was no IT training available in their rural areas. I23 (Feb 20, 2006) explained:

> Our IT literacy is limited. In rural areas, even if we want to learn IT, we do not have much chance. There is no IT training in my district but only in Hanoi and Bacninh [1 hour away from his commune - the author]. How can I arrange my time to study there, and to work here, as well as to look after my children? We have the same problem with English.

11.6 Main insights from Chapter 11

Insight 11A: Two kinds of computer and Internet users were identified in the study – "general users" comprising a group of people under the age of 35 years with diverse occupations and relatively high educational levels, and "gamers" a group of school and university students aged 21 years and under, who played online computer games as their primary form of computer use. The five main places where interviewees used a computer were Internet cafés, homes, workplaces, school, and friends' houses and PTC3s. Internet cafés were the most popular places where users could come to play games, chat, surf the Internet, and make international phone calls at low rates.

Insight 11B: Almost all of the ICT users had at least an email account, all with Yahoo; however activities on which users said they spent most of their time were chatting, playing online multiplayer games, searching for news or information, downloading online music, downloading educational materials, playing solo games, work, other entertainment, creating online content, making friends and relaxing. Bachelor's degree holders spent more time on work, study, news and information, and less time on chat and games. University students were the most active online users who exploited many functionalities of computers and the Internet, with relatively high usage rates for email, chat, and multiplayer games, followed by study, online music and searching news and information. University students also created online content.

Insight 11C: The Internet and computers have radically changed modes of communication, study, work and recreation. Technological advantages of the Internet were perceived by interviewees to include the availability of up to date information, convenient ways of processing and storing information, the availability of content and software at any time and place, the ease of communication, and the availability of different and challenging kinds of entertainment.

Insight 11D: Challenging issues relating to ICT use were seen to include strain injuries and other health problems, inappropriate content on the Internet, the lack of reliability of Internet information and often the costs of retrieving information, a shortfall in the quantity and quality of Vietnamese content, inadequacies of the infrastructure and concerns about Internet security, the role of government measures in controlling Internet use, the need for English language competency, IT literacy including shortfalls in Internet searching techniques and the role of educational institutions, and the disadvantages experienced by rural ICT users.

Chapter 12
Virtual worlds

12.1 Scope of the chapter

Draws on documentary sources and interviews conducted with ICT users, and particularly seven game players, to explore positive and negative aspects of online gaming and virtual communities. The chapter provides further background on the context of ICT use in Vietnam which helps to set the scene when considering the role of ICTs in relation to economic sustainability in the country.

12.2 Computer gaming and the emergence of multiplayer games

Computer games appeared in Vietnam in the late 1990s but only became a "fever" in 2005 when multiplayer or role playing games (RPG) were introduced. Unlike solo games which only allow a solo gamer to play against a computer, RPGs are games that allow large numbers of users to interact together via the Internet in a virtual world by assuming ongoing roles or characters with different features and distinct characters.

Millions of young gamers have devoted extensive amounts of time playing especially Volamtruyenky, MU, PTV or MU online which were developed in South Korea or China and distributed by Vietnamese companies.

12.2.1 Typical games that launched the "game fever"

(1) "Vo lam truyen ky" (VLTK) – "The First Myth", according to Vinagame, the Vietnamese distributor of "Vo lam truyen ky" (VLTK), this game was a Vietnamese version of "the First Myth", an MMORPG (massively multiplayer online role playing game) developed by Kingsoft in China. The First Myth was a fantasy simulation based on the classical Chinese myth "Romance of the Gods" (Kingsoft, 2006) which happened in the 12th century when the Southern Song dynasty was resisting the Liao invasion of China. That game has attracted many subscribers because of its clever design and plot that allows gamers to act as a favorite character in the story, either a just or unjust kung-fu knight on the Liao or Song side, to go on adventures, fight against, make friends with, or even fall in love with, other characters (Vinagame, n.d.).

(2) "MU Online" is a 3D combat-based fantasy MMORPG in which a large number of players interact with one another in a virtual world. MU Online features seven different worlds to explore.

(3) "PTV – Priston Tale Vietnam" is also a 3D fantasy MMORPG that centers on action-based role-playing. The game was created by Triglow Pictures Inc. (now known as Priston Inc.)

Mu Online and PTV were two South Korean 3D MMORPGs that were purchased and distributed in Vietnam by FPT Communication Corporation in 2005 (VietnamnetBridge, 2005). The games attracted many players thanks to beautiful design and graphics (I09, interview, Jan 24, 2006).

(4) Online and offline solo games – Although online solo games, such as Yahoo's "Bounce Out", "Collapse" and "Pop 'n Drop", and offline games like "Solitary" and "Star Wars" were seen as less attractive by some, they were preferred by female users because they did not involve making contact with strangers (I12, interview, January 25, 2006); and by rural players because they did not require a good Internet connection (I22, interview, Feb 20, 2006).

12.2.2 Games bringing people together

Interviewees explained that persons who played RPGs often chat or communicate with each other in games forums developed either by the game distributors or themselves.

Such follow-on action from game playing gave rise to many virtual communities in the Internet in Vietnam, such as:

- http://forums.gamevn.com/ and http://www.volam.com.vn/ for Volamtruyenky
- http://www.muonline.vn/ for MU online. In these forums gamers could share their opinions about the games as well as other hobbies (I08, Jan 24, 2006; I45, Apr 5, 2006; or I34, Mar 6, 2006).

12.2.3 Games forums as catalysts of virtual community

The games forums became places in which players, besides discussing and sharing information about the games, also explored other interests together – for example study, music, and searching for news and information. In other words, computer games acted as an intermediary motive to attract people to enter the Internet for broader purposes.

12.2.4 Games forums and learning

Gamers argued that they could learn by playing. I33, a junior player in grade 7, reported that when he did not play, he studied computer applications, such as Microsoft Office or tackled mathematical problems online (interview, March 6, 2006). Rural villagers found the Internet to be a new channel for accessing education opportunities which were often very limited in the countryside, as I22 stated: "If I enter the Internet, I can obtain more information about examinations and schools (interview, Feb 20, 2006)". Game playing also helped players to learn English. "My English has improved, and I learn more about computers through playing", said I08, a grade 12 student in Hanoi (interview, Jan 24, 2006).

12.2.5 From virtual communication to physical co-presence

The games forums helped people to go beyond the virtual boundary as well. The relationship of members of a game forum might develop further when gamers decide to meet face to face: such activities are called "offline" meetings. I34 in Hai Phong (interview, Mar 3, 2006) revealed that sometimes parties were held with gamers and their families from Hanoi. Similarly, I45 in Hochiminh city recalled hundreds of members of his AC Milan club gathered together to have an offline party (interview, Apr 5, 2006).

12.2.6 Online gaming as an enterprise opportunity

Online games have provided commercial opportunities for a range of enterprises, including game providers, game producers, game hosting companies and thousands of Internet cafés.

Online games were expected to be the hen that laid golden eggs. VINASA – the Vietnamese Software Association – dreamed of a market size of USD 150 million by 2010 (Bui, 2005). Several domestic products based on Vietnamese traditional culture have been developed (H. Nguyen, 2006b), such as:

- "Thời loạn" (wartime) and "Tintun" by the FPT group,
- "Thiên địa dư chí" (Heaven and Earth geography) by the 3DVN company, and
- "Thiên hà đại chiến" (Galaxy great war).

However imported games dominated the market. Competition became intense among big domestic software companies, notably FPT, Vinagame and VASC, seeking to gain market share and build profits by hosting the games and offering added services, on the basis of advertising or subscription fees.

FPT paid USD2,9 million to buy Vietnam distribution rights for MU Online from Korean Webzen in 2005 (T.V., 2005). Vinagame bought the license for VLTK from Chinese Kingsoft, and as a result became the top Vietnamese online game distributor in 2006 (The Phong, 2006). VASC spent about USD 2 million to buy American games RYL, HERROT and DARCANIA but failed in the market (I20, interview, Feb 18, 2006).

Online games are not only a business for software corporations, but also for some top-class gamers. These gamers play games not only for entertainment purposes, but also for profit. They become "gold farmers" on the net who seize in-game currency in a MMORPG by collecting virtual properties, such as in-game money or armor, through defeating rivals within the game. They then sell the collected virtual items to other players for real money (wowwiki.com, n.d.). A virtual sword was sold at VND 4 million (USD 266 equivalent) and virtual gold armor was bought at a record price of VND 30 million (USD 2,000).

12.2.7 Concerns about online gaming

Online games have become a new "fever" amongst youth around the world. It was estimated that about 25 to 37 million people were playing online games in China alone in 2005 and that this figure would reach 80 million by 2010 (Burns, 2006). In Vietnam, no entertainment could have gained millions of users in several years like RPGs. For the three top games only, PTV, VLTK and MU Online, about two million had subscribed by August 2005 (Vietnamnet, 26/08/2005). "Laodong", the Labor newspaper, estimated that the majority of gamers were aged 15 to 35 years, and that 35% were aged 14 to 18 years (Laodong, 2005). As online games were emerging and becoming lucrative for major distributors and thousands of Internet cafés and game dealers, many intellectuals, the government and millions of parents expressed concerns about the possible side effects.

The theme of violence

Dr. Nguyen Huu Le, Chairman of TMA Solutions RPGs, wondered if the development of the games industry as projected by VINASA was a proper strategy for the national IT industry. He was anxious that violence, with much fighting and blood, was overwhelmingly the theme in almost all games available in Vietnam. He questioned whether making or maintaining such products was the most effective way of using Vietnamese ICT expertise (Vietnamnet, 25/10/2008). Mr Pham Hong Hai, Director General of Telecom Department, urged that the government should work out some effective means of measurement to minimize the social negative impact of online games (Hien Tram, 2006).

Addiction

When gamers realized that excessive gaming was doing them harm, they nevertheless found it very difficult to stop playing games. I08 (interview, Jan 24, 2006), a student in grade 12 explained the way it feels: "I thought that I'd better cease playing. But I don't know why I can't stop it". In all the Internet cafés where the author and his assistant interviewed, dozens of teenagers forgot themselves in the virtual world with invisible rivals and imagined romance. With those who spent 10 hours or more a day at a computer, games became a new drug. I07 (interview, Jan 24, 2006) who later reduced the time he spent playing, described his friend, a game addict:

> But not many people think like me. My friend, a university student, has been playing since grade 1. When he comes back home from school, he does nothing but games. Computer gets itself into his blood. He claimed that he could leave anything but the online games.

Some parents thought that it would be the lesser evil to let children play computer gamers than to let them tramp the streets and expose themselves to even more severe dangers like drugs (I32, interview, Mar 6, 2006). However, it would be reasonable for parents to be almost equally concerned about online gaming if they heard the following story by I19, a university student:

> It costs a great of time, money and health for VLTK. Many gamers become addicts, crazily gazing at the screen all the time, even over nights. When they play like that, their scholastic results are badly affected. Several schoolmates of mine who got bad marks due to playing too much, then dropped out of school ... The players at high rank often played overnight. To keep on playing night long, we needed to smoke and remain sleepless. I had no time to study (interview, Feb 16, 2006).

Several gamers had reduced the time spent in gaming after realizing the bad effects they were having. I07, for example, claimed:

> I could have played 24 hours per day before. Now I only play 1 hour or two because I realized that such play is not beneficial to me, thus I'd better reduce the time. In addition, playing is harmful to the health or eyes. When playing

> much, one does not think of eating or drinking. Playing overnight made me thinner (interview, Jan 24, 2006).

Online gaming as a two-edged sword

The evolution of online games in Vietnam is only at an initial stage, but its impacts have already been remarkable – both positive and negative. Online games have introduced millions of young people to ICT use. It has helped them build skills and social networks both online and offline. It has also been a commercial opportunity for a range of enterprises. However the abuse of gaming has unquestionably become a dangerous social issue that needs to be dealt with in the context of Vietnam's development as a knowledge economy.

12.3 Virtual communities

Virtual communities or online communities were a relatively early development in cyberspace – Howard Rheingold's pioneering work "The virtual community" was first published in 1993 (Donath, 1999). Yet it is still a concept whose definition is controversial among scholars. For the purpose of this study, a virtual community is formed when people within a group interact together online, and also possibly offline, using, as a channel, the Internet, mobile phones, and/or other ICT means. Community members may share one or multiple interests or purposes, and some norms or values. However, where charging is involved, some members (such as VIP members or managers) may have to pay more than the others.

From an economic point of view, a virtual community is a type of commons-based peer production system where members share knowledge (for example about hobbies, music or their professions) and/or facilities (such as host servers). A strong manifestation of this kind of community is the free software or open source movement (Benkler, 2002).

The strength of virtual communities is that they allow individuals to present their knowledge and experience on the web, and to learn from, or enjoy the contributions of, others. Within virtual communities, relatively speaking, relationships can be characterized as strong or bonding ties, whereas between virtual communities there may emerge weak or bridging ties as individuals in one community make contact with individuals in others (Schauder, 2006).

Interviewee I04 captured the essence of virtual communities by saying that they "make people more connected and easy to meet" (interview, Jan, 20, 2006). In the case of virtual communities that inhabit a simulated virtual world, as in the kinds of online games described above, people became virtual residents because "there are no limits in relation to space (no physical organization or arrangement needed), time, personal background (for example social status and/or profession), race or geography" (I48, interview, Apr 5, 2006).

12.3.1 Participation by interviewees in virtual communities

The 22 ICT user interviewees were found to be engaged in virtual communities at one or more of three possible levels: (1) they stayed in touch with their real friends using email or chat to communicate, (2) they became members of an online forum, and also (3) an individual might serve as a manager of an online forum, either as a web administrator (system architect) or as a moderator.

Interviewee I06 said that although the current conditions of their lives meant that she and her friends could "meet by face less, they could "still help each other. By chatting, I can talk with my friends via the net" (interview, Jan 24, 2006). In saying this she might not have been conscious that she was acting in part as a "virtual person" to her friends.

12.3.2 Types of virtual communities

Beside one-to-one contact by email or chat, users participated in one or more of the following types of virtual communities:

- Personal blogs (P07[11], interview, Dec 20, 2005),
- Listservs - email mailing lists through which members can send mails to each other, or News groups - Internet based bulletin board (I06, interview, Jan, 24, 2006),
- Group online forums: such as http://vn99.net (S69 and S70, interviews, May 19, 2005) or http://www.acmilanvn.com (I45, interview, Apr 5, 2006),
- Corporate online forums; such as http://vn.vinagames.com/ of Vinagame, http://www.volam.com.vn/ of Fpt, and
- Public or governmental online forums: such as the portal of the government http://www.vietnam.gov.vn/portal/page?_pageid=33,127908&_dad=portal&_schema=PORTAL or http://vnexpress.net/Vietnam/Ban-doc-viet/

12.3.3 Purposes of virtual communities

The purposes of virtual communities reported by interviewees were very diverse and focused on services or topics including:

- public forums (S49, interview, Dec 14, 2004)
- sport fans (I45, interview, April 5, 2006)
- online games (I08, interview, Jan 24, 2006)
- free training (I48, interview, Apr 5, 2006)
- motherhood (I12, interview, January, 25, 2006)
- research (I01, interview, January 19, 2006)
- music (S69, interview, May 19, 2005)
- photos and video clips - even providing their collected videos to television (I45, interview, Apr, 5, 2006).

The list of purposes would be much longer if the sample size was larger. In the US, the top 10 purposes were: trade/professional, hobby, fans/sports, fans/entertainment, local groups, health related, shared beliefs, political, religious and sports teams (Horrigan & Rainie, 2001).

Factors which affect the success or failure of online communities have been discussed in the international research literature. These include trust, online time con-

[11] Respondent P07 was engaged in a pilot interview for the second fieldwork.

tribution, the ownership and maintenance of the virtual space, and the necessary equipment. Online communities do not require shared physical space, regular face to face encounters, or physical artifacts (Baym, 2003, p. 1015).

On the topic of volunteer-based websites like Vn99.net, S69 and S70, two moderators said in an online interview that the total devotion of the moderator and administrator, and participation by the members of the site were the key elements of its success. Members participated because young people are so fond of music and romance (interview, May 19, 2005).

Online newspapers are in varying measure virtual communities, with readers playing a special role. According to S49, a leading participant in Vnexpress.net, most online newspapers have a page for readers' feedback which functions as a virtual forum. Audience members can comment on a topic of interest or communicate with the editorial board. A thread posted by a reader stimulates others' reaction and a chain of discussions among readers ensues in the e-forum. Though discussions in official forums are often edited, corporate and public forums are popular with many people. These forums are usually well-funded by the government, corporate sponsorship, or advertising (interview, Dec 14, 2004).

12.3.4 Personal blogs, personal websites and virtual communities

In 2006 Vnexpress.net conducted an online poll with the question "In the Internet world, how do you often share your information and feelings?" Amongst 2,130 respondents, 53.7% joined forums, 14.2% created websites with a private domain name and 19.4% opened a free personal web log (Nguyên Hải. & Hằng Ngọc., 2006).

New versions of blog tools, such as 360° of Yahoo! or YouTube of Google have added a range on new capabilities for virtual community building – especially for the establishment of bridging ties. It is becoming ever easier to access resources in other virtual communities, and draw them to the attention of fellow-members of your primary community or communities.

Free and open blogs

Virtual communities often form around attractive personal blogs.

Personal blogs can be accessed by anyone, without intermediation, so popular blogs with interesting pictures or video clips attract a great deal of interest. Dung Tien suggested that many Vietnamese bloggers do not have a clear topic for their sites. Rather bloggers might create a resource about a topic with a sudden inspiration: through curiosity, imitation of others, or as part of an online diary.

He said that in relation to blogs created or used by interviewees in the present study, contents were usually about music, movies or books, or comments on them. "Hot" blogs might receive a great deal of attention, such as Cuong Oz's personal 360° blog that presents a unique investigation of the good and evil aspects of computer gaming in Hanoi, which receives more than one million views in a month (Tiến, 2006).

Specialized websites

Specialized websites can also attract a following which forms a kind of virtual community. In contrast to free personal blogs, some individuals buy a domain name to own a private "fiefdom" in cyber space for sharing "high-quality" information and experiences. An example was www.vninvest.com created by S44, a Doctor of Economics. This site presents a collection of resources, enabling surfers to find useful documents on markets, trade, banking, law, etc.

For such sites international domains with extensions .com, .net, .info, or .org were preferred to Vietnamese domains with .vn because the former were much faster and cheaper to establish (USD 10 vs. 60). The estimated total cost for building a personal website including domain name and host hire was about USD 300. A personal website might include some profit making functions like marketing or introduction of the host's image.

12.3.5 Challenges in relation to virtual communities

Whether offline or online, communities are characterized by identity formation, newcomer confusion, leadership roles and etiquette. Like real communities, virtual groups are faced with cultural issues of which trust is the foremost. In answer to the question "Who are you?" Sartre answered "I am my body to the extent that I am" (Sartre, 1992). In the online environment identity has to be established without bodily presence.

Identity issues and relationship building

In relation to identity and honesty among virtual contacts on the Internet, some users interviewed, both male and female, were very cautious in communicating with strangers. Interviewee I07 would only communicate with others known by face (interview, Jan 24, 2006). Interviewee I04 would only chat with old friends (interview, Jan 20, 2006). Both I08 (interview, Jan 24, 2006) and I01 (interview, Jan 19, 2006) thought it was hard to know the real name, face, age and gender of the virtual friend. Interviewee I08 (interview, Jan 24, 2006) had experienced identity deception, in this case a boy showing the face of a girl on his webcam. I10 (interview, Jan 24, 2006) argued that an online forum, if not controlled properly, could be a place for dramatic or extremist behaviors, as is the case of a website that encouraged Japanese and Korean teenagers to commit suicide. Thus, I48 (interview, Apr 5, 2006) urged that illegal online activities needed to be well policed.

Not everybody was as defensive as the interviewees cited above. Some people felt freer to make friends online. I07 (interview, Jan 24, 2006), though well aware of the unreliability of identity in the online world, claimed that he had "uncountable" virtual friends. However, he added, "the closest 10 friends came from his school." I45 (interview, May 5, 2006) explained:

> Chatting does not unite virtual chatters because they do not trust. Participating in online forums makes members friendly and trustful. Members meet not only online but also hundreds of them actually meet each other offline [actual face-to-face, added by the author], such as in the case of the forum www.acmilanvn.com.

In other words ICTs can link people together, but such linkage is still very weak and temporary. Face-to-face encounters help to strengthen virtual relationships.

Differences in attitudes and cultures

Different attitudes and cultures was another issue. I45, a super-moderator, had witnessed conflicts and quarrels among the online members of forums and said that such heated arguments needed to be settled (interview, Apr 5, 2006). I48, an experienced online resident advised: "There are different cultures of those who join in a forum. [People should] just quit if [they] cannot compromise" (interview, Apr 5, 2006). People join online communities with the expectation of sharing and gaining something, especially assistance from others. However, I34 revealed "Virtual friends do not help much" (interview, Mar 6, 2006) – except for co-gamers who often help the others how to play (I15, interview, February 15, 2006).

Costs of online virtual communities

When people participate in an online world they have to spend considerable time and effort to interact with the other members and contribute opinions and/or even money to the commons of the community. There is a kind of hierarchy in online societies in which the majority of participants are relatively passive exploiters. The main contributors are often the few key managers of virtual communities, web administrators and web moderators who might contribute their own funds to maintain equipment and host space, and who give their time to supervise the community activity without significant revenue, for example from advertisements. The motive of a main contributor is often identity and self fulfillment. When and if this key person becomes exhausted, or if the person has to stop the role for some other reason, the community collapses. Vn99.net, an online music forum with 150,000 members, was once one of the best Vietnamese online group forums, with a high ranking of 9,596 by alexa.com on August 1, 2005 (see Chapter 4). In August 2006, exactly a year later, the student-run-website was closed down because of a shortage of money and resources to maintain it. I70, one of the web managers called for fundraising from former members to revitalize the forum under a new name, 1vn.com. However the contribution of members never seemed to be adequate to revitalize the forum (1vn.com, 2006).

The reliability of virtual communities in times of need

In many real (non-virtual) communities, mutual cooperation, support and shared responsibilities link members together. The ties in "real" communities are often strong bonds. According to a study of community strength by the Department for Victorian Communities in 2004, about 92% Victorians believed they could get help from friends, family and neighbors when they needed it, and about 80% could raise A$2,000 in an emergency (Strategic Policy and Research Division, 2004).

A virtual culture can also be a source of support and assistance. I08 said:

> … in a virtual community…I have many relationships with strangers, know more friends and get help. If I have a problem and someone can help, they will do so (I08, interview, Jan 24, 2006).

Another interviewee, I09 said that virtual communities provide the opportunity to "...communicate with many and diverse people even at home, and make friends with many new people" (I09, interview, Jan 24, 2006). Virtual communities offer the benefit of being a source of mutual cooperation and support. But the bonds of such communities are often not as strong as those of a "real" community. Key managers and administrators can quit at any time. The collapse of Vn99.net mentioned above highlights the voluntary nature of these roles. I34 stopped managing his forum when he got married (interview, Mar 6, 2006), others because they did not have time, or because they changed to another hobby. This vulnerability needs to be something that is understood by those who come to be dependent on a particular virtual community.

12.4 Main insights from Chapter 12

Insight 12A: Positive aspects of online gaming include its role in bringing people together and in attracting people to enter the Internet for broader purposes; the capacity of games to provide useful learning experiences; the opportunity to improve English language skills when playing games in the English language; and the use of games forums as a vehicle for meeting other gamers face to face. Negative aspects of online gaming include the social impacts of the common theme of violence, and also the possibility of addiction.

Insight 12B: While online gaming provides different kinds of business opportunities in Vietnam, in recent times large Vietnamese gaming software companies have been faced with the difficulty of competing with overseas suppliers, and with each other.

Insight 12C: Key benefits of virtual communities are that they allow individuals to present their knowledge and experience on the web, and to learn from, or enjoy the contributions of others; connection between a community's participants is not limited by physical space limitations, time, personal background, race or geography; and a diverse range of services, activities and topics are available to meet different needs.

Insight 12D: Challenges in relation to virtual communities include the issue of trustworthiness and integrity in the creation of virtual relationships; the dangers posed by illegal online activities and extremists; the real costs of virtual communities to those who act as managers and hosts; and the extent to which virtual community participants can be relied upon as sources of assistance to individuals.

Chapter 13
The Vietnamese ICT industry

13.1 Scope of the chapter

In considering the extent to which Vietnam is able to achieve economic sustainability through human made capital it is necessary to obtain a micro-level overview of issues in Vietnam relating to the ICT industry and ICT providers, in both the private and public sectors. More specifically this chapter

- Describes key characteristics of the interviewees in the two study groups of relevance to this chapter: (1) ICT providers from the business sector and (2) CPCPs and PTC3s, two public service agencies providing postal and communications services to remote and communal residents
- Reports on issues identified by the ICT provider study group
- Reports on issues identified by the CPCP and PCT3 study group
- Discusses the role of government in supporting the ICT industry, with specific reference to management capacity, government regulations, telecom connections government websites and examples of effective uses of ICTs by government
- Discusses the role played by Vietnam's online newspapers, and of non-government forums and websites.

13.2 The balance between locally produced and imported ICT in the context of sustainability

In Chapter 5 a snapshot was presented of the IT industry of Vietnam based on data from official reports and available research. This "macro" overview highlighted the rapid growth that has been a remarkable feature of the ICT sector in Vietnam.

Sustainability at the national level is the ability of the current generation to reserve the resources that nature grants for the use of future generations. This can be done by keeping the consumption of physical resources at site to a minimum, and importing physical consumables from the global market. This has been termed an "importing sustainability" approach (Pearce, Markandya, & Barbier, 1989, p. 47).

In considering whether Vietnam can achieve economic sustainability by producing human made capital locally that is greater in value than the value of imported ICTs, it is necessary to ask whether Vietnam's domestic ICT industry is able to satisfy the needs of its growing local market. This chapter considers this question and reports findings from in-depth interviews with ICT providers in Vietnam to provide a "micro" level overview of the challenges they have experienced.

13.3 Description of the study groups

The interviews aimed to gather views from a diversified group of people working within the areas of telecommunications, hardware, software and services to provide a kind of "think-tank" about key issues relating to the industry:

Material for this chapter was drawn from the second and third fieldwork rounds consisting of interviews in Vietnam with ICT industry personnel. The second round was undertaken in the two major cities, Hanoi and Hochiminh, from January to April 2006. The third round was undertaken in rural areas in December 2006. Each of the interviewees was given a numbered code Ixx. The chapter also includes several views of the interviewees from round 1 of the fieldwork (encoded Sxx).

In the second round, together with an assistant, the author had discussions with 23 ICT business managers in Vietnam comprising: one American and 22 Vietnamese. 19 (82.6%) interviewees were in big cities (12 in Hanoi, 6 in Hochiminh city and one in Haiphong) and four (17.4%) in the countryside. Four of them were women (17.4%). The sizes of the company varied from one-staff-Internet-cafe (I32, interview, Mar 6, 2006) to 600 employees in VASC (I20, interview, Feb 18, 2006), and from tens of millions to 100 billion VND worth of turnover. Three interviewees were managers of the 10 top ICT companies in Vietnam, namely, FPT Telecom, VASC and Vnexpress.net.

In the third round, the author explored how communal posts and cultural points (CPCPs) provide services for communes, and how posts and telecom companies level 3 (PTC3) provide services for the districts and inter-communes where they are located. Two provinces in the north of Vietnam were selected, namely Hatay, 30km west of Hanoi and Haiphong, 120 km east of Hanoi. In each province, the author observed and interviewed one employee of a CPCP, a manager of a PTC3 and an Internet user at PTC3. The questions focused on the services, strengths and weaknesses of the posts and communications system in delivering information services to communal residents (see Appendix 6).

13.3.1 Positions and experience

Table 13.1 Position titles of the interviewees

Position	Number of interviewees in each role
General Director	3
Director, Manager	14
Salesman	2
Internet café owner	2
CPCP operator	2
Total	23

Of the 23 interviewees, the majority were key persons in their organizations: 8.7% were entrepreneurs, 13% general directors and 61% directors or managers who played the overall management role in their company such as setting up strategic plan, managing the business and so on. Classified by ownership, one was an American general director of a software consulting company and two worked in foreign companies, three worked in state owned companies and 19 were in private ICT companies.

Regarding the years of experience of interviewees within the ICT sector, 22% were of entry level - 1 year or less; 13% from 2 to 5 years, 26% were experienced with 6-10 years and 39% very experienced – over 10 years.

13.3.2 Businesses and services

While falling within four broad types: telecommunications, hardware, software and services, most of the ICT providers interviewed also specialized in one or more specific areas as shown in Table 13.2.

Table 13.2 Businesses and Services

Businesses / services	**Number of interviewees who offered each specialization (nomination of multiple specializations allowed)**
Internet café services	6
Copyrighted Software	5
Open source software	4
Postal & telecom	4
IT Consulting	3
E-content	3
Hardware retailer	3
Training	2
Cultural services	2
Games	1
Network & cable	1

N = 23

Internet cafes provided the most popular Internet-related services such as chat, computer games, information searching, and message exchanging or international

VoIP calls. In the countryside, PTC3s, (the affiliates of VNPT at the district level) besides Internet café type services, also provided postal and telecom services, while CPCPs provided 12 postal and telecom services, as regulated by VNPT (see more about CPCP and PTC4 in Chapter 5 and the section about CPCP and PTC below).

ICT consulting companies, in contrast, assisted clients in finding solutions to applying ICT to improve their businesses and organizations (I11, interview, Jan 25, 2006).

The business of many companies consisted of the prevailing tradition in the ICT industry, namely software production, hardware sales and telecommunications, though at a more advanced level than in the past. Software production was based on either copyrighted third-party vendor platforms (22%) or open source platforms (17%). The latter (though less diversified) grew quickly in Vietnam in the number of providers and their products (see section about open source below).

Apart from software, the sale of computers and hardware was the main business of 13% of the ICT companies, such as the company of I14 in Hanoi which sold PC servers and laptop computers (I14, interview, Feb 15, 2006), and that of I47 in Hochiminh city what traded all kinds of hardware from computers to peripherals or storage equipment (I47, interview, Apr 15, 2006).

Several companies, not limiting themselves to purely selling hardware, supplied value added services like network installation or Internet services. I17's company, for instance, specialized in setting up networking systems for industrial zones or foreign companies (I17, interview, Feb 16, 2006). I47's company delivered integrated solutions to customers (I47, interview, Apr 5, 2006), and the giant FPT Telecom delivered ADSL connections to households and individuals and leased line services to corporate customers (I40, interview, Mar 14, 2006).

In considering human-made capital, an important finding was that 16% of companies had moved from software production to e-content provision, a service that was of more practical value to end-users and more profitable to companies. The list of e-content services included an e-commerce pilot project from E-Commerce Department (I16 interview, Feb 16, 2006), news-online such as echip.com, online TV, online promotion and SMS via mobile phone for youth served by VASC (I20, interview, Feb 18, 2006), and online advertisement, music, games, news and VoIP of FPT Telecom (I40, interview, Mar 14, 2006).

In summary, the sample of 23 interviewees were diversified in their business, products, services, market, locality and business size as well as their seniority within the ICT sector.

13.3.3 Clients of the interviewees

The evolving of segments and specializations in services and products is a feature of the Vietnamese ICT sector. Clients described by the interviewees could be grouped into 3 types: 1) large corporations and organizations, 2) small and medium businesses and 3) households and individuals. Interviewees were requested to name their major customers. Some listed one single type of customer, for example, individuals in the case of an Internet cafe. Others responded to more than one category, for example, foreign company and large business. Their answers are shown in Table 13.3.

While large ICT companies often focused on large clients, such as government ministries and large corporations or foreign companies, small companies gave their attention to small shops and businesses. A glass dealer, for example, might need a small tailored software program to manage its materials which should not exceed a cost of 7 million VND (about $500). Larger companies might not accept such a project because the offer was too low in terms of unit value and the unlikelihood of additional income from duplication, whereas small companies would be happy to do so. A small IT company set up by I14 (interview, Feb 2006), an IT undergraduate, could satisfy such a customer's demand with very low cost.

Table 13.3 Clients

Clients	**Number of interviewees who nominated each client category** *(nomination of multiple client categories allowed)*
Individual (student)	10
Household	1
Internet cafe[12]	1
Small business	1
Medium business	1
Large business	2
Foreign company in Vietnam	2
Foreign market	4
Local government	4
Central government	6
University	2

N = 23

Some large companies also targeted end-users, for example VASC which delivers SMS via mobile phone to youngsters (I20, interview, Feb 18, 2006); FPT Telecom (I40, interview, Mar 14, 2006) which delivers ADSL connections to households; and Vnexpress.net (S48 and S49, interview, Feb 14, 2004) which delivers cyber-news. Internet cafes and hardware retailers were the closest suppliers to individual end-users in the cities, while CPCPs, PTC3 and Internet cafes provided services in rural villages.

In addition to their major role within the domestic market, Vietnamese companies are switching their focus to foreign markets by undertaking outsourcing contracts.

[12] One company which provided ADSL connection to Internet cafes was FPT Telecom.

13.4 Issues identified by the ICT provider sample

Some worldwide risks such as security attacks (for example, Trojan), viruses and information hacking were not perceived as major problems for Vietnamese business (I11, interview, Jan 25, 2006; I02, interview, Jan 20, 2006), but other areas, such as shortages of finance, infrastructure and human resources were seen as key challenges.

13.4.1 Finance

Most ICT companies in Vietnam were small and medium enterprises that were often short of money to expand their businesses and had difficulties in accessing funding. I13, a director of software development of MISA group in Hanoi (interview, Feb 13, 2006) remarked "there are still few investors with venture capital in the software sector and there is no incentive policy on mobilization of funds for ICT companies". I50, a vice director of a hardware retail company in Hochiminh city, said:

> The government should support funds for SMEs in the IT sector by giving loans. We never know where and how to get investment loans from the government. At the moment, there is only a mechanism to provide soft loans to software companies (interview, Apr 5, 2006).

The ICT business community expressed its concerns about the implementation of financial policies. Government regulations indicate a willingness to assist ICT business in seeking funding for their activities, for example Article 5 of the 2006 Law of Information Technology stating that the government

> encourages investments in information technology; allocates a government fund for IT utilization in essential sectors, for the establishment of the IT industry and for the development of human resource (Government of Vietnam, 2006c).

However data from this study suggests that the law is still far away from practice. It was felt that there is a need to review the effectiveness of current incentive policies such as IT law and policy on financial support, so that practical guidelines could be issued and implemented by both ICT business and government authorities.

Another concern for knowledge-based enterprises in particular was mechanisms for the unlocking of funds. For ICT businesses, knowledge is their main capital and property, but unlike physical enterprises whose properties are plant, machine or land, knowledge capital is intangible and hard to evaluate when applying to borrow funds or obtain loans from banks or financial institutions. I55, CEO of an IT consulting company (interview, Apr 11, 2006) suggested the need for a mechanism to facilitate the accessing of funding sources for ICT businesses, software, ICT training and consulting.

A suggestion for full institutionalization of electronic payment was given by another interviewee, I16, a chief of an e-commerce project (interview, Feb 16, 2006), as it would lay the foundation for ICT businesses to make online transactions. The law of electronic transaction was promulgated on March 1, 2006 but could not be implemented due to the lack of related regulations.

In terms of readiness for e-commerce in Vietnam, I57 – a senior official of the Department of Telecom (interview, Apr 12, 2006) explained that 5 government decrees in relevant areas were being drafted, namely e-commerce, e-banking, e-finance, digital signature and certification and e-transaction in public agencies. The implementation of these decrees will require increased capacity of related governmental agencies and the economy to realize e-commerce in Vietnam.

13.4.2 Infrastructure and Internet access

As noted in Chapter 5, the network of telecommunications and Internet has developed very rapidly in Vietnam in the last 5 years, laying down infrastructure for the development of ICT companies which are not only providers, but also consumers of telecommunication services. Nevertheless some interviewees felt that there was much more to be achieved. I24, a development manager of an American game company in Hanoi stated:

> ...(though) the telecom tariff, especially ADSL, has reduced a lot, nevertheless it is still high. The renting of leased lines is still very expensive; not many enterprises can afford this" (interview, Feb 22, 2006).

The government, both national and provincial, built software parks or IT centers, such as Hoalac Hi-tech park in Hatay province in the North, Quangtrung Software park in Hochiminh city in the South, and Danang IT Center in the Center to provide good facilities for ICT companies to hire. However, it was still very expensive for local software enterprises to hire space in software parks constructed by the government or municipal authorities, as remarked by I13, a manager of a software company (interview, Feb 13, 2006).

I47, a chief operating officer of Informatics Service Providers Inc. in Hochiminh city considered that "infrastructure had improved and telecom tariffs had decreased. Many new services such as overseas VOIP have become available and convenient" (interview, Apr 5, 2006).

However, not all of the interviewees were equally satisfied. The quality of Internet connectivity was complained about by many interviewees, such as I02 (interview, Jan 20, 2006); an Internet café entrepreneur in Hanoi (interview, Feb 20, 2006; and I43, a counterpart in Hochiminh city (interview, Apr 5, 2006) who said: "... connection provided by VDC, FPT or Viettel is not reliable." Interviewees recorded connection disruptions such as drops or jams (I11, interview, Jan 25, 2006; I14, interview, Feb 15, 2006; I43, interview, Apr 5, 2006) or lags in case of online games (I45), or no ADSL connection in rural areas (I64).

Even in large cities, foreign business personnel working in Vietnam had similar frustrations. I17, Director of a company which specialized in building network infrastructure for industry zones and foreign enterprises, commented that

> Large or foreign companies complain much about the quality of telecom infrastructure as they want to have high speed connections such as ISDN that is only available in Hanoi and Haiphong (interview, Feb 16, 2006).

I11, an American who was a general director of the software consulting firm mentioned the unreliability of Internet access in Vietnam:

> I really hoped that it would be more improved. I have been working here for 9 years. Now I still have the same problems as I had 5 or 6 years ago. Line drops, [and connections which are] unavailable too often. It is the problem of a long time monopoly (interview, Jan 25, 2006).

In provinces other than Hanoi and Hochiminh, the telecom infrastructure, especially Internet connection, is poorer. I13 revealed that Internet was bad in small towns and not available in rural areas,

> Regarding telecom infrastructure, there is no network connection to remote provinces. We want to provide online assistance to clients but cannot provide the service, nor even send emails. To deliver supporting services attached to our products, [such as] a financial software application, requires linking to online banking that no bank in Vietnam offers (interview, Feb 13, 2006).

I40 from FPT Telecom, the second biggest Internet exchange provider in Vietnam, gave reasons in the interview,

> In the coming years, we will invest much in cable network infrastructure in Hanoi and Hochiminh city... because the market develops quickly in those cities. We won't do it in other provinces as the telecom network does not well develop there. And you know, telecom infrastructure follows the cable of VNPT. We don't have telecom cable of our own thus we can't compete with but have to take a lease from them. The cable network in big cities is quickly developed. But if we hire cable in provinces from VNPT, the leasing cost will be higher than the in-house price that VNPT charges its affiliates (interview, Feb 14, 2006).

A monopoly and profit-making have been suggested as reasons preventing ICT companies from expanding their markets outside large business centers. It was argued that limited competition in the telecom sector made products and services more affordable to most inhabitants in key business centers of Vietnam (AT&T, 2006). In 2006, ADSL was the most popular broadband service at a cost of USD 12.80 to 64.10 a month.

However, with the development of the telecommunication sector recently, especially the openness to limited competition in 2003, telecom infrastructure and the Internet in Vietnam have quickly made great strides. In fact, the monopoly apparently seemed not to be the main issue in Vietnam where there were at least 10 IXPs (Internet eXchange Providers) and ISPs (Internet Service Providers) delivering Internet services. The market share of the long-time monopoly of VNPT has been taken away by new competitors, as shown in Table 5.8 in Chapter 5. The deregulation in the telecom sector has created a dramatic growth in infrastructure and quickly brought down prices, thus bringing ICT services and knowledge resources closer to the people. In 10 years from 1995 to 2005, telecom and Internet tariffs have declined 9 times, and that trend is continuing (see Chapter 5).

The unsatisfactory nature of the quality of Internet connection lies in the argument that carriers increase the numbers of their customers without increasing bandwidth and quality sufficiently. A rough comparison below shows that bandwidth has grown more slowly than demand for use. Bandwidth growth was 51.7% in 2005 from 2.5 Gbps to 4 Gbps, compared with 86% of usage growth from 7.1 million to 12.9 million. Cairncross suggested that increases in the complexity of usage behaviors (multimedia rather than textual data) consumes more bandwidth (Cairncross, 1997, p. 115). As a result, national connecting points become bottlenecks in the system. Responding to the growth in demand for connectivity with the world and inside the country is very important. At the same time, the protection of the existing cable network is also a crucial measure in the event of any sabotage of the network[13].

13.4.3 Markets – Customers – Competition

The quality of infrastructure directly affects the behavior of customers. Internet café managers complained that if the connection was not reliable (slow or often disrupted) customers would leave (I43, interview, Apr 5, 2006). While ICT companies mainly focused their business in cities where they could gain profit quickly, the needs of rural and remote areas were left to be covered by public services delivered by subsidized organizations, such as PTCs and CPCPs.

Local ICT companies have greatly improved their capacity to provide services to customers, to the point of succeeding in dominating the domestic market (I13, interview, Feb 13, 2006). I11, remarked:

> I started my work in Vietnam 9 years ago. I saw national IT companies and many foreign companies appearing in Vietnam. We competed. National IT projects always chose foreign companies. Now there are almost no longer IT foreign companies, only Vietnamese ones. All small and medium IT projects are undertaken by Vietnamese. They provide much lower cost. I think they have overcome the problems associated with the capacity to provide ICT services. Just only very high level and complex projects have to go overseas. Basic IT - all is covered by Vietnamese providers (interview, Jan 25, 2006).

This perception of success in domination of the domestic market suggests that Vietnam is making great progress towards becoming knowledge based economy.

If the monopoly is a visible problem in the telecom market, unfair and monopolist control of the governmental market by state-owned companies was also found in the IT sector. I42, a general director in Hanoi made this criticism:

> These companies create monopolized advantage by their closed relationship with governmental agencies, or by offering very low prices to win IT bids. Due to the un-recoverable cheap prices, they win, but often provide low quality products to customers that are incomplete and thus useless. To cover the problem, they use money rather than technical solutions to corrupt governmental IT managers. Finally, governmental institutions receive what they can't use but have paid for (interview, Mar 23, 2006).

I46, a general director in Hochiminh city, argued further:

> This is partially because of the lack of independent IT consulting and supervision of construction projects. Fifty percent of IT bids in the past 5 years have faced that problem. For example, in a project to set up a computer based library in a university, while other bidders quoted \$300,000, company X quoted only \$200,000 and won. However their final product did not have the full number of functions for which the other competitors had quoted. To pass the

[13] In May 2007, about 100km optic cable of two out of eight submarine Internet cables laid under the Vietnam sea, namely, TVH and ACPN, were stolen as scrap by fishermen owing to the permission of a provincial authority. This severe violation cost about US\$2.6 million and half of the Vietnam Internet bandwidth. Another cable, SMW3, was in great risk .

> checking stage before acceptance, they kicked back $20,000 to the evaluation panel (interview, Apr 5, 2006).

While large customers were largely served by foreign ICT companies and medium customers, by domestic ICT companies, the market of some Vietnamese ICT companies, such as I14's company focused on small shops and retail enterprises (I14, interview, Feb 15, 2006). This segmentation of the IT market in Vietnam was noted earlier. This kind of specialization brought down the cost and also diversified the availability of ICT services to various customers.

In addition to serving the domestic market, local companies also provided services to overseas markets. Vietsoftware undertook outsourced work for IBM (I41, interview, Mar 21, 2006), Punch Entertainment used Vietnamese employees to produce games on mobile phone for the North American market (I24, interview, Feb 22, 2006), and IES Co. Ltd. exported software to Nokia and General Electric (I46, interview, Apr 5, 2006). Efforts to obtain outsourced work for offshore clients were strongly encouraged by the government as an initial step towards exporting software. I14 (interview, Feb 15, 2006) claimed that this approach only exploited the cheap labor of Vietnamese and did not take advantage of Vietnamese's creativity. However, in the long run, the link to Nokia or General Electronics may bring local benefits by way of investment and knowledge transfer for Vietnam and the advantage of low labor costs to the investors. In talking about the USD$1 billion investment to increase the capacity of the assembly and test facility of Intel in Hochiminh city, Brian Krzanich, vice president and general manager of Assembly and TestIntel, said:

> By expanding the planned size of this facility we expect to gain greater efficiency to improve our ability to meet our customers' requirements (…) The Vietnam facility will be the model for larger, more efficient assembly and test facilities that will make Intel even more competitive (Intel, 2006).

Another point made was that the low IT skill and understanding levels of clients impacted negatively on the success of ICT projects. I13 claimed that "clients in government administration areas do not have good ICT awareness and skills, thus the application of IT and other technologies is low" (interview, Feb 13, 2006). I17 (interview, Feb 16, 2006) argued "if they have high awareness, the market will be larger." I55, a managing director of an IT consulting company, reported his difficulty with his clients "they simply think that IT is equal to hardware, and are not aware of the value of after-sale services, so they are not willing to pay for the services". I47 said: "… investment in IT often follows fashionable trends, rather than being based on cost-effectiveness considerations" (interview, Apr 5, 2006).

In the public sector, I11 explained his concerns:

> Internal issues of clients block the chance to give good advice and services by providers: hence delays and no clear decisions... E-government should not be for its own sake, but for re-engineering processes of the organization for the sake of improvements. The application of IT in business and organization without conducting business re-engineering is not effective (interview, Jan 25, 2006).

Additionally, the providers' unfamiliarity about clients' culture could hinder the abilities of ICT companies to expand their businesses, as in the case of southern company of I49 (interview, Apr 6, 2006) which did not open its business in Hanoi (the North) because of its difference in culture.

13.4.4 Human resources and training

The problems of the education system were preliminarily discussed in Chapters 4 and 5. In 2006, the phenomenon emerged in Vietnam of increased competition to find and retain good ICT employees. ICT companies tried to take qualified staff, especially top experts, from their competitors, and offered attractive remuneration to keep their staff, as revealed by I20 (interview, Feb 18, 2006) of VASC Software and Media Co. who explained that it was not easy to get satisfactory personnel from the ICT human resources market.

At the low and medium levels, it would not be difficult to find good people, I11 noted, because the Vietnamese people are very dynamic and intelligent and have good education (interview, Jan 25, 2006). However, at the senior level, it was very rare to find Vietnamese applicants, and thus this area was an advantageous niche for foreign competitors (I50). At Hochiminh city, I47 (interview, Apr 5, 2006) complained that it was hard to find professional and hardworking IT people. In Hanoi I13 (interview, Feb 13, 2006) revealed that about 5-10% of applicants were suitable for selection, while the rate in Hochiminh city was 30 out of 3,000 candidates (I46, interview, Apr 5, 2006). Recruitment of a capable employee is challenging but retaining her is also not easy. Enns, Ferratt, & Arasad (2006) suggested that not all young IT professionals belong to the high maintenance stereotype, namely that they "provide significant value to their organization, but they expect their organizations to meet their many needs", such as "high pay, benefits, interesting work, recognition, and opportunities for growth and development."

Many ICT managers interviewed believed that a large source of the problem was linked to aspects of the formal IT training system in Vietnamese universities, which lacked practical applicability. The managers said that important knowledge and "soft" skills, such as communication skills (I24, interview, Feb 22, 2006), up to date technology and fundamental knowledge (I46, interview, Apr 5, 2006) were not taught, therefore those who were good at IT might not understand business processes and vice versa (I50). Regarding informal IT training, I46 claimed that vocational training centers, such as FPT Aptec, only provided training in programming languages like C++ or Java.

Similarly, I14, a former practical engineer from Hanoi University of Technology assessed his training program:

> We were taught in school to be technicians in hardware retailing who repair or assemble computers. We had to learn software mainly by ourselves. We had to study and improve on our knowledge ourselves, and asked our teachers at the school to teach us more. When I started my work in computer science, I found that the knowledge and social communication capacity I had acquired was not satisfactory. I then came back to the university to study 3 years more to get a bachelor's degree (interview, Feb 15, 2006).

Newly recruited employees often need to be trained usually for a period of 6 months to a year to meet requirements (I46, interview, Apr 5, 2006; I13, interview, Feb 13, 2006). I24 said that sending new employees to train overseas was expensive, while local training institutions were not capable of providing required training (interview, Feb 22, 2006).

Some domestic companies were able to provide supplementary training for new graduates. I41 (interview, Mar 21, 2006) said that his company had set up an in-house training center to provide supplementary training for new staff. According to him, given that new graduates did not have adequate experience and expertise to start work in the business environment, in the last year before graduation they should learn technological knowledge rather than much theoretical knowledge. I41's company was offering collaboration with public universities to provide free training for about 70 IT students in the 6 months before their graduation. The company's training center was expected to fill the practical gap by teaching students, for instance, computer languages like Java, and by using that knowledge to work on real IT projects. However, to implement that initiative, the company needed the government and public universities to share the cost of facilities, computers and space, as well as incentives to pay the students. At the time of the interview, that company had made joint commitments with several universities, such as the Hanoi Technology University, the Open University, Transportation College and Thanglong University to provide IT training.

From another perspective, I24 (interview, Feb 22, 2006) and I42 (interview, Mar 23, 2006) criticized the high degree of importance attached to the role of ICT by the government and society which resulted in software programmers mistakenly thinking that they were of a superior rank to other professions, and required much higher payment (USD500 per month for entry level while a senior engineer of other professions is only paid at USD100-150). Even within the IT sector, software programming work ranked more highly than positions such as sales, testing, document writing, and technical support. This increased the number of universities and IT training institutions that provided software training as well as the wages level of programmers. In other words, it also decreased the likely competitiveness of the Vietnamese people through low labor cost advantages. The government, with the assistance of the Canadian funded IT Policy Implementation Assistance project (of which the present author was a vice director) made some attempts to set up an IT Titles Classification for Vietnam, together with criteria and a treatment regime for IT public and civil servants. However at the time of writing this book, such classification was still not available, thus, no data are available about IT roles within the industry with respect to occupations, such as tester, designer, programmer, etc.

13.4.5 Intellectual property issues

In developing the software industry, the protection of intellectual property (such as software copyright and trade secrets) is very critical as it is the most important asset of software companies. The 2006 BSA and IDC's report on piracy of global software (BSA & IDC, 2006) claimed that Vietnam was the country with the highest rate of piracy (90%). In this study, information on IP gathered in the interviews fell into three main areas: the protection of trade secrets and software copyright, IP violation and socio-economic conditions for IP protection. These are discussed in turn below.

Protection of trade secrets and software copyright

Interviewees expressed concern about how to keep business secrets within the company. I24's anxiety was:

> Legal binding with employees is not tight, thus the risk of IP loss is high. When an employee quits the company, no legal tool binds him to guard our secrets and knowledge of products (interview, Feb 22, 2006).

And to I13 at the time of the interview the protection of software copyright was in fact based very inappropriately on copyright law relating to literature:

Firstly the author of a literary work is an individual, but with software products, authors could be a group of people or a company. Secondly, the modification of a literary work is not allowed but of software, it is essential. Thirdly, the similarity of two literary works is visible, but the similarity of two software programs is much more complicated to assess, requiring an appraisal by a professional organization.

Therefore, I13 argued that it was hard to apply the current law on copyright in software sector (interview, Feb 13, 2006).

IP violation

I13 (interview, Feb 13, 2006) stated that usually, when software copyright was violated, source code was stolen and it was hard to sue the offending party. Since the revenue of software companies mainly comes from the selling of their software products, the violation of IP protection endangers the sustainability of companies. I48 endorsed this point: "... without IP protection, the development of the software industry is difficult or impossible" (interview, Apr 5, 2006).

I24 of Punch Entertainment mentioned likely problems in an environment where IP is highly violated:

> There is unfair competition between those that comply, and those that don't comply with IP protection;
>
> Companies have to spend money to protect their products by technical measures rather than legal measures;
>
> Companies can't share knowledge with partners to the extent that they wish because of being concerns about the leaking of trade secrets;
>
> Companies can't outsource even though they might wish to do so;
>
> Employees can violate copyright due to their low level of awareness of IP (I24, interview, Feb 22, 2006).

To counter the problem, software companies in Vietnam on the one hand used copyrighted software tools or open source tools to avoid possible disputes with the tool developers, and on the other hand protected themselves from piracy undertaken by users. I14's company, for instance, focused on tailored products thus reducing anxiety about piracy – with tailored products the buyer received the exclusive right and ownership of the tailored product, not the developer (I14, interview, Feb 15, 2006). I17's group developed software embedded in systems or hardware so that cloning would be almost impossible (I17, interview, Feb 16, 2006).

Socio-economic conditions for IP protection

Some interviewees had strong views against the extreme protection of copyright and believed that this protection was inappropriate in the current social and economic conditions in Vietnam. I11, an American entrepreneur working in Vietnam, argued:

> Only foreigners are talking about IP. It has to be reasonable for developing countries to have the right to copy, and it is not a criminal activity. If something is reliable, attractive and easy to copy, you copy it. It's very natural and normal. But the western companies try to maximize their profits, making that

> very normal activity a criminal activity so that they can charge much money. But it is stupid. If they try to make it criminal and always charge for pharmaceutical, software and many other products, people in the developing countries will not pay for or use their software programs (interview, Jan 25, 2006).

I39 claimed that "piracy in the view of ICT users is beneficial. They don't have to pay a lot of money to possess a high quality and valued product" (interview, Mar 14, 2006). I02 was afraid that if Internet cafes had to buy software, it would be very hard to get back their investments (interview, Jan 20, 2006). The Ministry of Posts & Telematics (now Ministry of Information and Communications) does not allow the charging of a fee for the use of a computer in a Internet café over the ceiling limit of VND10,000 per hour. I48 recommended

> It is impossible for low income Vietnamese to buy legal software at the moment. More relaxed IP policy for developing countries like Vietnam should be allowed to give them a chance to catch up with world technology and knowledge standards (interview, Apr 5, 2006).

I14 (interview, Feb 15, 2006) argued that in pursuing proper IP policy, the government should consider social and economic factors such as the low level of income, economic standards and IP awareness. Thailand and China have successfully negotiated with Microsoft to have a reasonable price policy.

Similarly, I47 suggested that

> If Microsoft wants to enter the Vietnamese market, it should modify its price policy that is very high to suit most Vietnamese business and normal users today. The IP issues mainly relate to Microsoft because 90% of IT users are utilizing MS products (interview, Apr 5, 2006).

In China, cheap software developed by domestic companies, such as Kingsoft's Office, helped the country to reduce the piracy rate. Kingsoft's package is now the official document management software in the Chinese Government. In 2005 the government bought 11,143 copies, making Kingsoft Office the standard software of 57 ministries and sectors in 31 provinces and cities, and that Microsoft-like product accounted for 56% of the office software market of China. In May 2007 Kingsoft introduced the English and Vietnamese versions of its Office 2007 to the Vietnam market (Government of Vietnam, 2004a). It was still too early to know if this Chinese company would be accepted in the Vietnam market.

In summary, though opinions on IP issues were diverse, there was a common theme: that domestic intellectual property arising from the Vietnamese people, should be protected to nourish the young industry, and that the protection of software copyright, especially Microsoft software, should take into account the socio-economic conditions of Vietnam as a less developed economy.

As an alternative, open source software was considered.

13.4.6 Open source software and ICT users

Open source software (OSS) emerged as an alternative solution to the problem of IP. In 2004, the government issued a decision to use and develop OSS as a solution in order to be "compliant with IPP, reduce the cost of software purchase and promote the IT industry in general and software sector in particular (Nhan Dan, 2007).

Data from the interviews summarized below reflect OSS in Vietnam from the perspective of two communities: (1) OSS developers and (2) OSS users (including ICT companies and the 22 ICT individual users described in Chapter 11).

Open source software developers

Four companies (17%) chose OSS as a business strategy. I42's company used Zope, an open source application server for building content management systems to develop TVIS, a web based information system for the Ministry of Foreign Affairs and its representative offices overseas (I42, interview, Mar 23, 2006). I41's company focused on Compiere, an OSS ERP&CRM tool to develop ERP applications. I39's group integrated OSS components from various developers for a particular application in Vietnam which none of the single OSS solutions could satisfy. I46's company delivered consultation on OSS solutions in Hochiminh city (I46, interview, Apr 5, 2006).

Strengths of the OSS solutions

Optimistic interviewees believed that OSS should be encouraged as a good alternative to proprietary software because of its low cost (I12, interview, Jan 25, 2006), and the transparency and ease of modifiability for the developer (I13, interview, Feb 13, 2006). I42 argued that OSS had many features that were as good as copyrighted software (interview, Mar 23, 2006). With OSS, software developers could save the costs of production, avoid IP violation and more importantly, could get strong and quick support from the OSS community when they had difficulties. According to I39, resources of OSS were abundant on the web with hundreds of thousands of projects which, if properly exploited, could enable Vietnamese IT companies to "stand on the shoulders of giant's" (interview, Mar 14, 2006).

Weaknesses of the OSS solution

Although OSS is an alternative solution for Vietnamese users and developers, many interviewees believed that OSS could not compete with commercial software. For those who did not use OSS as a tool, the reasons were:

> No long term development orientation, thus low reliability; Responsibility and accountability of OSS developers are not clear (I11, interview, Jan 25, 2006; I13, interview, Feb 13, 2006)
>
> The professional level of OS is often lower than that of commercial (I13, interview, Feb 13, 2006); the quality is not high and guaranteed (I24, interview, Feb 22, 2006);
>
> The relationship in relation to commercial products is mutual profitability between the developer and the user, whereas members of the OSS community unilaterally exploit the fruits of OSS projects. The core of a large OSS is often funded by large corporations, e.g. Open Office funded by Sun or Mozilla by Netscape (I13, interview, Feb 13, 2006).
>
> The total ownership cost of an OSS solution (including the cost of training, maintenance, the required capacity of users, the quality of the product, etc.) could be even greater than that of an equivalent commercial solution (I11, interview, Jan 25, 2006)

> Software company only use commercial tools to develop software because it is more convenient to get technical assistance (I13, interview, Feb 13, 2006)
>
> Installation of OSS is more challenging since technical support is not adequate, and supporting documents and training programs for OSS are limited or not available (I13, interview, Feb 13, 2006; I24, interview, Feb 22, 2006)
>
> OSS is more suitable for professional developers than for normal users, except for very well-funded products like Open Office or Mozilla FireFox(I13, interview, Feb 13, 2006).
>
> OS conditions may cause a developer to lose the right to hold IP rights for herself (I24, interview, Feb 22, 2006); commercial companies can steal ideas from OSS and work more efficiently (I12, interview, Jan 25, 2006).

For those who were currently developing OSS, they expressed the following concerns:

> It is not acceptable to sell OSS based products at high prices, even though the labor-cost to create the products is not low in comparison with vendor-based platforms (I46, interview, Apr 5, 2006).
>
> The sale value of commercial products is the cost of development, while that of an OSS is often after-sales services or value added services (I39, interview, Mar 14, 2006).
>
> OSS developers are volunteers; there are few full-time contributions (I39, interview, Mar 14, 2006)
>
> GPL (General Public License) allows the free use of OSS but also does not regulate the accountability of the author, thus does not guarantee any possible loss that end-users may encounter (I39, interview, Mar 14, 2006)

The OSS market

Among the ICT customers interviewed (whose details have been described in Chapter 11), only a few individuals were found to have had some experience with open source, for example, I07 – an IT student (interview, Jan 24, 2006), or I12 – an IT specialist of the Ministry of Posts and Telematics. I12 said that she had tried Redhead or Apache but added that she still preferred vendor software, although to her vendor software and open source software were not so much different (interview, Jan 24, 2006).

The utilization of OSS was also very rare in governmental agencies. According to one interviewee, the Ministry of Science and Technology, the focal point of OSS in the government, "only makes OSS on paper, but actually does not use OSS for itself" (I46, interview, Apr 5, 2006).

Even among ICT companies, very few domestic ICT companies actually used Linux or Open Office. The reason was firstly because there is no actual pressure to use OSS because IP protection is not respected in Vietnam. Software companies all agreed in interview that they would use more OS general software to replace MS Windows or MS Office if copyright was tightly enforced in Vietnam (I13, interview, Feb 13, 2006; I07, interview, Jan 24, 2006; I47, interview, Apr 5, 2006). Along similar lines, Internet café manager I02 claimed:

> … business is based on profit making. At the moment, almost 100% software are pirated and OSS is not a must. We will use OSS if IP protection is tightly enforced, OSS awareness is raised, and OSS is easy for users (interview, Jan 20, 2006).

At the moment, they encountered the following difficulties when using OSS:

> Open source software, like Linux, becomes more friendly and easy to users but lacks guiding documents. OS can save money if the in-house technical staff is strong [i.e. highly skilled] (I13, interview, Feb 13, 2006).

> MNCs, such as BP, did not have time to build new software applications, thus they often buy off-the-self packages available in the market. Furthermore, MNCs do not accept the high risk of OSS when they need to upgrade or expand a system (I48, interview, Apr 5, 2006)

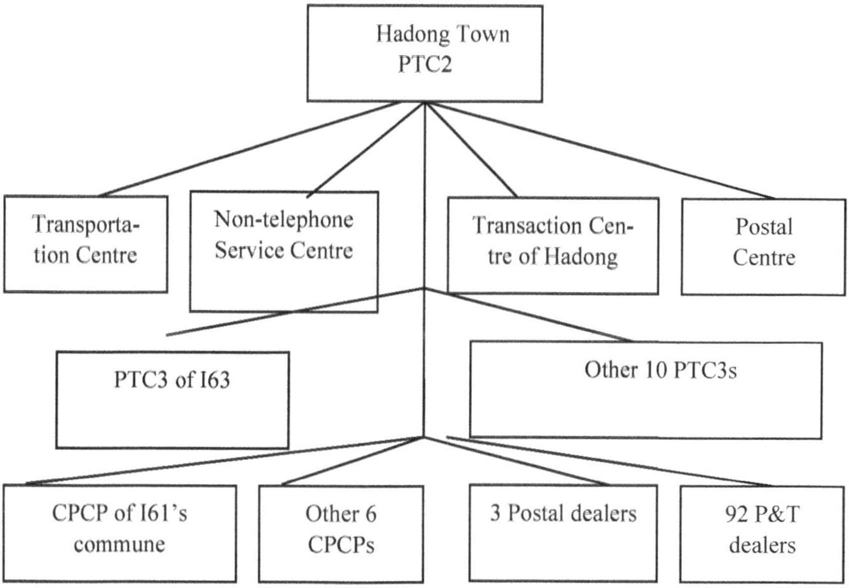

Figure 13.1 Organizational structure of Hadong PTC2 and its affiliates

IMNCs like Punch Entertainment could use general OSS applications like Firefox or Apache as an end-user, but would not want to use OSS as development tools due to the worry of business secrets possibly leaking under the conditions of general public license (I24, interview, Feb 22, 2006).

Finally, three years after the introduction of an OSS plan, during the visit of Microsoft CEO Steve Ballmer, a software deal to use licensed programs in government desktop computers was signed on May 21, 2007 between the government of Vietnam and the giant software company (Hung, 2007). On the one hand this action showed the commitment of the government to reduce rampant software piracy, but on the other hand it was clear evidence of the failure of the 2004 government's plan on "Use and development of open source software in Vietnam in 2004-2008" (Chapter 5) and left only a narrow door for OSS in the public sector.

In summary, although the requirement to use licensed software is likely to become more compulsory in Vietnam, at the moment, the potential for using OSS as an alternative to commercial software is not obvious due to a series of obstacles in the areas of technical issues and ease of use of the OSS and far more importantly, the real market demand and the willingness of customers at the moment. It seems that unmanaged and non-commercial non-profit making movements like OSS have not been able to compete to the same ability with commercial and profit making software like Microsoft Windows.

However, the emergence of alternative competition like OSS can give more choice to people to fight against corporate monopolists, and efforts on OSS need to be further encouraged.

13.5 Issues identified by ICT providers at communal and remote areas

n rural and remote areas, ICT providers were thinly scattered and VNPT affiliates in the form of posts and telecom companies and their dealers, such as CPCPs were sometimes the only providers for the local residents. The government subsidizes VNPT to develop a ubiquitous network of CPCPs and PTCs that covers almost all communes in the country (see Chapter 5 about the status of CPCP in Vietnam). In Hatay, one of two provinces the book author visited, the structure of PCTs and CPCPs was as shown in Figure 13.1.

Figure 13.2 The front of a CPCP. Photo: the author

13.5.1 CPCPs

According to I61 and I63, CPCPs were well presented in almost all the communes where there were no PTC3. They all included a separate concrete building; often next door to the communal people committee's headquarters. A picture of one of them is shown in Figures 13.2 and 13.3. They were equipped with a telephone connection, a bookcase with hundreds to a thousand books on agriculture, cultivation, breeding techniques and legal subjects.

In front of the CPCP or in a clear position in the CPCP, customers could see large posters with information about the services and tariffs. The main services available at I61's CPCP in Hatay province, as shown on its notice board, were

- Receiving and transmitting domestic and international postal mails or parcels;
- Transferring domestic money order;
- Telegramming and fax;
- Domestic and international telephone;
- Consulting over the 108 telephone number;
- Collecting telecom subscription fees;
- Providing free book and newspaper reading;
- Selling envelopes, stamps and phone cards;
- Subscribing to period or daily newspapers.

The main customers were local residents, people passing-by, local students and staff of the neighboring communal people's committee. Yet, CPCPs were often deserted, only 10 to 15 customers per day compared with crowded private Internet cafes where teenagers often have to wait for their turn. We observed only one client coming to send mail in I61's case, and none in I62's CPCP in Haiphong during our one hour visit. Once a quarter, I62's CPCP held a talk to encourage local people and children to read books and newspapers, and get comother basic services, receiving and transmitting money was visibly essential to the local people in the absence of banking and financial systems in the communes.

The provision in communes of books, magazines and newspapers by CPCPs (under the guidelines of the Ministry of Posts and Telematics) in many communes duplicated services provided by communal libraries (under the silo management of the Ministry of Information and Culture existed). With the merginments from them. It also exchanged books with the communal library.

Given that mobile and fixed line phones were not well presented in the country side, CPCPs, despite their low frequency of use, played an important role as a bridge to connect local people with the rest of the world. Among g of the Ministry of Posts and Telematics and the Ministry of Information and Culture into one single agency (i.e. Ministry of Information and Communications) it is unknown whether this redundancy has been removed or not.

Each CPCP was run by an employee who was a young local person. Since the number of customers was not large, the revenue of a CPCP was not high, only 400,000 to a million VND per month (from USD 25 to 62.5). Income for employees, even with the compensation for overnight guarding of the CPCP facility was thus quite low, about VND 400,000.

Besides low revenue, CPCPs were not well known by the residents. I61 (interview, Dec 12, 2006) claimed that her CPCP was not located near the school or main path of the commune, therefore few people understood what service she could provide to them via CPCP, even though a large notice board was laid at the front door. In addition, the appearance of the CPCP was not really inviting (damaged fence and rough entrance).

Not many CPCPs had an Internet connection. According to I62 (interview, Dec 18, 2006), before the time of interview, his CPCP of Doanxa in Haiphong had been equipped by PTC3 with 2 computers with Internet connections. However, two computers had not been enough for customers who often came in groups of four to five persons and each all demanded a computer. When the request had not been satisfied, they went to other Internet cafes. As a consequence, few people had wanted to come to his CPCP for Internet purposes. Finally, as revenue from Internet service was not high, instead of providing more computers, the two computers had been moved to another place.

We noticed that several kilometers away from these CPCPs, private Internet cafes and communal post offices with Internet connection located in a larger road and more populous areas were often full of Internet users.

Figure 13.3 A CPCP employee serving a customer. Photo: the author

In sum, CPCPs provide basic postal and communication services to, and stimulate the information of, local residents. However, the effectiveness and efficiency of this top-down investment from the government via VNPT, in the author's observation and opinions from the interviewees, was not as high as widely described by the mass media. Actual needs of local people will gradually increase along with the progress of the economy and the urbanization process.

13.5.2 Post and Telecom Company level 3 (PTC3)

Unlike CPCPs which are run by private contractors, PCT3s are state companies totally owned by the VNPT group. Figure 13.4 was a picture of a PCT3 Internet chatroom in Haiphong which was not much different from a private Internet café (Chapter 11). They provide a full range of services of P&T plus money transfer, newspaper and magazine delivery and Internet services (Chapter 5). Some, such as the I64's PCT3, delivered librarianship services but these had ceased due to low use. In this study, the author only focused on the Internet services that PTC3s were providing.

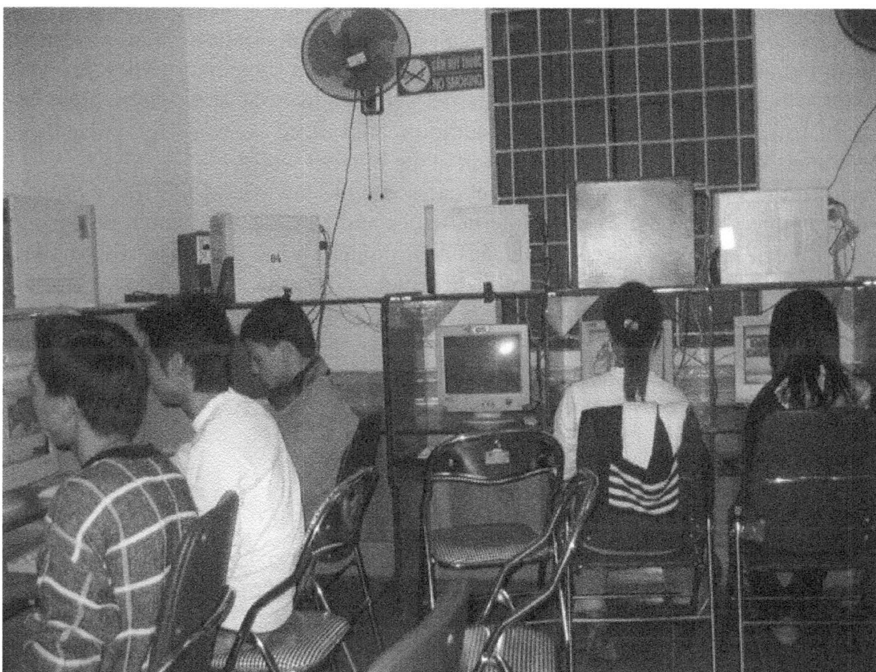

Figure 13.4 Inside an Internet room of a PTC3. Photo: the author

In each of the two PCT3s visited, there was an Internet room with 10 PCs separated from other functional rooms. Main customers were students of various levels and workers. The registration book showed that the number of customers of PCT3 was much higher than that of CPCPs, 78 on the day before in I63's PTC3 (Hatay province) and 51 in I64's PTC3 (Haiphong city). Revenue from the Internet varied from 2 to 4 million VND.

Compared with private Internet cafés in the communes, customers claimed that they preferred to come to PTC3s because they could enjoy good services, new computers with better Internet connection in a comfortable air-conditioned atmosphere. Many persons used the Internet along the way with other P&T services. However, the study did not find much difference between PTC3s and private cafés. The state run and private run Internet cafes existed concurrently and delivered similar Internet services at the same location. The function of PTC3s in these locations as an Internet bridge to local people was not outstanding.

However, in poorer areas where the private Internet café is absent, the delivery of ICT services by PCT3s and CPCPs, though modest, is important to bring opportunities to the unwired regions in the villages. Incentive and subsidiary policy for businesses which provide telecom and Internet services in rural and remote areas are crucial to reduce the digital gap. In addition, subsidized schemes for villagers in these areas would encourage them to join in the Internet world.

Given the success of private Internet cafes and the low use of CPCPs, a private-public partnership in the form of Internet-cafés and CPCP hybrid could be encouraged in needy and poor areas. That business model on the one hand would allow needy people to enjoy the benefits of the Internet, and on the other hand would allow business managers to get fees from other customers to cover the costs of operations.

13.6 The role of government in supporting the ICT industry

13.6.1 Government as a customer of ICT providers

National and local government agencies are the biggest customers of the ICT providers (see Table 13.2). Government's decisions on ICT procurement strongly impacts on the business of many Vietnamese ICT companies (43.5% of the interviewees). A 2007 deal with Microsoft on copyright in the public sector, valued at about USD 1 billion, may have a strong impact on the local ICT business. I17 (interview, Feb 16, 2006) expressed concern about the capacity of management in ICT government agencies and state-owned companies.

I46 complained that:

> The attention to IT in government remains more formal than practical. Ministries and sectors are buying IT solutions directly from foreign vendors at very expensive prices and give little attention to domestic companies (interview, Apr 5, 2006).

The government plays an administrative role, thus its management manner influences the development of the ICT industry. I46 (interview, Apr 5, 2006) said that if a project was difficult to manage, the governmental officers took an easy option of cancelling it. For example in 2007, the IT authority in Hochiminh city suspended the issuance of new licenses to Internet services of Internet cafes due to its limited capacity.

I49, manager of a hardware retailer in Hochiminh city, whose main market was the south-east provinces was concerned that the digital gap and unfairness emerged from uneven investment in IT in the cities and provinces: large cities received more money, therefore they had more chance of access to the Internet than others (I49, interview, Apr 6, 2006).

13.6.2 The ICT legal environment

In regards to the ICT legal environment, ICT business interviewees mentioned two points: government regulations on IT procurement in the public sector and IT tax. I11 claimed that ICT procurement in most organizations in Vietnam was not rational but based on a subjective relationship: an unspoken agreement between buyer and seller (interview, Jan 25, 2006). I46 (interview, Apr 5, 2006) explained this was because of the absence of appropriate policy and regulations on IT procurement and investment in the government sector. IT projects were currently treated as construction projects where the proportion of design was only 10% and there were no rules or norms on the proportion of expenditure in software, consultation and services in governmental IT projects. It was thus difficult for IT consulting or software firms to get projects (I55, interview, Apr 11, 2006).

Regulations on import taxes were also concerns. I24 from Punch Entertainment complained that sometimes his company had difficulty in getting permission to import tools and facilities for software production. He further recommended that "the equipment should enjoy as incentives taxes which were lower than that for general goods" (interview, Feb 22, 2006).

ICT businesses' dissatisfaction about transparent and fair treatment of the Government mainly turned around the government regulation on IT procurement in the government sectors and the capacity of the government to undertake a purchase fairly and efficiently. Six month later from the above interviews, on July 7, 2006, the Prime Minister issued a decision #169/2006/QĐ-TTg which provided a framework of key principles on IT procurement with government funding (Government of Vietnam, 2006c). That document, similar to the majority of other IT related documents mentioned in the chapter, is far from concrete to be followed. Detailed guidelines, particularly regulation on ICT consulting and services, need to be provided soon to meet the demand of the industry. Concurrently, there should be relevant programs on the one hand to enhance the capacity of ICT governmental managers in procurement, and on the other hand to tightly control ICT bids in government to avoid scandals like the Nguyen Lam Thai's case who bribed at least 37/66 provincial VNPT directors in 2006 (Pham & Anh, 2006).

Tran An (2006) summarized a survey undertaken by the Ministry of Science and Technology of 11,000 enterprises in 30 northern provinces of Vietnam which indicated that 39.6% of the enterprises needed information about government policy, 25.9% needed information about new technology and 22.6% needed information about markets and marketing. The report revealed that while information about new technology and markets are vital elements for foreign entrepreneurs, their Vietnamese counterparts are mainly concerned with information about government policy.

13.6.3 The provision of telecommunications infrastructure

In rural and remote areas, the number of telecom connections which are essential for ICT business to provide services, and for the local people to have access to the benefits of ICT, was abysmally low. The low teledensity - telephone lines per 100 inhabitants (International Telecommunication Union, 1998) - reveals the reverse side of the laissez-faire economy. Few businesses want to invest in areas where the beginning stage often requires high investment costs but a low rate of revenue due to low numbers of users.

In this case, the government has undertaken strong intervention. The Vietnam Public Telecom Service Fund was established by the Prime Minister in 2004 "to support the development and delivery of public telecom services in areas where the market mechanism of business's revenue can't recover costs" (Government of Vietnam, 2004b). The fund is financed by the government and from compulsory contributions from telecom service providers.

Organizations and business have also joined in the effort, for example, FPT-Lead, CMS and Intel Vietnam launched a low price computer program for farmers in the same year (H. Nguyen, 2006a) and the Youth Union had a program to send 20 million young volunteers to the countryside to educate farmers about access to ICTs. I57 (interview, Apr 12, 2006), vice general director of Telecom Department and I58 (interview, Apr 14, 2006), senior officer, IT Industry Department, MPT, believed these activities should be undertaken and supported with the partnership of the government, ICT business, associations and the villagers.

However, these attempts to bring ICTs to rural areas are not always as successful as expected. If CPCPs are in low use, projects to produce cheap computers for peasants, such as the G6 computer of the alliance of 6 IT companies in Hanoi, namely Trần Anh, Mai Hoàng, Phúc Anh, Vĩnh Trinh, Máy tính Hà Nội, and Ben; and Thanhgiong computer of FPT and CMC all died in silence (H. Nguyen, 2004).

In the "Strategy of information and telecommunication to 2010 and vision to 2020", the government has committed to increasing its capacity and effectiveness in the management of ICT:

> Strengthening the ICT government management system at the national and local levels to ensure the motto "management capacity must keep pace with the development (Government of Vietnam, 2005c).

It is hoped that the findings in this chapter will provide helpful inputs to that development.

13.6.4 Government websites

A survey by the Ministry of Science and Technology of 11,000 enterprises in northern Vietnam found that while 60.2% of enterprises used computers, only 11.55 had LANs and 2.1% had a website, and almost no company transactions were conducted using e-commerce applications (Tran, 2006).

By 2007, about 502 websites were managed by governmental agencies. Around 30 central agencies and 64 local governments have been developing their online presence with varying degrees of sophistication. In relation to some of the websites there have been expressions of concern about unfriendly design and usability difficulties, with end-users reporting difficulties in finding the information they needed and low speeds of website access (see Chapter 5).

An official website on sustainable development (va21.org) was established in 2004 by the Sustainable Development Office at the Ministry of Planning and Investment. The site covers law, policy, action programs and many materials and research documents on the topic in both the Vietnamese and English languages. Compared with personal blogs like Cuong oz's, this site was less popular, with an automatic count by the website of 43,293 views by December 2006.

13.7 Tourism: an example of ICT industry success

During the fieldwork phase of this book project, the author noted many examples in which ICTs were used effectively at different levels, even though interviewees might not have been satisfied from a cost-effectiveness perspective. ICT applications in the tourism sector, for instance, were very impressive. S29, CIO of the Vietnam National Administration of Tourism said that its main outcomes were as follows: 1) ICT awareness among the leadership was upgraded and 2) the number of ICT applications increased, with 70% of travel companies and 25-30% of hotels establishing websites. The IT center at the headquarters of the agency aimed to maintain two websites containing information about 300 travel companies, advertisements and travel programs to support tourist companies and an online hotel booking system. This IT center was also providing IT training for tourist staff so that they could update their information to the web. S29 revealed that he encountered problems in a number of areas such as the legal requirements for e-payment of hotel bookings and the small amount of interest by the agency's leadership who were usually over 40 (S29, interview, Nov 18, 2004).

13.8 E-content and online news media

The vital role of knowledge and ICTs in the struggle for sustainable development incorporates two aims: to change human behavior to be more environmentally friendly, i.e. being less harmful to nature, and to create new substitutions for natural resources, diminishing global warming. In this struggle, e-content and online newspapers play crucial roles. During the first fieldwork round, 4 journalists were interviewed. They were S11 of Vietnampost newspaper (interview, Nov 2, 2004), S26 of PC World Vietnam magazine (interview, Nov 10, 2004) and S48 (interview, Dec 9, 2004) and S49 (interview, Dec 14, 2004) of Vnexpress news-online. The electronic form of newspaper is called "news-online" because, like Vnexpress, it is paperless.

Progress towards greater openness and democratic participation have brought a new vitality to Vietnam's mass media which allows the appearance of hundreds of news-online websites today (S49). Journalists are given more freedom and openness, and news-online websites are working for a better society, and against evils such as corruption (S11). Chronic corruption is generally viewed as a serious threat to development and a major deterrent to investment in Vietnam (World Bank, 2004, p. 90). Great resources wasted by corruption can severely jeopardize the sustainability of future generations.

Mass media, especially news-online like Vnexpress.net or Thanhnien.com.vn revealed many crimes on the Internet. Hundreds of high ranking officials in the trade, oil and gas, and transport sectors and even in the Government Office have been arrested for their involvement in corruption, thanks to the uncovering of corrupt practices by the mass media and the dissemination of the news to the public which have required prompt action by the authorities. Via the Internet and mass media, individuals can now raise their voices, though through the censorship and monitoring of the enterprises' editors, to the highest leader of, or any official in the administration. In other words, democracy brought about by ICTs and a mediated press have contributed much in relation to the goals of growth and sustainability.

News-online surpasses previous media in being a gigantic speedy and two-way channel of information for the 12 million Vietnamese people online. Vnexpress.net maintained by FPT Corporation, for instance, was the first Vietnamese inclusion in the list of 500 most accessed websites in the world in 2005 (T. T. Nguyen & Johanson, 2006). On December 3, 2007, its rank in Alexa.com jumped up to 81 with more than 2.7 million page-views per day. Readers can find, from this huge store of the hottest news and information, a political scandal or a soccer result only some seconds after it has occurred. And often, if someone is unluckily labeled with a bad reputation on such a site, he or she is virtually sentenced. The "hammer and axe" of public opinion is so fierce that the relevant authorities have little choice but to take action. On the positive side, this is evidence of the power of democracy to the people given by ICTs. On the negative side, however, individual freedom – called in some legal systems the presumption of innocence – is seriously violated. The person targeted has not yet been sentenced by a court.

In online news media information on sustainability such as global warming, water scarcity or a new technological invention by a farmer, though not in a separate column, is also frequent. S49 revealed that the main revenue of the publication came from advertisements. In this case, the money received from profit making business had been used smartly in the diffusion of information, including information about sustainable development, to the society.

Internet press has more advantageous technical features than other media. To attract audiences, information provided must be extremely "hot" – hence the 24 hour news-online services are pre-eminent. Moreover, news-online services have multimedia functions such as text, audio, graphics, animation and video to inform or entertain the audience, and can redirect people to limitless links on the Internet. And finally,

> the interactivity on the forums between editors and readers, and among readers is the "super-strength" of this electronic media. Readers, by exchanging information and identity (not necessarily giving their real names) with others, receive needed esteem and self-actualization. News-online, in its turn, receives popularity and information from the audience (S11, interview, Nov 2, 2004).

Maslow developed a hierarchy of needs which consists of five levels: the four lower levels are grouped together as "deficiency needs" associated with physiological needs, while the top level is termed growth needs associated with psychological needs. Deficiency needs must be met first. Once these are met, seeking to satisfy growth needs drives personal development. The higher needs in this hierarchy only come into focus once all the lower needs in the pyramid are satisfied (Maslow, 1954). With active participation in online forums, people can achieve the highest level of satisfaction in Maslow's hierarchy of needs. In other word, besides economic motive, the spiritual recognition of others in the virtual world is one of the main drivers for individuals to participate in the virtual world and to contribute their intellectual creation.

Nevertheless, news-online is not a perfect solution without any drawbacks. If ICTs bring more power to journalists, they induce new pressures too. S11 (interview, Nov 2, 2004) was in charge of an ICT column in Vietnampost. In August 2004, he went to Danang city where he had an interview with some local people and transmitted news, photos and the papers of correspondents to the office. Without computer access and the Internet, he said, he could not deliver such work at a speed that would satisfy customer expectation. This journalist spoke from personal experience as a reader of his own newspaper – during the last two days before his interview with the author, his network connection had been down and he could only read the newspaper as a "normal" reader and found this highly frustrating. Clearly, ICTs have become indispensable factors for journalists like S11 and to news-online services and many of their readers. In response to readers' preferences reporters become technology dependent. Reporters, mainly those who are in their later years, and are not able to adapt to the new technologies, become obsolescent in the networking environment.

To some degree, the government's orientation towards and control of the press limits the openness and democracy of mass media. S49, a manager and editor of an online news service, argued

> The press appears to be more of a one-way propaganda [mechanism] by subjective writers rather than allowing readers' self assessment and thinking. Readers have few chances to express their opinions on news-online. Even news-online are censored by the authority. The legal environment should give more democracy and freedom to the press. Journalists should provide readers with the most objective and multi-directional information and facts, and should avoid subjective assessments, and should give the right of comment to the readers (S49, interview, Dec 14, 2004).

And news-online is not for all, at least at the present time. S49 claimed that while urban inhabitants have much more incentive to be online "... rural counterparts without connectivity, knowledge and money, are often out of the game".

13.8.1 Non-governmental forums and websites

Online forums are administered by a group of people or an organization. They contain collections of knowledge, experiences and opinions. Personal blogs and private websites are also kinds of online forums.

Some websites are purely forums. They are usually a common place for thousands of members to view, chat and/or search for information. There are also online places and clubs where people from a particular province or with a particular interest can meet. www9.ttvnol.com (Online Vietnamese Heart) is a non-profit and non-political forum covering multi-faceted, broad topics such as culture, history, astrology, music, and economics. The forum has been maintained by Nguyễn Xuân Mai and several former members of Trí Tuệ Việt Nam Online (Online Vietnamese brainpower) since 2002 (M. X. Nguyen, n.d.).

Other websites are created to disseminate information. http://www.vacne.org.vn/ is a website of the Vietnam Association for the Conservation of Nature and Environment which publishes the chapters of a book entitled "Vietnam - Environment and Life" in both Vietnamese and English languages. Another site with similar purposes is http://www.vusta.org.vn/ which is published by the Vietnam Union of Science and Technology Associations, and provides scientific and technological information and activities.

According to Alexa.com's statistical data, http://www.vacne.org.vn and http://www.vusta.org.vn, websites without forums receive much lower online traffic than ttvnol.com, the online forum, or http://www19.dantri.com.vn, the website that provides both information and forum services. Hybrid website-forums may draw more attention because they satisfy two important needs of Internet users: seeking and sharing information. The website http://www19.dantri.com.vn ("People's Intellectual") of the Vietnam Society for Study Promotion and the electronic newspaper of Study Promotion is such a combination. It attracts access from a large number of students and intellectuals.

13.9 Main insights from Chapter 13

Insight 13A: The ICT industry ranges from Internet café micro-businesses to large foreign and state-owned corporations. In addition to Internet cafes, strategy for rural service provision is largely based on CPCPs (contracted out to micro-business private contractors) and PTC3s (directly operated by government). There remains scope for government policy to facilitate the closing of gaps through public-private sector initiatives.

Insight 13B: Local ICT companies have greatly improved their capacity to provide services to customers, to the point of dominating the market – all small and medium IT projects are undertaken by Vietnamese companies. Local companies are also providing some services to overseas markets. Nevertheless there are still issues concerning state monopolies, project development expertise and due diligence. Also access to capital for start-up and expansion of enterprises lacks clear guidelines.

Insight 13C: There is a major move towards content based products and services (including e-news media and e-forums), although the original focus of the Vietnamese ICT industry on software development, hardware sales and infrastructure services remains strong. Major issues affecting the progress of software and content development especially relate to open source versus proprietary business models and related legislative frameworks. Outsourcing is becoming an ever-more prominent feature of ICT service provision.

Insight 13D: The deregulation in the telecommunications sector has facilitated growth in infrastructure and brought down prices. However there is a wide difference between the availability and reliability of high bandwidth communication in the major cities compared to rural Vietnam.

Insight 13E: Different views were reported in regard to levels of human resources available in the IT sector, the quality of formal education, and existing projects which provided supplementary training for new graduates.

A summary of challenges and recommendations is given in Appendices 9 and 10.

Chapter 14
Sustainable development or permanent scarcity?

14.1 Scope of the chapter

- This is the first of three chapters that use insights gained from the earlier chapters of the book inspire a range of visualizations or theoretical models.
- These theorizations aim to capture, in whole or in part, the complex interplays of factors that are discernable in the accounts contained in the earlier chapters, whether based on literature or case material collected during fieldwork in Vietnam. Additional literature and data are also introduced to assist the theorization process.
- In this chapter the author uses formalized logic to argue the proposition that the information of an organic system or a community of organic systems tends to increase over time to infinite volume to resist the Law of Mass and Energy Conservation and the Law of Entropy.
- The effect of this interaction, he suggests, is that by accumulating information, organisms can sustainably replicate themselves and maintain their appearance in the world. He calls this perspective on sustainability the "Law of Information Increment in Organic Systems".
- Applying the perspective both to economic history and contemporary trends, he offers an illumination of the concepts of natural and human-made capital and their relationship to information and knowledge in economic development.

14.2 Modeling the relationship between development, negative aspects of the knowledge economy concept, and sustainability

14.2.1 Law of Mass and Energy Maximum, Law of Information Increment in Organic Systems and Law of the Economic Value Maximum

One of the most influential paleontologists of the twentieth century George Gaylord Simpson in his book "The Meaning of Evolution" in 1949 concluded that evolution is "the Problem of the Problems" and added "Evolution is neither wholly orderly nor wholly disorderly" (Simpson, 1967, p. 185). Georgescu-Roegen, in explaining Simpson's statement, warned in "The Entropy Law and the Economic Process":

> If evolution of large organizations and, especially, of our own species seems somewhat of a mystery, it is only for two reasons: first, not all natural phenomena follow the pattern of the rhythm, and second, the condition of mankind is such that we can observe nature only once, or more exactly, only in part. That is the only fundamental reason why evolution is "the Problem of Problems" as G.G. Simpson labels it – and why men can hope only to grasp it in a very imperfect way. Even set in this modest manner (as it should be), the task is no less than titanic (Georgescu-Roegen, 1971, p. 208).

Similarly with the concept of development, scholars have fervently desired to discern "the rules" governing development, and have devoted extensive energy to the quest.

In the case of evolution Georgescu-Roegen, for example, attempted to use the second law of thermodynamics to interpret the evolution of human species and its associated economic processed using the second law of thermodynamics. This law, also known as the Entropy Law, was discovered by R. Clausies in 1865 and states that the total entropy of any isolated thermodynamic system like the universe tends to increase over time, approaching a maximum value. In the course of this process, there is a continuous and irrevocable qualitative degradation of free or available energy into bound or latent energy. The final outcome is a state where all energy is latent, the Heat Death (Georgescu-Roegen, 1971, p. 129). By the Law of Entropy, Georgescu-Roegen argued that in a definite time human beings will use up all the natural resources that the earth grants as a free gift.

The present author acknowledges the explanatory strength of Georgescu-Roegen's formulation. However in relation to the real-world case of economic development he suggests that human activity works to induce a "death" state well ahead of the Heat Death of the universe as a whole. He would argue that the transformation of materials in the universe following the second thermodynamics law evolves very slowly and it may take many billions of years before all energy in the universe becomes totally latent at the Heat Death. It is the human species which, with accelerating profligacy, has been using up the natural resources of the earth and has brought the planet to the verge of a human-caused analogue of the heat death, that we call ecosystem death.

Although a powerful analogy, the argument of the Heat Death as applied to development can be misleading. The Heat Death of the universe is, to the best of human knowledge, ultimately inevitable: no human action can alter a process of such cosmic scale. If this were true of economic development no human action would be necessary. Redclift wrote: "Once they are regarded as 'natural' such laws tend to be confused with the social and political implications that follow from their adoption" (Redclift, 2006, p. 79). *However in the author's perspective, the demise of the global economy, if it happens, will only be blamed as a consequence of human-made action, not nature.*

The universe and all of its elements, including all organisms, follow the Second Law of Thermodynamics and two other universal and classical laws of thermodynamics: the Law of Conservation of Mass and the Law of Conservation of Energy. The Law of Conservation of Mass or the Lomonosov-Lavoisier law states that the mass of a closed system of substances will remain constant, regardless of the processes acting inside the system. The Law of Conservation of Energy indicates that the total amount of energy in an isolated system remains constant, although it may change forms, e.g. friction turns kinetic energy into thermal energy. Overall, the total quantity of matter and energy in an isolated system is not altered.

From these laws of nature as axioms, the present author will propose a model of development as follows.

An organic system, in its development and expansion, tends to accumulate mass and energy, thus, its combined mass and energy grow to a maximum value. If these organic bodies gather in a community, such as a community of mushrooms or bacteria, that community also tends to accumulate mass and energy approaching a maximum value.

In general, *we can safely state that the mass and energy of an organic system or a community of organic systems tends to increase over time to approach a maximum value.* This is the *Law of Mass and Energy Maximum in organic systems.*

However, in organic systems evolution "puts up a fight" against entropy. The increase of mass and energy in an organic system, to a certain degree, resists the Law of Entropy, deferring the point at which the system decays and perishes. Brooks and Wiley argued:

> Evolution is a process that slows down the entropic decay of lineages, minimizing their entropy increases. This suggests that, as the interplay of information and cohesion, biological evolution should exhibit an intrinsic tendency towards efficiency or parsimony which in turn should relate to the principle of minimum entropy production (Brooks & Wiley, 1988, p. 106).

To resist the law of entropy, during its existence, beside mass and energy, the organic system also amasses experience in the form of information. Genes and DNA accumulated in bacteria, plants and animals; and memories in animals are passed from generation to generation. The storage of information in organic systems requires the minimum volume of mass and energy, and thus attracts the lowest level of entropy. Maupertuis stated it in his Principle of Least Action that "If there occurs some change in Nature, the amount of action necessary for this change must be as small as possible" (Barrow, 1988, pp. 80-81). This principle is somehow similar to the first paragraph that Laozi enunciates about nature in Tao Te Ching:

> Is the action of nature not unlike drawing a bow? What is higher is pulled down, and what is lower is raised up; What is taller is shortened, and what is thinner is broadened. Nature's motion decreases those who have more than they need; And increases those who need more than they have (Laozi & Henricks, 1989, p. 77; Laozi & Merel, 1995).

Generally, we can assume that *the information of an organic system or a community of organic systems tends to increase over time to infinite volume to resist the Law of Mass and Energy Conservation and the Law of Entropy.* By accumulating information, an organism can replicate itself then maintain its appearance in the world. The present author calls this phenomenon the *Law of Information Increment in Organic Beings.*

Like any organic being, the development of human beings, firstly follows the Law of Mass and Energy Maximum and the Law of Information Increment. Not only accumulating mass, energy and information, humankind by its labor and social organization for all forms of human production, creates and accumulates capital value, a purely artificial concept which was never created before by Nature. Within capital value, economic value is commonly regarded as the most important property of the production processes of human beings, even higher than the mass and energy that might be seen as the initiating or original inputs of human production systems.

Debates about economic value can be traced in works of economists since Adam Smith or even the physiocrats before Him.

Therefore, we can state that human capital comprises two value components: natural capital and human-made capital.

This allows enunciation of what this author calls the *Law of the Economic Value Maximum: the accumulated aggregate of human-made capital derived from natural capital tends to be hoarded increasingly to a maximum volume.*

The accumulation of value, especially economic value, in reality is the appropriation of mass and energy that used to belong to nature proprietary possessions of individuals or groups. However, that accumulation is not unconditional. As bounded by the Law of Conservation of Mass and Energy, when energy and mass are used up by human consumption, all or part of them convert from a usable to a useless form.

14.3 Development of human species – the human-made / natural capital model

14.3.1 Some definitions

The following definitions are helpful as they assist this author to propose a further level of explanation that links closely to the mainstream international discourses on development.

Natural capital (NC) is the un-appropriated mass and energy in the planet or the total stock of natural resources, land and environmental quality (United Nations. Statistical Division & United Nations Environment Programme. Economics and Trade Unit. Division of Technology Industry and Economics, 2000, p. 5).

Human-made capital (HC) is the mass and energy that human beings have exploited and used up so far, and human-made capital is "a stock that yields a flow of valuable goods and services over time" (Costanza & Daly, 1992).

Environment quality (EQ) includes the sink functions and service functions. *Sink functions* are the function of the environment in absorbing the unwanted by-products of productions and consumption. *Service functions* are the habitat functions of nature, such as land, air, water, and landscape for humankind and all living beings. (United Nations, European Commission, International Monetary Fund, Organization for Economic Co-operation and Development, & World Bank, 2003, p. 5)

Total planet capital (PC) is the total of mass and energy in any form on planet Earth, including its surrounding atmosphere. PC is constant.

By the Law of Conservation of Mass and Energy, we may write symbolically:

$$NC + HC = PC.$$

When the author says "symbolically" because in this equation (as well as in those of the same nature that he shall write hereafter) the mathematical signs should not taken in the strict sense, but rather as convenient signs for summarizing the imponderable elements that constitute, in a positive or negative way, the entity represented on the left of the equality sign.

With a certain level of knowledge, technology and other human-made capital such as labor and financial accumulation, humankind can exploit the planet capital to a certain level of *accessible resources* (AR). With a higher level of knowledge and capital accumulation, people can find new resources (e.g. discover new mines) and thus increase the proved reserve of accessible resources.

Yet, human beings must have a minimum level of exploitable natural resources required for a positive growth of capital production and consumption per capita. Let's call that level the *existence threshold* (ET) for positive development. Below that level, HC cannot increase.

As the existence threshold is affected by the volume of population (Pn), per capita consumption (C) and technology (T), then ET = ET(Pn,C,T). For example, the existence threshold of energy at the moment is the minimum stock of fossil fuel required for the human beings to maintain its normal production and consumption.

As these factors are variable by time (t), we have NC= NC(t), ET = ET(t), HC= HC(t), EQ = EQ(t), PC = constant and therefore:

$$NC(t) + HC(t) = TC \quad (1)$$

We note ΔHC(t) which is the difference of HC across time.

Then the condition of a development with a positive difference is:

$$\Delta HC(t) > 0 \text{ if } NC(t) > ET \quad (2)$$

And the condition of a development with a negative difference is:

$$\Delta HC(t) \leq 0 \text{ if } NC(t) \leq ET \quad (3)$$

If there can be no substitution for conventional resources, then as HC(t) increases, NC(t) declines in an inverse relationship. In other words when HC increases, NC decreases. In time the system would reach ET at a point that can be called permanent scarcity. After that point HC(t) would decline.

Figure 14.1 shows the relationship between human-made capital growth and natural capital degradation. Development starts from the right hand when human-made capital is zero and natural capital is at maximum abundance as an Accessible Resources (AR). When human made capital grows, natural capital declines correspondingly. During that period, development is in a stable expansion phase. Growth reaches its peak when the remaining natural capital is equal to the lowest level that development requires in order to maintain its positive growth. We have called this level ET - the existence threshold. After this crisis point, the production and consumption of the mass/energy drops in turbulent destructive cycles to zero. Regarding natural resources in this analysis, this author discerns four phases: Abundant, Pre-Scarcity, Permanent Scarcity and Post–Scarcity.

The difference of AR and ET, AR – ET, is the limit of growth that humans can achieve under conditions of stable expansion. When the remaining stock of AR reaches ET or AR-ET = 0, human beings reach a condition of scarcity of natural resources. In the extreme case when all natural capital is used up, this becomes *a permanent scarcity of natural resources.* The author would call it the permanent scarcity to differentiate it from short-term scarcity in conventional theories of economics which is caused by fluctuation of supply at a certain time, for example, the temporary scarcity of food in the Second World War or the temporary shortage of oil in the time of the Yom Kippur/Ramadan war in the Gulf in October 1973 (Klare, 2004). Permanent and temporary scarcities in nature are different but have similar impacts on human beings. Under temporary scarcity, when the causes are removed, the impacts are diminished. In contrast the permanent scarcity of mass and/or energy is similar to the extinction of a species. When the number of individuals of a species is below a minimum amount, this species vanishes and there would be no conventional way to revitalize it. The phrase "conventional way" is used deliberately, because in a later section we shall discuss the impact of unconventional ways made available through new knowledge.

Often, temporary scarcity has a quick impact on the supply of the mass/energy and consequently when the supply does not match demand, price soars. During the Yom Kippur/Ramadan war for instance crude oil rose by over 400 percent in less than a year (Rogers, 2004, p. 147).

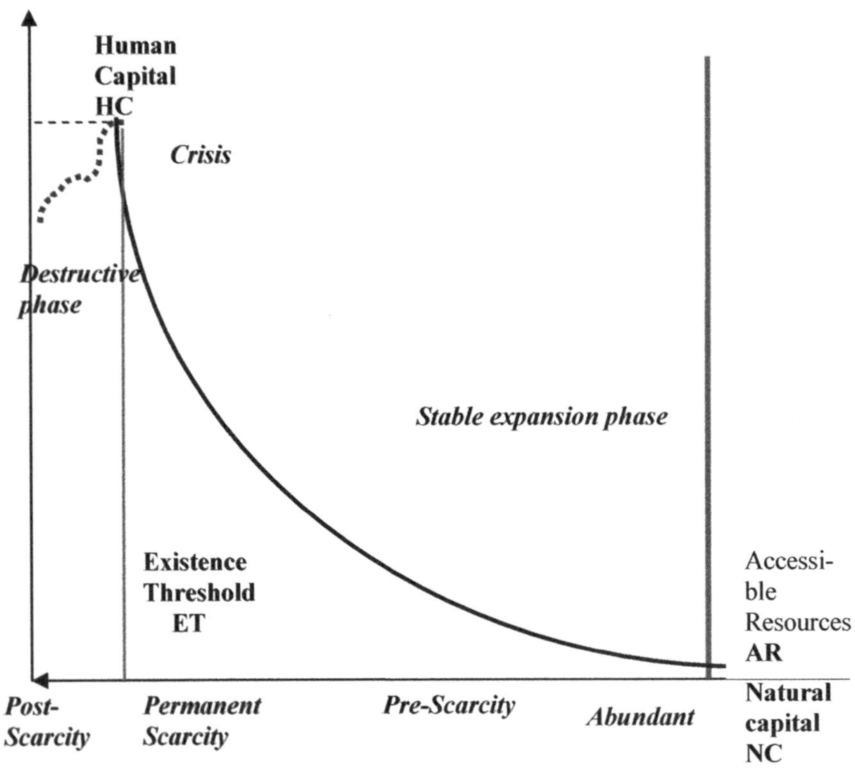

Figure 14.1 Development cycle in the Human-made /Natural capital model

Classical economists like Marx argued that the change of price caused by short term scarcity is a fluctuation that changes the value of mass as commodity (Marx & Engels, 1954, pp. 112, 124-128). By current economic theories, when the supply of a commodity decreases, its price increases. And as consequence, its producer tends to produce more to gain the profit increment afforded by the price increase. Finally, the market tends to balance the shortage. However, under a condition of permanent scarcity, there would be no availably adequate supply of the mass and/or energy that could be found from nature, and the balance could in no way be restored.

14.4 A long term view of cycles of development of the global economy

Wallerstein, founder of World System Theory which he used to analyze large-scale social change over long periods of historical time, often used Kondratieff cycles to explain swings in the world economy (Elliott & Armson, 1986). For instance, the period 1945 to 2000, according to him, is a Kondratieff cycle of the world capitalist economy which consists of two phases. The upward swing or the A-phase went from 1945 to 1967/1973 when economic development grew positively. The downward swing or the B-phase went from 1967/1973 until 2000 when the contraction of the world economy occurred (Wallerstein, 1998a). But Wallerstein suggested that there is another conjunctural wave that is longer-term than the Kondratieff. This is the cycle of hegemonies that could trace back centuries before the present decline of US hegemony (Wallerstein, 1998b). Mensch (1979) used a metamorphosis model of

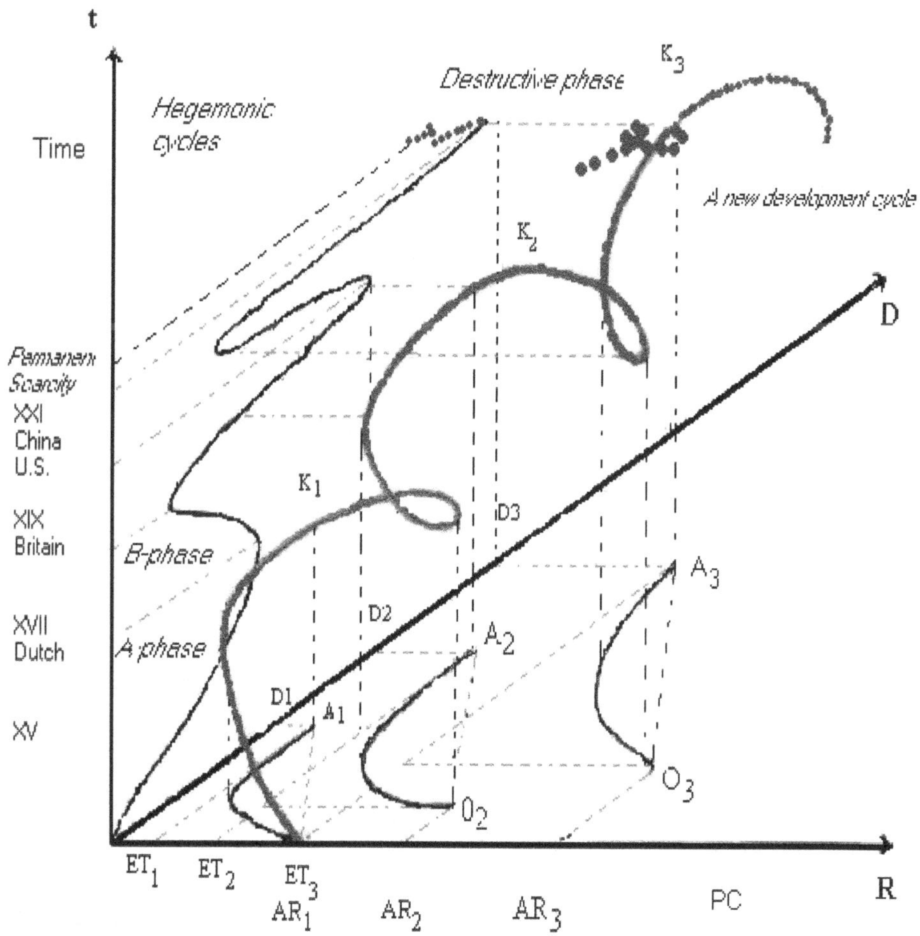

Figure 14.2 Spiral cycles of development and resources

socio-economic development by which a national economy or an industry experiences phases of stable growth, crisis, restructuring and turbulence.

Taking the capitalist world-economy as a whole, Arrighi shows that it also experiences cycles of development with phases of material expansion (A-phase) and financial expansion (B-phase) which ends with discontinuity when reaching its limits (Arrighi, 1994). Arrighi and Silver studied as examples of hegemonic waves three hegemonies - Dutch, British and the U.S. – that emerged since the fifteenth century dominated then broke down in the world economy.

The first hegemony, Dutch, grew from the United Provinces of Holland as the successor of the Genoese merchant bankers that had accumulated capital since the latter half of the fourteenth century to attain Genoa's peak in the seventeenth century with the 1648 Treaty of Westphalia. By this agreement, the Dutch established their monopolistic control over Iberian seaborne traffic and Baltic supplies as well as the supplies of silver brought to Europe from the Americas. Together with military superiority over the Spanish enemy, the Dutch become the focal commercial and financial entrepôt of the European-centered world economy (Arrighi & Silver, 1999, pp. 39-41). When the British seized supremacy from Holland by the Treaty of Utrecht in 1713, the latter gradually lost its control over supplies of material and labor to the winning power. At its decline, accumulated high finance became the main capital for the Dutch to maintain their wealth and power. This ended around 1772 with the collapse of an English banking house which had been entrusted with extensive Dutch assets (Arrighi & Silver, 1999, p. 53).

After victory over the Dutch, Britain emerged as a new hegemony. It controlled the supply of labor and materials across the world, as Barrat wrote:

> Most of the British Empire had already been established by 1850 – not only in Canada, and the Caribbean, Madras, Bombay and the Cape Coast from the seventeenth century, but in Gibraltar, Bengal, Ceylon, the Cape, Botany Bay, Penang, Guiana, Guiana and Trinidad by the end of the eighteenth; and to these were added by 1850 virtually the whole of India, plus Hong Kong, Australia, New Zealand, Natal. Further increments, then, were almost entirely on the African continent (Barrat, 1988, p. 32).

Like Dutch, British hegemony was a commercial entrepôt where one-third of world exports went to Britain in the middle of the nineteenth century. Moreover, Britain was also the industrial entrepôt or the "workshop of the world" for approximately a century. However, while the Dutch managed to maintain their domination for about one hundred and fifty years, Britain effectively lost its role as the world governmental power to the U.S. within half of that time, around the end of the nineteenth century. During the breakdown, the British Empire went through financial expansion, depression, turbulence, war and the disintegration of Britain's world order. The growth of the U.S as the world's major imperial power started from the Crimean War in the latter half of the nineteenth century and reached its height in the Second World War when the U.S. assumed the leadership role over the world economy except for the former socialist bloc (Arrighi & Silver, 1999, pp. 58-74).

Figure 14.2 sketches a view of spiral cycles of development in three dimensions: time (t), resources (R) and development (D). The swing of the three hegemonies is metaphorically sketched by a blue line in the left side of the figure. Now let us come back to the human-made/natural capital development model. The world production and consumption of the Dutch hegemony was characterized by agrarian productions like grain, salt, meat, wood (Braudel, 1984), a class of production whose consumption of natural capital is low. That is why the global agricultural-based capitalism led by the Dutch could survive for almost three centuries, starting from a low accessible natural capital AR1 and slowly ending with a low ET1 in the curve O1A1.

The second curve O2A2 of the global economy led by British Empire is a new discontinuity of development which is based on industrial capitalist production and a larger accessible natural resource AR2 (AR2>AR1). This growth started from a level of development O2 higher than O1 but lower than A1 as the consequence of the previous phase. However, the new speed of industrial production and capital development required a higher existence threshold than the previous phase (ET2>ET1). Thus with a higher speed of consumption of natural capital, the world capitalist system got to its peak and turbulence phase sooner than in the first cycle.

The third curve O3A3 is the development of the contemporary global capitalist system which is based not only on the production of mass but also the production of mind (e.g. intellectual property) for which the US hegemony is taking the lead. The boundary of accessible natural resource AR3 is greater than ever but it is also closer to the total planet capital PC than ever. With a destructive pace of expansion, the world's accelerating capitalist system also requires a greater than ever level for the existence threshold (ET3>ET2) and uses up the accessible resources in a shorter time. With the speeding-up of the world capitalist system through the progress of technology, space and time have been compressed quickly during the last five centuries. Dicken showed that in three centuries of the Dutch age, the best average speed of horse-drawn coaches and sailing ships was 10 m.p.h; in the next eighty years of British age, steam locomotives and steam ships ran faster at about 36-65 m.p.h; and the jet passenger now can fly at 700 m.p.h. (Dicken, 1992, p. 104). The present author would add that with the Internet speed, time and space seem to be shrunk into almost zero.

Development cycles O1A2, O2A2 and O3A3 seem to be a discontinuity as claimed by scholars like Drucker (1969) or Boisot (1995), however, in a three-dimensional space whose axes are time, resources and development, development is a continuous spiral of upswings and downswings as depicted by the red line in Figure 14.2. This figure also helps us to predict three scenarios for the global future. The pessimistic scenario (Destructive cycles) is when natural capital declines to ET3, and in the absence of any adequate substitution, the growth of the world economy would reach its global crisis and fall in turbulence with violence, war, and massive environment-induced emigration at the global scale as many authors predict (Giddens, 1990; Hammond, 1998; Hirst, 2001; P. J. Taylor, 1996). The positive scenario (new development cycle) is based on the assumption that humankind can break through the limit AR3 to a new, higher AR4: as technology optimist Georgescu-Roegen wrote "as surprising as it may seems, the entire stock of natural resources is not worth more than a few days of sunlight" (Georgescu-Roegen, 1971, p. 21). This scenario will be extended in the next section. The third scenario is somewhere between the positive and negative ones.

Lenin and Luxemberg pointed out that the imperialism of the 19th century was characterized by export of high finance (Lenin, 1974) and capitalism's need for a non-capitalist environment (Luxemburg, 1951). Nowadays, new physical and virtual non-capitalist markets are targets of capitalism's to expand, and export of human-made capital is the main mode of exploitation. New physical markets cover the former USSR and socialist Eastern Europe, China, Vietnam and the Arab world. However, these initially non-capitalist lands take a relatively short time to become new capitalist competitors who for their own existence also need both material supply and market. Therefore, the growing global capitalist system seeks yet further unconventional markets and sources of supply. New physical sources targeted for appropriation include global resources formerly seen as a "commons", such as water, the bio-system, the melting Antarctic and Artic, oceans, the atmosphere, and space. New virtual markets are the less knowledge-intensive economies, cyberspace and the knowledge "commons".

The shift of industrial economies to the so called "post-industrial" or "knowledge economy" is nothing but a new form of double-surplus exploitation of both positive and negative surplus values (see Chapter 15) in the time of pre-scarcity. Instead of exploiting directly their own natural resources, they exploit those from developing countries of inferior position, by the means of price intervention. If Hobson defined classical imperialism "desirable, or even necessary, in order to absorb and utilize the surplus of our ever-growing population" (Hobson, 1938, p. 41), the author of this book argues that knowledge economy is a form of imperialism in the pre-scarcity phase. In such a context, class struggle is replaced by a nation/state struggle for growth and survival. Cooperation and competition become two parallel counter-efforts for development and "denaturalization" (i.e. appropriation from natural to human-made capital). More importantly, the switching to a knowledge economy is also to the best way to expand the knowledge accumulation in the capitalist economy to buy time and options for the future. As the eclipse time from now to ET3 is shorter, whoever can save more natural capital would most likely gain more time and opportunity to contrive a means of escape.

The alienation, not substitution, of natural capital through which developed countries suck sustainability in the fashionable name of knowledge economy from the future generations in less-developed economies is illustrated below in Figure 14.3. In the last three decades increasing physical resource consumption and decreasing production in the core or developed countries, as the consequence of a new international division of labor, has shifted the physical-based economy from the core to the periphery. The disappearance of many forms of physical production from the developed countries perhaps feeds a myth of an abundant knowledge economy from their perspective. However the accelerating utilization of physical commodities by both old and new capitalist economies places inexorably increasing pressure on both renewable and non-renewable natural resources, jeopardizing biodiversity and environmental quality and threatening the survival of human beings, with developing nations being the first in jeopardy.

14.5 Sustainable development and the positive role of knowledge

14.5.1 A quick review of literature on conditions for sustainability

Knowledge economy and sustainability at national level

If we assume that the population of a country is constant, so-called weak sustainable development for that country can be achieved by maintaining no decrease in its total capital (TC) or "non-declining per capita national wealth by replacing or conserving the sources of that wealth; that is, stocks of produced, human, social and natural capital" (United Nations et al., 2003, p. 4).

The total capital, as mentioned earlier, is the sum of natural capital and human-made capital. Human-made capital flow can be measured either by the total supply of goods and services or the total demand. The total supply (TSP) can be delivered by either the domestic production (DP) or by suppliers from the rest of the world as imports (IM). The total demand (TD) is a sum of consumption (CS), capital formation (CF) and exports (EX). That is:

$$TSP = DP + IM$$

and is equal to

$$TD = CS + CF + EX$$

If population Pn grows at ∂Pn, the following condition is required over time, to maintain a weak sustainability in the long run:

$$\frac{\partial TSP(t)}{\partial Pn(t)} \geq 0 \qquad (4)$$

Given that

1. At present natural capital – the input to human-made production - is often treated in the world market as being fully substitutable by other forms of capital, for example financial capital or knowledge capital. In fact this may not always be possible (e.g. the world's fossil fuel resources are finite and for some purposes non-substitutable).
2. Natural capital, the input to human-made production, is declining and limited. Therefore the production of physical commodities becomes less sustainable. Knowledge capital, on the other hand, which is an increasingly important component of human capital, is limitless.

Thus, a shift from traditional production of physical commodities (for example minerals or agricultural goods) to immaterial production (for example education service, banking service or software production) would be more sustainable for a nation.

A weak sustainability of a country only can be achieved under the following conditions:

- International regimes which ensure an acceptable ratio of exchange value between human-made capital and natural capital (the H/N ratio) so that the substitution between natural capital and human capital can happen;
- The supply of imported goods and services, especially material goods, from the rest of the world is sufficient to balance the deficit between the domestic demand and domestic production;
- The exports of the country have to realize a monetary value no less than the value of imported goods and products;
- The population growth of the country must be under a level so that the consumption per capita does not outstrip affordable supply.

Knowledge economy and sustainable development

In the short run and at a national scale, an economy that shifts from a traditionally material-based economy to a non-physical economy, such as a knowledge economy, satisfies the above conditions, therefore most likely enjoys a weak sustainability. This county may gain advantages:

- A lower level of depletion of natural resources, because it does not directly exploit, and thereby destroy, its original natural capital.
- International regimes that distort price towards a favorable H/N ratio to create superior surplus for knowledge countries
- As argued above, buying time to discover options for the future.

Clearly countries that are superior in human-made capital (e.g. financial capital, human capital or military superiority) can enjoy a better sustainability than others. But can they be better off without making some others worse off? The answer can be "yes" for a single nation or several nations but is impossible for the globe as a whole. Perce et al. discuss importing and exporting sustainability. A country can secure a sustainable development path by importing raw materials while leaving waste discharge associated with extraction of the raw materials in the vendor country. Using its comparative advantage in technology and human skills it can convert the raw materials to final product, then export the final product to the less advantaged economies. Because the value it adds is converted into money, it can then further import all its mass/energy needs, with waste discharge being taken care of either by the vendor countries or the natural "commons" (e.g. the atmosphere). Thus the nation in question may keep its natural capital stock intact, but the nations from which it imports may well be experiencing a decline in their natural capital stock because it is being exported" (Pearce et al., 1989, p. 45)

Furthermore, nations that specialize in "clean" industries such as financial or knowledge production can maintain a good environment and ecosystem while those specializing in material based production suffer hazardous problems affecting both their natural resource stocks and their ecosystems. In the long run, the current generation of a more advanced countries arguably can leave a same amount of natural resources (including environment quality) to its next generation, in other word, achieve sustainable development within their national boundaries. However the children in developing countries will inherit only remnant of the natural resource stock plus a huge amount of accumulated waste residuals left by their predecessors in the environment.

In other words, the sustainability that a country with a knowledge economy may achieve does not come from its future generation, but from the future generation of the periphery. The flow of sustainability runs from the future generation of the less advantaged country to the future generation of the more advanced economy as shown in the Figure 14.3.

So far, we assume each country is an individual open system. It means that a country can reach out beyond its border. A country with higher human-made capital and a superior position to manipulate price interventions will have a greater chance to be richer, and to consume more material, than a less developed one.

For countries such as the U.S., 80% of the economy is human-made capital based and only 20% natural capital based. How can this giant country consume resources on such a vast scale and expect this consumption pattern to continue indefinitely? Where are the deficits of physical resources to come from? The continuity of such an economy depends on how long the U.S. can substitute the resource deficit by importation from the rest of the world. Through importation it will avoid over-exploitation of its natural resources as well as the depletion of its environment. In a defined context of space (inside the border of the U.S.) and time (as long as it can exchange human-made capital for natural capital) that country is apparently sustainable.

Globally we are trading off a reduced future environmental stock for an increased stock of manufactured assets. If just considering GDP, it seems that human beings are becoming richer. However, human-made capital is not truly a substitute which can replace natural capital but *only a supplement* to it.

Figure 14.3 illustrates how a human-based national economy H (such as in the core) can survive without damaging its environment. H can change its human-made capital (i.e. knowledge, finance or political power) into money, then buy foods, and other agricultural and materials needs from a material based economy M (such as in the periphery). The green arrow indicates that natural capital flow comes from the environment to M, then to H. The orange arrow denotes that human-capital flow runs in the reverse. The big orange arrow indicates that country M directly destroys its local resource capital (through exploiting it) not only for now but also for the future, while H can avoid that problem.

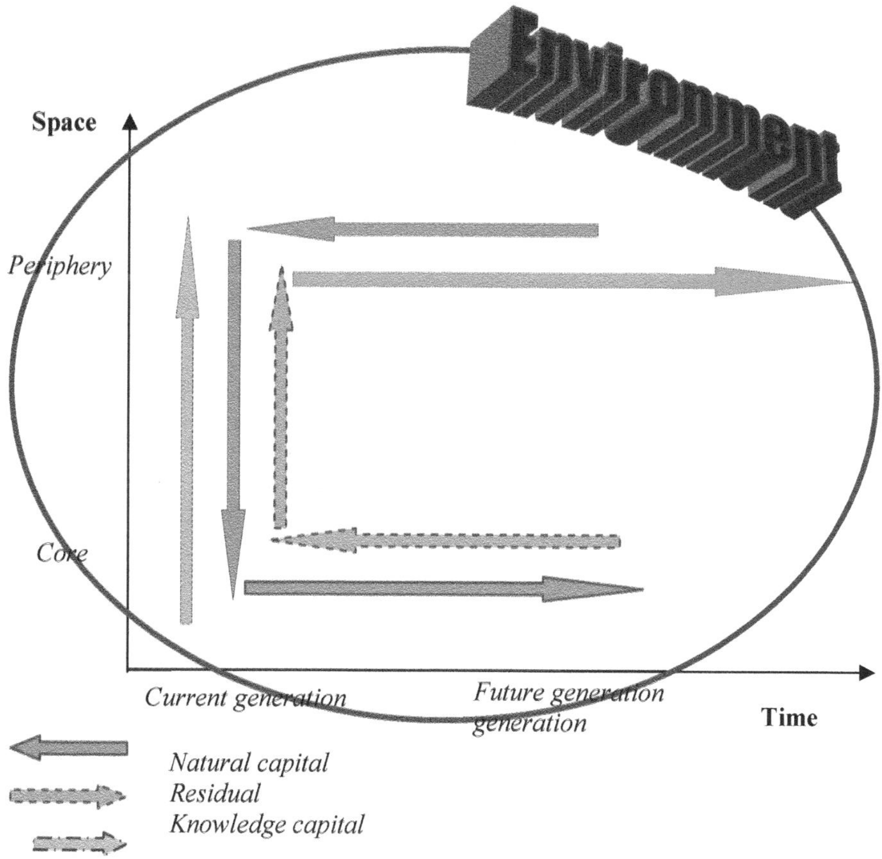

Figure 14.3 Flows of sustainability from the core to the periphery by generations

A metaphor of this situation is the case of two persons H and M in an isolated boat in the sea and the boat is made of bread. H is the ruler and N is the rower. H gains his bread by selling his command to N. N gets the bread by digging it from the boat. They both eat. (They both use their hats to trap drinking-water from rain.) As a ruler H can exchange his command for a great amount of food. He can thus accumulate food in reserve. N is less savvy, and uses up all his food both for survival and paying for the purchase of H's command. At a first glance, it seems that H is in a better position and becoming "richer" by his reserve. However, we don't know when they will both die by eating up the boat. The richness of neither H or N can substitute for the loss that they cause to the boat and their lives. Can we say as Keynes did, "In the long run, we're all dead" or can we expect that H and N may find other alternative, for example, catching fish from the sea with their hats or recycling their bodily wastes to make bread?

14.5.2 Components of Natural Capital and Human-Made Capital – the role of new knowledge

The presentation so far is based on classical laws of thermodynamics which seem to show no hope of sustainable development for human beings, even in quite a short-term future. So where is the technological hope? With the discovery of special relativity by Albert Einstein, the conservation of mass does not hold true in certain circumstances, for instance, a bounded system has a mass less than the sum of its parts. The difference, called a mass defect, is a measure of the binding energy — the strength of the bond holding together the parts (in other words, the energy needed to break them apart). Otherwise, mass can turn into massless energy according to Einstein's famous formula $E0= mc^2$ whereas E0 is rest energy, m is the mass of a body and c is the speed of light (the author remarks that the correct formula of Einstein is not $E= mc^2$ as often wrongly cited) (Einstein, 1951; Okun, 1989; E. F. Taylor & Wheeler, 1992).

The conservation of mass is not applicable for information and knowledge, too. In a closed system that contains living beings, information, such as knowledge created by human beings or DNA in biological creatures, is not created on the basis of mass-energy degradation. The generating of information in this case creates new value with a much lower level of mass/energy and entropy, and thus is more sustainable than conventional production.

Thus the author suggests that knowledge and new mass/energy (e.g. websites) created through new ways of digital production or consumption slows entropy and should be regarded as unconventional capital.

Under this light, Planet Capital (PC) now consists of conventional earth resources (CER), Unconventional earth resources (UCER) and human made capital (HC).

Conventional Earth Resources (CER) comprises of stock earth resources (SEC), renewable or reusable earth resources (RER) and environment quality (EQ). The stock of non-renewable earth resources (SER) are resources which only exist in a finite quantity, which once used will no longer be available for future economic activity, e.g. fossil fuels. The new technology or knowledge of use can reduce the rate of loss, thus extend the lifetime of resources. Renewable/reusable Earth Resources (RER) have a natural capacity, within defined physical limits, to replace that proportion of the stock which is lost through economic harvesting or natural decay. Minerals or metals ores are examples of reusable resources whereas land and ecosystem including fresh air, water, timber and fisheries are examples of renewable resources.

To the author, Unconventional Earth Resources (UCER) are resources which are or will be made available for human use because this is made a preferable option by new technologies, and/or because of the urgency of demand in the post-scarcity period. Examples of UCER are wind, tidal or geothermic energy.

Besides Earth resources, the Earth - not being a closed system - receives energy originated from space, such as light energy from the sun or tidal energy from the moon which could be called Space Resources (SR).

Finally, planet capital includes also human-made capital (HC), the human-made stock of material and non-material capital accumulated during the growth of humankind.

By introducing these elements, the accessible natural capital AR is as follows

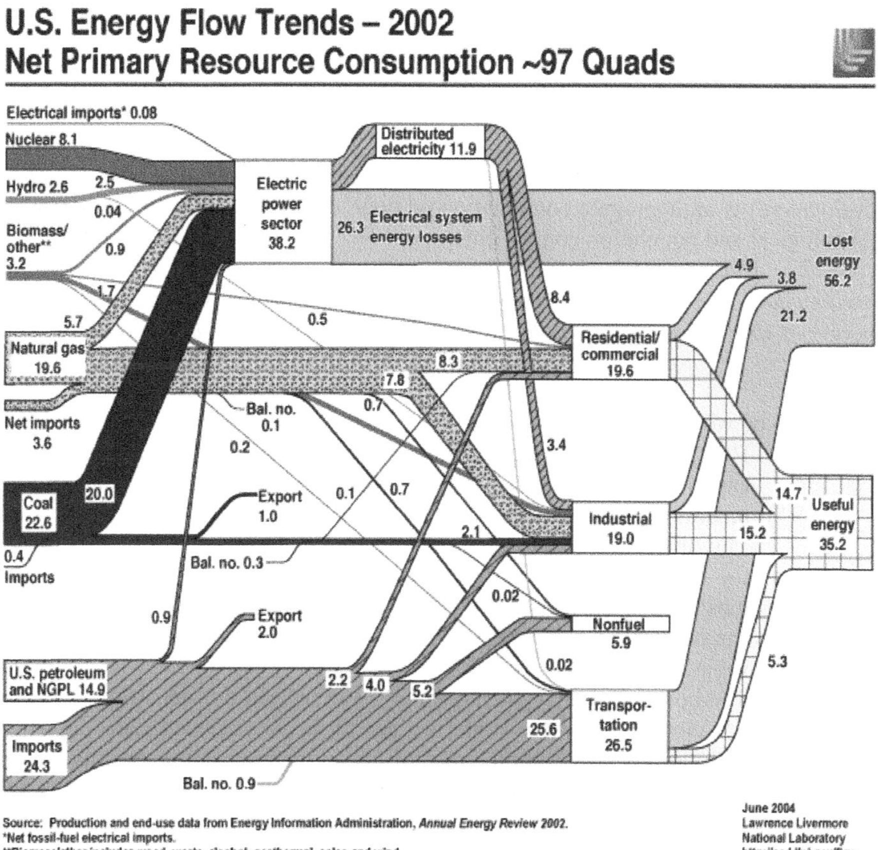

Figure 14.4 Permanent scarcity and the U.S. energy. Source: Adapted from Energy and Environment Directorate (2002)

(note that t stands for time, and T for technological level):

$$AR(t,T) = EQ(t) + ER(t,T) \tag{5}$$

Because $ER(t,T) = SER(t,T) + RER(t,T) + UCER(t,T)$ (6)

we have AR(t,T) = EQ(t) + SER(t,T) + RER(t,T) + UCER(t,T) + SR(t,T) (7)

The total capital (TC) that mankind can exploit at time t and technological level of T is:

TC(t) = EQ(t) + SER(t,T) + RER(t,T) + UCER(t,T) + SR(t,T) + HC(t) (8)

Because non-renewable resources will be used up by time, thus, at a certain time tET, SER(tET ,T) = ET (the existence threshold) then will decline to an economic zero.

Therefore, after that, if:

TC(t) = RER(t,T) + UCER(t,T) + SR(t,T) + EQ(t) + HC(t), when t> tET (9)

the formula for positive development would be:

$$\frac{\partial HC(t)}{\partial Pn(t)} \geq 0 \qquad \text{, all t} \qquad (10)$$

Conditions (note that C stands for consumption level):

RER(t,T) + UCER(t,T) + SR(t,T) + EQ (t) > ET(Pn,C,T) (11)

To summaries, the condition for *positive economic development*, (*not sustainable development*) is an aggregate combination of time, socio-economic, environmental, technological and population components.

Positive development is *development that can maintain positive growth of total human-made capital per capita and the natural capital that is essential for the production of the human-made capital, mainly consisting of renewable resources, unconventional resources and space resources, at a level no less than the existence threshold.*

It would be worthwhile to remark that:

- Human-made capital, after all, can be converted into the unit of natural capital because to produce the human-made capital, a certain amount of natural capital (e.g. water, iron ore, coal, oxygen) has been consumed. The difference of the two forms of capital, if any, is the ownership of the capital, whether to some person or no person. Human labor, the most important factor to create value in Marx's theory, after all, only can be created in the basis of food, fresh water and other physical consumption by the laborers.
- The boundary between renewable resources and non-renewable ones is not clear. Renewable resources are those whose stocks are depleted with use but which can be replenished by nature such as fresh water or timber; or through human intervention such as recycling or reuse of metals. However, the replenishment by nature often takes time and if human use exceeds the absorbing capacity of nature, renewable resources, such as fresh air, water or timber can be depleted and totally used up. The decline of sea fish production in Vietnam due to fishermen trying to take the maximum yield is another example showing that the concept of renewable resources is relative. What seem to be renewable resources can be converted into non-renewable resources due to carelessness in human use. Furthermore, the extension of a resource's life expectancy by recycling methods is not endless because we cannot

recycle 100 per cent of the material. By the end of that life, the material is totally used up.

- When we increase AR, the accessible resources, we can extend the lifetime of development. But at the same time, we shift further to PC, the planet capital or the final limit of development. It is still unknown how far away it is but scientific reports, such as of water reserves (Barlow & Clarke, 2002; WWF, 2006), ecosystem (Millenniumassessment.org, 2005) or global warming at the global scale (Chapter 3) or at regional scale as the case of Vietnam (Chapter 4) suggest that many essential natural resources are depleting towards the existence threshold. If only one of these essentials reaches that critical limit, the total growth of the global economic system will be in turbulence.

The energy production and consumption of the US will be used as an example to explain the model.

First, let's look at the left hand of Figure 14.4[14]. The natural energy of the US consists of non-renewable Earth stock resources energy (coal, natural gas and petroleum), renewable resources (biogas, hydro-power), unconventional energy (wind, biomass, nuclear) and space resources (solar and other).

At 2002, the majority of that energy comes from stock resources. The scarcity of oil and energy accelerates as the U.S. administration on the one hand increases production at home through "the use of more efficient drilling methods and by exploiting untapped reserves in protected wilderness areas" and on the other hand by "removing obstacles – political, economic, legal, and logistical – to increased procurement of foreign oil by the United States" (Klare, 2004, pp. 168-169). As a result, U.S. intervention in Iraq, Iran and other oilfields of the world is inevitable.

The reluctance to use nuclear power as an alternative source of energy comes from the public concern about the safety of nuclear energy therefore the last completed plant in the U.S. was commissioned in 1973 (United States. National Energy Policy Development Group & Bush, 2001).

As consequence, the proportion of renewable, unconventional and space energy consumed is very modest, even in the U.S. - the world leader in technology.

Secondly, let's look at the right hand of the chart where we can notice that more than half of the produced power output was lost. To maintain sustainable development does not only mean to increase production output. It also means careful and efficient consumption. Saving of lost consumption is also critical in giving nature a chance to maintain the healthy quality of environment that is vital to the existence of human beings.

14.6 Human/Natural Ratio, the Efficiency of Natural Capital and the Principle of Least Action

Economists tend to focus on produced capital such as the formation of wealth and therefore income. Inputs from nature, such as natural resources, are considered as free gifts of nature which are supposedly in limitless supply. Changes in the total stock of nature do not appear in the production costs. The material costs associated

---

[14] A quad is one quadrillion (10^{15}) BTU's – British Thermal Units

with production do not fully internalize so-called intangible material costs, environmental quality costs, inefficiency costs and the possible effect of depleting natural stock on the future value of material. All those costs become external costs, which are not directly born by either producers or buyers but by the whole of society and by nature. In such a situation, this author argues that a greater quantity of goods is produced and sold than is optimum; and therefore a greater quantity of natural resources is consumed than is optimum. The producing country suffers an environmental external cost in the form of a welfare loss when the activities of one economic activity cause negative externality effects on another, and no compensation is paid by the activity causing this externality (Waud, 1996, p. 487).

The Human-Made Capital to Natural Capital Ratio (the H/N ratio) is defined as:

$$\frac{H}{N} x \frac{PI_h}{PI_n} \qquad \text{(see Chapter 15)}$$

where

H – Human-made Input

N – Natural input

PI_h – Price intervention upon human-made capital

PI_n – Price intervention upon natural capital.

To maximize the quantity of material based products that a developed country can exchange with a unit of human capital from developing countries, the former would try to get a maximum from this ratio, that is,

$$\frac{H}{N}\frac{PI_h}{PI_n} \rightarrow \text{max.}$$

However, if H/N is maximized, the inverse of H/N ration will be minimized, that is,

$$\frac{N}{H}\frac{PI_n}{PI_h} \rightarrow \text{min.}$$

The $\frac{N}{H}\frac{PI_n}{PI_h}$ expression is called the Natural/Human-Made Capital Ratio or the N/H ratio which indicates how much human capital can be produced or exchanged from a unit of natural capital. In other words, N/H ratio indicates the efficiency of natural capital value. For example if a company from the US like Microsoft can exchange a unit of human made capital, such as software, for the highest amount of physical goods from Vietnam, the latter will be at the lowest efficiency of natural capital value. In other word, the current global knowledge economy that maximizes the human-made value also is minimizing the value of Nature and accelerating the destruction of Nature at global scale.

If N is a primary good for export and H is a non-primary good for import, then the N/H ratio becomes the terms of trade that refer to the relative prices of a country's exports and imports.

To survive and to grow, humankind must accede to Maupertuis' Principle of Least Action that was stated at the beginning of the chapter. That is "If there occurs some change in Nature, the amount of action necessary for this change must be as small as possible" (Barrow, 1988, pp. 80-81). ICT, with its networking capacity, could help humankind to learn and apply that principle.

14.7 Main insights from Chapter 14

Insight 14A: If declining sustainability is conceptualized as increasing entropy (i.e. increasing disorder or randomness in a system), it is possible to demonstrate through logical argument that in organic beings and communities information and knowledge acts to reduce the rate of entropy and increase sustainability.

Insight 14B: From the perspective of developed countries it may seem that a knowledge economy relieves exploitation of the physical environment, but in fact environmental damage is shifted to less powerful developing economies through WTO-style price interventions. Nevertheless switching to a knowledge economy is the best way to expand the knowledge accumulation in all capitalist economies to buy time and options for the future (retard entropy in the economic sphere).

Insight 14C: Substitution – whether of knowledge resources for non-renewable physical resources (e.g. use feedback monitoring to vastly increase efficiency of use), or of renewable resources for non-renewable resources – is the core challenge in sustainable economic development. The pessimistic scenario is that substitution fails and the world descends into turbulence with violence, war, and massive environment-induced emigration at the global scale. The positive scenario is characterized by the triumph of substitution, captured in Georgescu-Roegen words "the entire stock of natural resources is not worth more than a few days of sunlight" (Georgescu-Roegen, 1971, p. 21).

Insight 14D: While the problem of environmental sustainability is global, each nation needs both collaborative and competitive strategies to achieve an optimal national trajectory on sustainable economic development. This is because powerful nations can more easily influence the price-intervention settings in the world trade order governing the exchange of renewable knowledge resources for non-renewable physical resources.

Chapter 15
The exchange of mind for mass

15.1 Scope of the chapter

- Explores various approaches to the establishment of pricing regimes for the exchange of knowledge goods (e.g. license to produce a US patented pharmaceutical drug in Vietnam) for physical goods (e.g. tons of shrimp, an aquaculture crop that pollutes vast amounts of water).
- Examines selected models of the creation of surplus capital, from Marx and Ricardo until the current "green economics", and how these relate to the transformation of natural capital into human-made capital.
- Focuses on the concepts "era of pre-scarcity" and "permanent exhaustion" in relation to natural resources, using crude oil as a major example.
- Uses formalized logic to illustrate how the current global terms of trade result in negative surplus value for natural capital and positive surplus value for human-made capital.

15.2 Pricing a commodity

For a very long time, natural resources (natural capital) have been considered a "free gift" from nature – considered to be so abundant that no price, or a very low price, is ascribed to them.

In Marxist economics, raw materials are not seen to contribute any surplus value to production which is generally argued to be attributable only to the extent of the labor component (Marx, 1915). In capitalist economics the System of National Accounts (SNA) prior to 1993 assumed that natural resources are limitless and no change in their value could thus be treated as income (United Nations et al., 2003). Schumacher criticized,

> We are estranged from reality and inclined to treat as valueless everything that we have not made ourselves. Even the great Dr. Marx fell into this devastating error when he formulated the so-called 'labour theory of value' (Schumacher, 1973, p. 12).

Economists generally measure a commodity by its economic costs and economic value (or price) in monetary units. Economic costs are normally divided into two groups: explicit costs and implicit costs. In the short run, economic costs can be defined by fixed and variable costs. Demand and supply are two major factors, besides economic costs, that elasticize the price of a commodity. Other factors are government policies (indirect measures such as tax, subsidies or direct price control, such as quotas, licenses, price ceilings or price floors, rationing), technology improvement, and the economies of scale (Waud, 1996).

15.2.1 Pricing of knowledge products

Kibirige proposed three approaches to pricing information commodities. The first approach is cost-based pricing which entails setting prices mainly on the basis of cost of producing the items. The second method is demand-based-pricing that looks at the intensity of demand for a product or service: the higher the demand the higher the price. The last method is competition based pricing (Kibirige, 1983). However, Shy argued that the approach of cost-based-pricing is not appropriate for software because it is difficult in

> predicting software prices as it implies that marginal-cost pricing associated with competitive markets cannot prevail in markets for software simply because marginal-cost pricing implies a loss to the software developer (Shy, 2001, p. 54).

According to Shapiro and Varian, competition among sellers of information commodity pushes prices to zero. They argued that information is priced by its value, not its cost, hence: i) Production of information is costly but reproduction is cheap ii) When the first copy of an information commodity has been made, most costs are sunk iii) Multiple copies can be produced at almost constant (and very low) per-unit costs and iv) There is no natural capacity limit for additional copies (Shapiro & Varian, 1999).

15.2.2 Green economy – the pricing of natural capital

"Green economy" or environmental accounting theories were proposed by Pearce, Markandya and Barbier (1989) (1992) to measure the effect of environmental changes on welfare. Costanza and Daly (1992) applied these theories to the definitions of capital and production. They argued that as "capital is a stock that yields a flow of valuable goods and services into the future" the difference as to whether that stock is created by humans or by nature is merely a distinction between kinds of capital. If the stock is made by humans they called it conventional capital, and if it is made by nature they termed it natural capital. For example, a stock of fish in the sea provides an annual yield of fish which can be presented as valuable goods. Natural capital, thus, is "made up of all the natural resources and environmental functions that are essential for human life and economic activity" (Costanza & Daly, 2001, p. 14). As a result, the System of Integrated Environmental and Economic Accounting (SEEA), an adjunct to the 1993 SNA, was introduced to redefine the value of a stock of a natural resource terms of the net present value (NPV) of the stream of future resource rents. The principle underlying this approach is based on the standard method for valuing produced assets. As non-renewable natural resources provide a stream of services to the economy, their value as capital declines. This decline is called the consumption of natural capital, parallel to the consumption of conventional capital.

According to SEEA, the value of a stock of a non-renewable natural resource (RV) is calculated as the summing function of the resource rent (RR), the years the resources is expected to last until exhaustion (n) and a discount rate (r):

$$\text{RV} = RR\sum_{k=1}^{n}\frac{1}{(1+r)^k} = \text{RR}\left[\frac{(1+r)^n-1}{r(1+r)^n}\right]$$

If Pm is the market price and C is the cost of exploitation, NPV across time of a natural resource is:

$$\text{NPV} = (Pm-C)\sum_{k=1}^{n}\frac{1}{(1+r)^k}$$

(United Nations et al., 2003, p. 284)

The most controversial confusion of the SEEA perhaps is the Mexican case. In 1990-1991, the United Nations Statistical Division (UNSTAT), the World Bank and the Mexican Instituto Nacional de Estadistica, Geografia e Informatica (INEGI) applied alternate methods, one of which was SEEA, to revaluate natural resources in order to decide what policy Mexico should adopt. According to Vu Xuan Nguyet Hong (Vu, 2004, pp. 200-203), two methods were applied.

The first method calculated the net price of the natural resource. By this method, the environmental-accounting value of timber was:

Vw = Pm – C = 21.5 peso/m^3

where Pm is the market price of wood and C is the cost of exploitation.

The second method followed the method of estimating income from extraction. El Serafy developed the method in 1979 to calculate the value of oil as a depletable resource (El Serafy, 1979, 1981). This is the SEEA method mentioned above. By this method:

$$\text{NPV} = (Pm-C)\sum_{k=1}^{n}\frac{1}{(1+r)^k} = 1.46\ \text{peso/m}^3$$

Where, r is the discount rate of wood and n is the years timber is expected to last until exhaustion.

As shown above, the problem of SEEA calculation is that did not favor the price of natural resources as expected but undervalued it much lower than the market price. It presented a bewildering dilemma to Mexican policy makers. Should they reserve natural resources for their children or sell them quickly? Because if this theory was true, timber would become so cheap as to be nearly free in the future (Pearce et al., 1989, p. 137; Vu, 2004, pp. 200-203).

15.2.3 Price distortion

In reality, price is often distorted and does not represent exactly the value of a commodity or service. Jenkins identified three categories for interventions into price: 1) Prices affecting foreign trade - protection of agriculture, protection of industry and exchange rate 2) Prices affecting the factors of production - interest rates, wages 3) Prices affecting domestic prices - tariffs and inflation. Neo-liberals claim that price distortion is more serious in developing countries where there exists no "market economy" (Jenkins, 1992, p. 159). However, consideration of the terms of trade support an opposite argument, that price is even more distorted in "free trade" economies where market protectionism is erected, whether deliberately or not, through political, economic and technological superiority (see Chapter 2).

The terms of trade refer to the relative prices of a country's exports and imports. It is calculated by dividing the index of export prices by the index of import prices. Presbisch (1962) and Singer (1950) assumed that there is a tendency for the terms of trade of developing countries to decline over time in dealing with developed countries. According to the Presbisch-Singer hypobook, developing countries whose economies are based largely on export of physical products suffer more disadvantages than developed countries whose economies are based largely on processed products and services. Michael Porter's theory on competitive advantage suggests that those who produce goods and services with higher information intensity have more competitive advantage than those who do not (Porter, 1985). In the regard to price intervention, higher intensity of information very likely implies higher intervention.

15.3 A theory of exchange between human-made capital and natural capital

Economics plays a central role as it decides what has more market value, and what has less. In the market place, innumerable distinct qualities which are of vital importance for human and society must somehow be boiled down to the common denominator of price. Schumacher exclaimed:

> Thus the reign of quantity celebrates its greatest triumphs in 'The Market'. Everything is equated with everything else. To equate things means to give them a price and thus to make them exchangeable (Schumacher, 1973, p. 41).

Paul Hawken wrote that: "The single most important trend to understand is the changing ratio between mass and information in goods and services" (Hawken, 1983, p. 11). James Heskett observed that "Manufacturing is substituting information for assets. Nearly every program to reduce inventories has this character" (Heskett, 1986, p. 160). Gilder, in "Microcosm: the quantum revolution in economics and technology", saw the information revolution as essentially one of "mind over matter" (Gilder, 1989).

15.3.1 Modeling the exchange value of a commodity

In this section, a model is developed that seeks to explain how commodities are equated by price and thus exchanged. It also seeks to demonstrate how dominant countries specializing in knowledge commodities, such as consulting, financial, education services or high-tech products, can substitute human-made capital with natural capital from other countries.

Let consider a commodity Φ, for instance, a kilogram of rice, a financial service or a DVD copy of Microsoft Windows Vista. A commodity is

> an object outside us, a thing that by its properties satisfies human wants of some sort or another. The nature of such wants, whether, for instance, they spring from the stomach or from fancy, makes no difference. Neither are we here concerned to know how the object satisfies these wants, whether directly

as means of subsistence, or indirectly as means of production (Marx, 1915, p. Section 1).

Total actual cost - Components of real costs in a commodity

This section attempts to identify elements that contribute to the value of a commodity Φ.

The author will endeavor to bring Pearce, Markandya and Barbier's suggestion of *greening the world economy* (Pearce et al., 1989) into conjunction with Marx's theory of value. In that light, we need to consider this commodity over its entire life. This cycle starts from the moment humankind forms an intention to make this the commodity; then moves to the assigning of labor and the collecting of raw materials from nature; then producing and using the commodity; and finally the returning back of the commodity to nature (in form of waste, for instance).

The cost of producing that commodity consists of two elements: natural input and human-made input. The natural input N is defined in its most general sense, by which any natural resources and environment losses over the entire life of the commodity, as defined above, is accounted for. These inputs consist of more that just the direct physical inputs such as water, timber, oil, gas and mineral deposits required to produce the commodity and then discharge or process the wastes left at the end of the commodity's lifecycle. The inputs also include other indirect resources from the environment, typically the decline of environmental functions due to the production of the commodity (e.g. climate change) and the decline of habitat function in the environment (e.g. environmental toxicity).

The human-made input H refers to the total human labor inputs, either skilled or unskilled, that are necessary to convert the naturally-made input into the commodity. The total human labor includes not only the direct labor costs to pay for the knowledge and capabilities of the labor used in production, and the social organization (e.g. management) required to make the production possible (in the form of wages), but also other human-made capital such as – machines, roads, buildings, or investments that are necessary for the production of the commodity. This means that

$$TAC = N + H$$

where TAC stands for the total actual cost of the commodity Φ in its entire life, N for total natural inputs, and H for total human inputs.

By this definition, N is much broader than the constant capital c in Marx's theory because to input a fully accountable amount of material c into production, a higher amount of natural inputs are consumed, including M – the accountable natural input, ITM - the intangible natural inputs such as water, fresh air; and EQL, the environmental quality losses due to the entire life of the commodity. Thus

$$N = (M + ITM + EQL) / ec \qquad (1)$$

where ec is the efficiency coefficient of resources used in production, $0\% < ec < 100\%$.

Some explanation of each component is given below.

Intangible natural input (ITM) refers to non-costed environmental input to the commodity. An example of ITM is the water used in shrimp cultivation in Vietnam. According to Nguyen Quang Dang (Q. D. Nguyen, 2003, p. 42), shrimp production is a leading export area in Vietnam with 22,374 hectares for intensive farming in 1996 (compared with 2,681 ha of China). However, for the production of on-sand shrimp culture, 54,600 m3 of fresh underground water were estimated to be used in each hectare per crop (L. T. P. Nguyen, 2003, p. 34). This precious resource is not accounted into the cost of production and therefore in the price of exported shrimp. If this water were accounted as a cost of production, the price of shrimp would be much higher than the current one.

An example of the loss of environmental quality (EQL) is provided by the shrimp production above. Firstly, because water is not accounted as cost, no funds are returned to spend on water treatment. Thus water discharged from shrimp ponds without proper treatment. It contaminates the surrounding land and sea as well as the groundwater. Secondly, overexploitation of groundwater leads to the collapse of geological strata and increases the intrusion of salt from the sea. Thereby, the environmental quality deteriorates (L. T. P. Nguyen, 2003, pp. 34-35).

The efficiency coefficient of resource used in production – ec - depends on the level of technology and management. The better the level of technology and management, the higher the coefficient is. An instance of ec is the loss of gold in production in Vietnam. According to S54 (interview, Dec 21, 2004), due to backward technology, only about 30-40% of gold can be collected from gold ore in comparison with 92-97% extraction in developed countries like Australia. In addition, only ore with gold content above 10 gram/ton can be produced with profit while with advanced technology, ore with 0.5 gram/ton can be exploited. This means that with low technology, only 3 grams of gold out of 100 grams can be collected and all the ore with gold content below 10 grams/ton is discarded as waste into the environment. In this example, ecVietnam = 30% while ecAustralia = 92%.

Current pricing practices, such as the 1993 SNA, do not bring into account intangible material costs, environmental quality costs and the efficiency coefficient, and only tangible material costs are accounted as material costs (see for instance, United Nations et al., 2003). In other word, these practices presume that ec = 1, ITM = EQL = 0 and N = M.

15.3.2 Price and the total actual cost

This section builds a formula to formalize (but not quantify) the relation between price and the total actual cost which will serve as a basis for argument in coming sections.

In the marketplace, this commodity Φ has an exchange value P as its price, influenced by a price intervention PI that deviates price from the total actual cost. The relationship of P, PI and the total actual cost TAC is expressed in the following formula,

$$P = TAC \times PI = (N + H) \times PI \qquad (2)$$

Price intervention is defined as the deviation of the market price from the total actual costs of production of a commodity. Price intervention can be classified into four main groups, related to natural, human-made, supply and demand factors.

Price intervention of natural input (PIn) refers to that category of price interventions which affects the exchange value of the natural input of the commodity. Examples of this kind of intervention are subsidies or import/export taxes on agricultural products, ceiling-price policies for physical products or protectionism applied to fuel.

Price intervention of human-made input (PIh), refers to that category of price interventions which affects the exchange value of the human-made components of the commodity. Examples include protectionism applied to industry or intellectual property, exchange rates, bank interest, ownership interest, tariffs, inflation, profits, dividends and interest payments from foreign debt, direct or indirect foreign investment, loans and ownership. Another example is the difference of wages for labor in a developed country compared to a developing country though they may have same skills, experience and workload etc.

Price intervention of supply (PIs) or elasticity of supply refers to the degree of responsiveness of the level of total stock of natural input ingredients of a commodity to a change in price of that commodity. Where PIs is absent or low the less stock there is, the higher the price tends to be.

Price intervention of demand (PId) refers to the degree of responsiveness of quantity of a commodity demanded by humankind to changes in price of the commodity. Intervention of demand is affected by population, consumption behavior and capital formation. Intervention of demand is equivalent to elasticity of supply in macro economics.

Thus, price intervention is an array variable of four elements,

$$PI = A(PIn, PIh, , PIs, PId) \qquad (3)$$

Notice that PIn and PIh are mappings of social, economic and geo-political interventions on the price relating to natural and human-made inputs to the commodity whilst PIs and PId are interventions relating to the supply-demand rule of the market.

15.3.3 Transformation from human-made capital to natural capital

This section seeks to build a formula to explain how mind is exchanged with mass.

Let us construct a model to show how a country with a superior position in terms of human-made capital exchanges a smaller quantity of its commodity for a larger quantity of natural resources from a less advanced country. We will consider a two-country, two-commodity model.

For simplicity, first, we assume that the first country has a commodity Φ_1 of which the pure input is natural input and no human-made capital. Examples of Φ_1 are a deposit of crude oil or raw assets that are sold without any human-made process. The price P_1 of Φ_1 in formula (2) is,

$$P_1 = N \times PI_n \qquad (4)$$

where N is the cost of natural input, and PIn is the price intervention upon the price of that input N.

Similarly, the second country has a commodity Φ_2 which consists of no material but only human-made input. An example of Φ_2 is a service in consulting, education or finance. The price P_2 of Φ_2 is

$$P_2 = H \times PI_h \qquad (5)$$

where H is the cost of human-made input including direct labor cost (wages) and indirect labor cost (such as capital, interest, knowledge) and PIh is the price intervention upon the price of input H.

Suppose each country sells its product in the international market to buy the same exchange value of the other's product for its own consumption and vice versa, that is country 2 sells q_2 units of commodity Φ_2 to obtain money to buy q_1 units of commodity Φ_1. As the exchange values of two commodities are equal, then:

$$q_1 \times N \times PI_n = q_2 \times H \times PI_h \qquad (6)$$

From this equation, we have $q_1 = q_2 \dfrac{H}{N}\dfrac{PI_h}{PI_n}$

Let call $\dfrac{H}{N}\dfrac{PI_h}{PI_n}$ (7) - *the human / nature exchange ratio* or H/N ratio.

In this ratio, if $\dfrac{H}{N}$ depends on the level of the social forces of production, $\dfrac{PI_h}{PI_n}$ is a product of the social relations of production (for discussion about the social forces of production and social relations of production see Chapter 2). The H/N ratio tells us that what quantity of a human-made input can buy what quantity of a natural resource.

15.4 Natural capital

15.4.1 Absolute value and total economic value of natural capital

In this section, the attempt is made to define an absolute value of a commodity so that we can effectively assess the degree of scarcity of a natural resource in relation to its importance for human needs. Let us hypothesize a *principle of value for supply and demand* which assumes that a higher absolute value reflects more scarcity and/or high demand; and a lower absolute value reflects more abundance and/or low demand. This hypobook relates exclusively to exhaustible resources whose absolute value would, in economized utilization, stimulate increased exploration or technological substitution.

A simplistic method is to treat as constant the total value of a natural resource stock across time, because we assume direct inverse proportionality between the unit cost of the stock and its decline in quantity (for example when the amount of stock halves, the unit price doubles). This assumption preserves the total value of the stock at any time as a constant. Let us denote as TS the total stock of a non-renewable resource Φ1 in nature (for instance, oil). At first, let's assume that the demand for this material (or the rate of extraction of the resource by year) is constant, with annual consumption denoted as AC.

The lifespan of this deposit, denoted as n, is as follows:

$$n = \frac{TS}{AC}$$

The rate of consumption upon total stock (Rc) is:

$$Rc = \frac{AC}{TS} = \frac{1}{n}$$

Let denote as AV(0) the absolute value of an unit of TS at year t_0 – the assumed starting time that the resource is initially exploited by humankind.

The total stock at year 1, namely TS(1), is a deduction of AC from TS(0), that is:

$$TS(1) = TS(0) - AC$$

As the supply reduces by an amount of ΔTS(1) = TS(1) – TS(0) = AC, according to the proposed supply and demand value principle, the absolute unit value of this material now increases:

$$AV(1) = \frac{TS(0)}{TS(0) - AC} AV(0) = \frac{1}{1 - \frac{AC}{TS(0)}} AV(0) \qquad (8)$$

Replacing $\frac{AC}{TS}$ = Rc, we have AV(1) = $\frac{1}{1-R_c}$ AV(0)

Generally, at a given time t, we have,

$$AV(t) = AV(0) \frac{1}{1-tR_c} = AV(0) \frac{1}{1-t/n} \qquad (9)$$

An example of the application of formula (9) to estimate the AV of crude oil is illustrated below.

Absolute unit value of crude oil

Let assume that t(0) = 1880 – the year when oil was first used on a global scale and AV0 = $20 – the price of a barrel of crude oil at that time. Let is also assume that 2050 is the final year when oil would be totally exhausted as predicted by BP (2007) and other forecasters.

Therefore, n = 2050-1880 = 170, and

$$AV(2000) = \$20 \text{ x } \frac{1}{1-(2000-1880)/170} = \$68$$

$$AV(2007) = \$20 \text{ x } \frac{1}{1-(2007-1880)/170} = \$79.07$$

$$AV(2049) = \$20 \text{ x } \frac{1}{1-(2049-1880)/170} = \$3{,}400$$

Please note that $79.07, the 2007 absolute value of a barrel of crude oil, is derived as the theoretical absolute price calculated on a relative basis from the starting point of the 1880 price level. Convert it into the real-world 2007 price (by multiplying with the price index) and it would be equivalent to about $1,653.4, much higher than the real-world 2007 market price of about $100.

In general, the absolute unit value calculated at time t, is

$$AV^*(t) = \frac{\lambda(t)}{1-t/n} AV(0) \qquad (10)$$

Where λ(t) is the price index at t compared to t_0.

Now we consider the case when the actual demand is not constant, but incremental along with population growth, consumption behavior and the level of knowledge.

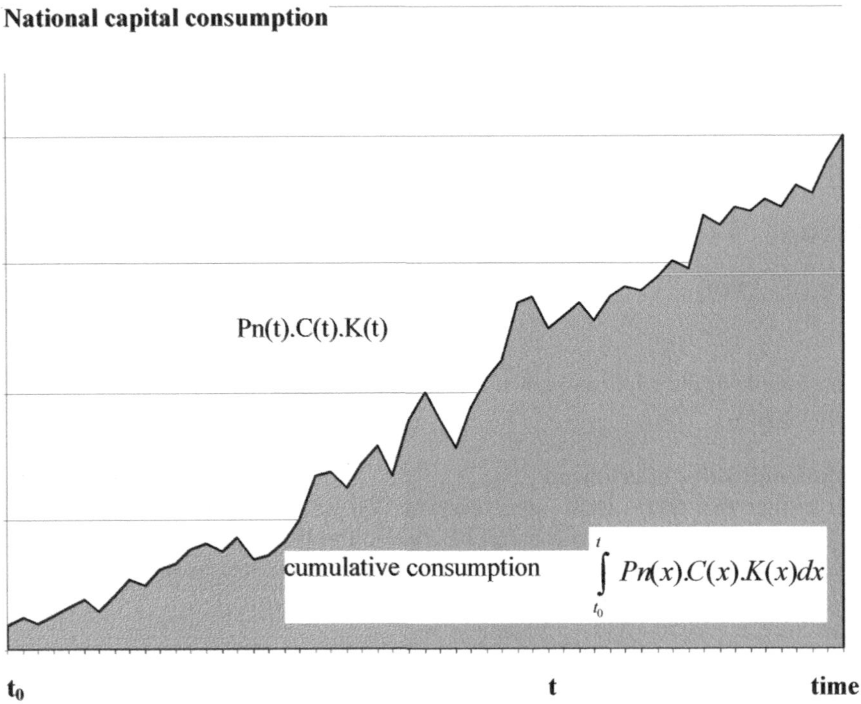

Figure 15.1 Cumulative consumption of national capital by time

It is argued that human impact on the environment can be measured by PnxCxT where Pn is population, C is the average consumption level and T is the level of technology development (Aplin, 2002, p. 12). Using this approach, the cumulative consumption of the natural input Φ1 up to time t can be measured by the integration of Pn(x)C(x)K(x) from t0 to t (see Figure 15.1), that is,

$$\int_{t_0}^{t} Pn(x).C(x).K(x)dx \qquad (11)$$

Where: Pn(x) is a function of global population, C(x) is a function of average consumption level, and K(x) a function of the level of knowledge development at the time x.

Combine (8), (10) and (11), the absolute unit value of the natural resource, expressed as the relative price at time t, is:

$$AV^{*}(t) = \frac{\lambda(t)}{1 - \frac{1}{TS}\int_{t_0}^{t} Pn(x).C(x).K(x)dx} AV(0) \qquad (12)$$

Now let extend (12) by bringing into the absolute unit value the concept of *Total Economic Value* (TEV). Pearce et al. suggested that TEV is a sum of *Actual value* (AV)- "the present use value of natural resources", Option value (OV) – "potential benefits in the future", and *Existence values* (EV) – values that "are closely bound up with the uniqueness of natural environment" (Pearce et al., 1991, pp. 81-82). Altogether, we have the future unit value of natural capital as:

$$\text{TEV(t)} = \frac{\lambda(t).\text{AV}(0)}{1-\frac{1}{TS}\int_{t_0}^{t_0} Pn(x).C(x).K(x)dx} + \text{OV} + \text{EV} \qquad (13)$$

It is noted that:

1. In reality, market prices are less likely to reflect the scarcity. Pearce argued that the flow of production is not significantly affected by gradual depletion of the total stock. Price may then rise only as the last units of the resources are extracted (Pearce et al., 1989).
2. The total loss of option value and existence value caused by the exploitation of a resource in t years is:

   $$(\text{OV} + \text{EV}) \quad \int_{t_0}^{t} Pn(x).C(x).K(x)dx$$

3. In this model, the total stock is assumed to be the proven reserves. If there are new discoveries, say, an amount of ND, the total stock at the time t is:

   $$\text{TS} + \text{ND} - \int_{t_0}^{t} Pn(x).C(x).K(x)dx$$

In this case, the absolute unit value would be then:

$$AV^{*}(t) = \text{AV}(0)\ \frac{\lambda(t)}{1-\frac{1}{TS+ND}\int_{t_0}^{t} Pn(x).C(x).K(x)dx} \qquad (14)$$

4. The annual consumption of Φ1 may decline if there are:
   - technology improvements which accommodate less consumption of that material;
   - change in human behavior towards economizing the scarce resource;
   - decline in population;
   - if there is a substitution of this material (e.g. new technology improvement, solar power or wind energy which replaces fossil fuels at economic scale) that is cheaper, consumers will shift to that new replacement, the economic value of the former will decline, similar to scenario 3 above.
   - In the worst-case scenario, as the resource approaches extinction in a permanent scarcity phase, the price would jump sharply to an *unbearable threshold* which could not be born by majority of the population. For instance, oil could jump to $3,400 per barrel against the relative price of 1880. In this case, consumption would drop quickly to a tragic

crisis as deposits become completely exhausted and no new deposits exist to take their place. The present author terms this scenario the ***Permanent Scarcity Era***. Under that condition, all prevailing economic theories would be no longer valid. The Second World War is an illustration of global temporary scarcity when all primary resources jumped much higher than processed products (e.g. the price of rubber in Germany). An example of present-day permanent scarcity is the near-extinction of tigers on the global scale and the price of tiger-derived raw materials on the black market. More discussion about this scenario is in Chapter 14 or discussion of Taylor (1996), Giddens (1990) or Wallerstein (1997).

15.4.2 Negative surplus value of natural capital

Figure 15.2 illustrates the difference between the market price and the absolute value of oil. The black line is the actual market price and the purple line is a symbolic estimation of the absolute value of crude oil as hypothesized by formula (13). As the total stock of oil declines by time the absolute value would increase accordingly. However, due to price intervention, the actual market price has fluctuated only marginally around $20 per barrel for more than one century from 1874 to 1999. The price of oil only surges up when the main supply of oil is threatened, for instance, in the Middle Eastern wars of 1973, 1991, and from 2005 up to present.

Let's call the zone between the price line and the absolute value line the *Negative Price Intervention Zone*, where the intervention of natural input (already discussed above) applies downward pressure on the market price, to levels much lower than the absolute value could be.

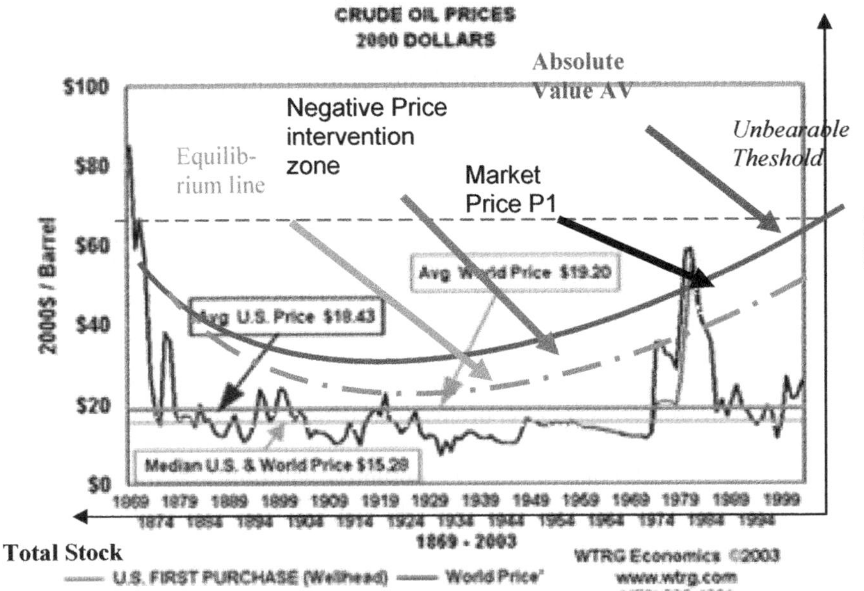

Figure 15.2 Absolute value, market price and total stock of crude oil. Source: The graph is created by the author with the background from WTRG Economics 2003, cited Williams (2004).

If AV(t) is the absolute value and P(t) is the market price of a physical asset, then the difference ΔP= P(t) - AV(t) is a *negative surplus value* of this commodity. This author calls it negative surplus value because if a producer sells the commodity at the price P(t), he loses surplus value ΔP and at the same time the buyer gains surplus ΔP. By extension, it means that *in the exchange with other commodities, such as knowledge capital, as long as the price is undervalued by the intervention of natural input (in the Negative Price Intervention Zone), the more the supplying country loses and the buying country gains.* It is widely known that consumers will concede a higher profit margin if they don't have to pay the external cost of production, thus unrecognized negative surplus value creates "externalities" (Pearce et al., 1989; Waud, 1996, p. 437).

Most economists since Ricardo (1821) generally believe that when the price of a commodity fluctuates, market forces with an "invisible hand" operate to adjust the price toward a static equilibrium level. Until the present, the concept of equilibrium has been well accepted as a key axiom of economics. However, from the model, this author argues that rather than a *static equilibrium point* in relation to which the price of a commodity fluctuates. Instead, there exists a *dynamic equilibrium line* (illustrated as an orange dash-dot-line in Figure 15.2) which tends to increase in relation to the absolute value with a time lag, and market price fluctuates around this dynamic equilibrium line.

15.4.3 Other disadvantages of natural capital production

In addition to negative surplus, producers who depend on natural capital like farmers face other critical disadvantages, namely:

- the limits of natural resource stock
- the increased cost of marginal production, and
- the decreased return on marginal production.

To maximize economic revenue in the production of physical commodity, the producer has to produce a maximized level of output. But this action does not bring more profit to the producer, because it faces the law of increasing cost and the law of diminishing returns. The law of increasing cost states that "when moving along the production possibilities frontier, the cost per additional good obtained measured in terms of the good sacrificed rises due to the different productivity of resources when used in different production process" and the law of diminishing returns states that "as more and more of a variable factor of production is used together with a fixed factor of production, beyond some point the marginal product attributable to each additional unit of the variable factor begins to decrease" (Waud, 1996, p. 186). In addition, the producer accelerates the depletion of the natural resource stock.

15.5 Human-made capital

15.5.1 Economic value of human-made capital

The production of a human-made capital, for example a software product, does not bear the disadvantages of physical production as argued above. Let consider a knowledge capital Φ2, such as a CD copy of Windows Vista software. This is non-rivalry capital, which means that after selling it, the supplier still holds the exclusive right to this intellectual property, as protected by TRIPS and other IPRs regime, and can make endless copies for on-going sale. In other words, the supply of this capital is infinite.

As illustrated in Figure 15.3, the supply of a knowledge capital is infinite, denoted as the red line SS. No matter what real-world price, a limitless supply would have an absolute unit value AV(t), equal to the almost-zero marginal cost of production. It is knowledge protection regimes, such as TRIPS, that restrict the use and dissemination of this product by people other than its owner; therefore, artificially create a limited supply S'S' of Φ2. At this level, a unit of knowledge capital has a market price P(t) greater than AV(t), as May wrote

> The construction of scarcity in knowledge is one of the chief aims of knowledge entrepreneurs. Only when a commodity is scarce and therefore treated as property can it be accorded commodity status, allowing it most importantly to command a price (May, 2000, p. 49).

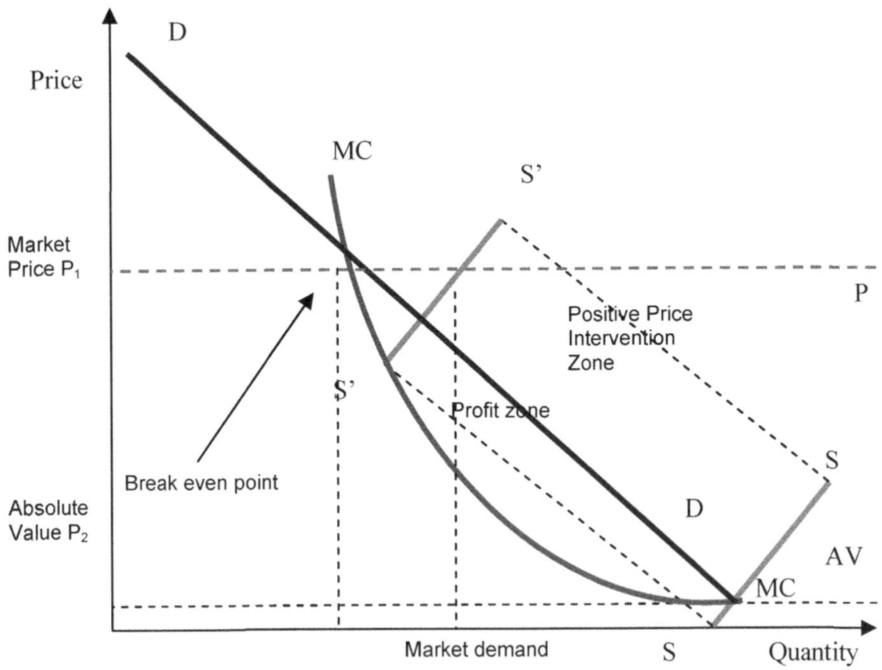

Figure 15.3 Market price and absolute value of a unit of knowledge capital

The zone between SS and S'S' is defined by the present author as the *Positive Price Intervention Zone.* In this zone "Intellectual property rights construct scarcity to ensure market relations in 'knowledge objects' can be undertaken" (May, 2000, p. 42).

The owner keeps on producing Φ2 as long as he can maintain a demand level (by the intervention of human-made input, such as marketing) that allows a P1 which is greater than marginal costs.

15.5.2 Positive surplus value of human-made capital

If AV(t) is the absolute unit value and P(t) is the market price of a human made commodity, then the difference between the values, $\Delta P = P(t) - AV(t)$, is called a positive surplus value of this commodity at time t.

Let's call it *positive surplus value* because when a producer sells the commodity at the market price P(t), he gains surplus ΔP and the buyer loses this same amount ΔP. And by extension, in the global market, it means *in the exchange with other commodities, such as natural capital, as long as the price is overvalued by the intervention on human-made input e.g. IPRs (in the Positive Price Intervention Zone), the more the supplying country gains and the buying country loses.*

Unlike the production of a physical commodity which faces critical limits as discussed above, the production of human-made capital holds prominent advantages. First, there is no limit in supply as long as the producers can maintain a market demand for their product they can always supply that amount. Secondly, there is no increasing cost of marginal production. Therefore, the production is not affected much by the law of increasing costs. Finally, because there is almost no marginal cost, production of such projects does not follow the law of decreasing return on marginal production.

In the formula of price, $P = (N + H) \times PI$ (bearing in mind that PI denotes price intervention) two major components of an economic value of a human-made capital are sunk cost and marginal cost. Sunk cost consists of basic costs like materials (e.g. plastic, machine and other physical facilities to produce a CD); financial costs, such as investment interest or ownership cost; knowledge value, including R&D investment; and wages to pay labor. Shapiro & Varian (1999, p. 21) showed that marginal costs drops sharply after the break-even-point to a low long-time cost arisen by the intervention on human made input SIh.

As Shapiro & Varian indicated, the most dominant cost in knowledge commodity production is the "first copy costs". Once the first copy of a knowledge product is created, the marginal cost of reproducing might be around a dollar in the case of CDs. Major marginal costs after the moment of production are the fee to protect IPRs, promotion and transactional cost. Shapiro and Varian argued that the dominant component of the fixed costs of producing knowledge are sunk costs, and these costs are not recoverable if production stops (Shapiro & Varian, 1999, p. 21).

When the producer recovers the sunk costs, in economic terms, he achieves the break even point, and will not pay any extra cost at all except the marginal cost. Thus, beyond the break even point, he will gain *a net profit equal to the product of positive surplus multiplied by annual average market size and the time of protection.* In Figure 15.3, the zone between the price pink line P1-P1 and the marginal cost MC-MC is the Net Profit Zone.

To maximize economic profit of a unit of knowledge capital (see Figure 15.3), there are several approaches which are widely applied in the software industry. The first approach is maximizing the length of life of IPRs protection to maintain the scarcity of the good. The term of interventions is regulated by TRIPS or bilateral trade agreements. They vary from 5 to 75 years. Not only TRIPS, but also bilateral trade agreements (BTAs) such as the US-Vietnam BTA also regulate that developing countries have must abide by IPR protection, for example:

> where the term of protection of a work is to be calculated on a basis other than the life of a natural person, the term shall be not less than 75 years from the end of the calendar year of the first authorized publication of the work or, failing such authorized publication within 25 years from the creation of the work, not less than 100 years from the end of the calendar year of the creation of the work (Government of Vietnam & Government of the United States of America, 1997).

The second approach are general marketing strategies such as: maximizing demand by the expanding market, gaining a monopolized position, offering a competitive price and building the attraction of the commodity that holds the "love" of consumers (Kibirige, 1983).

15.6 Main insights from Chapter 15

Insight 15A: The long-respected doctrines of Adam Smith and David Ricardo which hold that comparative advantage is best achieved if all countries specialize in exports that they can produce at the lowest cost appear to be inadequate in relation to the global knowledge economy and sustainable development. The present author argues that human-made, knowledge intensive products in fact appear to have a structural advantage over natural resource based products in the world trade order.

Insight 15B: The present author proposes a new mathematical formula (the Natural-Human ratio - N/H ratio) for terms of trade. According to this formula, terms of trade are composed of the trade protection not just of one country but both parties, together with the real values of human-made and natural components.

Insight 15C: Studies show that terms of trade for primary products have continued to deteriorate since 1870, except in rare periods of history (during the world wars or oil crises) when primary products reveal their real value with higher value of the N/H ratio. When oil demand of the world, especially from the U.S. and the growing Chinese economy, sharply surged in the chaos of the Iraq war, oil suddenly broke the artificially low human-made H/N ratio to partly reveal its real economic value e.g. rising from USD20 per barrel to USD54 on October 12, 2004 (Barbajosa, 2004) and up to about USD100 in 2007.

Insight 15D: The systemic, structural advantages of human-made resources over natural resources in the world trade order pose particular difficulties for developing countries such as Vietnam in achieving sustainable economic development, and lend credence to accusations that the global knowledge economy is a new form of imperialism.

Chapter 16
Metatypes for knowledge promotion and the enhancement of domestic value

16.1 Scope of the chapter

- Argues that an emphasis on knowledge promotion strategy, subsuming information and knowledge management, is likely to be the most empowering approach for the growing of a national knowledge economy in Vietnam.
- Systematically presents the Coconut, Orchid, Vine and Rain Forest analogies as the basis of theoretical metatypes helpful in the analysis of current situations in the knowledge economy, and in the formulation of strategy and plans.
- Draws on the content of this and previous chapters to develop a view on knowledge and the adding of value in the domestic national economy. Adapting Porter and Millar's information intensity matrix, uses two new concepts Social relations of production: price intervention and Social forces of production: human-made capital intensity to theorize the conceptual terrain within which knowledge promotion operates to add economic value.

16.2 For a national knowledge economy – manage or promote knowledge?

Many nations in the world have recognized the need to build a knowledge-based economy, and each has its own approach. While some issues for consideration are common to all nations, others are not the same. Common issues include the use of technology to create, obtain and share knowledge. They also include the extent of access to, and cost of, telecommunications. Issues where marked differences occur concern the extent to which countries focus on the marketplace rather than the public interest as the main driver of the knowledge economy; on the commercial value of knowledge and its consumption; and on copyright and the protection of intellectual property (Debowski, 2006, pp. 308-309).

A "knowledge management" approach to the formation of national knowledge policy was discussed by a number of authors and in respect of a range of countries, for instance, Ritter (2004), (Dahlman, Aubert, & World Bank Institute, 2001) or (Dahlman, Andersson, Organisation for Economic Co-operation and Development, & World Bank Institute, 2000), and was seen to offer some valuable concepts and techniques. However in the light of the fieldwork data collected in Vietnam during the course of this study, it seems clear that what might be termed core or conventional knowledge management, as developed in the context of formal organizations, needs to be extended and enriched to comprehend the complexities of knowledge policy development and implementation at a national level because:

a) The scope of a national policy is much wider and more complicated than that of organizations, the setting in which most knowledge management research and implementation has occurred.

Table 16.1 Extension of Knowledge Management to Knowledge Promotion

Elements	**Conventional KM Focus**	**Extension into KP Focus**
Purpose	Right and timely knowledge	Relevant, precise, sufficient and timely knowledge for development
Object	New knowledge	Existing and newly created knowledge, data, information and expertise
Scope	Organization	An entity, ranging from a single person, a small group, an organization, or a community to a country or the whole globe
Targeted users	Employers, employees and customers of an organization	All members within the entity. People outside the entity who have a common interest.
Who manages	Chief knowledge officer, Chief information officer	Leaders, managers and all members
Method	Management of knowledge, including tacit knowledge. Technology-centered	Promotion of environment (culture, human capabilities). Management of implicit and explicit knowledge. Human-centered
Approach	Top down, bottom up	Multi-directional
Intermediary	Computer network	Computer network, human contact and other channels

b) Full consideration needs to be given to indigenous knowledge like Bat Trang ceramic art, Dong Ho folk printing, Quan Ho folk songs (Chapter 9) and locally created knowledge like know-how about golf course design as cited by S62 or know-how about eco-tourism as discussed by S67 (Chapter 8). The ways in which indigenous and other endogenous knowledge is used and shared is an important intellectual property issue for Vietnam in building a knowledge economy.

c) Top-down or bottom-up relationships as typically mapped by organizational charts are not adequate for building a national knowledge economy, especially in societies where relationships among citizens are neither simply top-down nor bottom-up, but far more "organic", evoking the metaphor of a rain forest which is characterized by an intimate and interconnected tangle, woven over centuries, of vertical, horizontal and crisscross trees and climbing plants.

Thus the author argues that in order to build a national knowledge economy, the notion of managing knowledge is less important than the promotion of knowledge. If we define knowledge in the most general sense as those mental constructs of the world that are learned or created by individuals and held in their minds, then knowledge ranges across a seemingly endless series of spectrums including: from wrong to right, from unconscious to conscious (for instance, tacit knowledge or the roadside glimpse that a passer-by gains by chance); from slowly to quickly captured; from undiffusible to diffusible; from unintentional to purposive; and from valueless to valuable. The sources of knowledge can be from the external environment (by experiencing, learning, communicating, seeing, hearing, touching, feeling or even perhaps psychic sensing) or from self-creation (by imagining, reasoning or thinking). In a collectivity like a nation, the promotion of knowledge or *knowledge promotion* (KP), this author suggests, can be defined as: *a process to promote favorable conditions and culture in which old and new knowledge can be shared, exchanged, accessed, learned, used, created or re-created by each member of that society in a relevant, precise, sufficient and timely manner. The target of knowledge promotion is empowering and enabling people in different ways of knowing to gain new value.*

In Table 16.1 this author attempts to explain his concept of knowledge promotion as an extension of conventional understandings of knowledge management.

The rest of this chapter will attempt to theorize:

a) a series of Knowledge Promotion Metatypes that may assist policy development and research in Vietnam, building on insights gained through the "Foundations" and "Grounded Accounts" chapters of this book;
b) a general explanation of how increased information intensity resulting from effective deployment of these models can sustainably enhance domestic value.

16.3 Local knowledge and global value

Previous chapters have discussed the role of creativity and newly created knowledge, endogenous knowledge (that is domestically made such as indigenous heritage) and exogenous knowledge (such as imported ICT systems and content) in a knowledge economy. Arguably endogenous knowledge, either inherited or newly created, is the most important intellectual asset that public policy needs to protect and promote to produce value in the global market. Attaching too much importance to imported knowledge and technologies while ignoring self-generated internal capacity would seem to be unwise policy. The emerging knowledge economy of China is an example of success in promotion of its unique and rich indigenous culture by proper commercialization and globalization of *old cultural knowledge* to fit the desires of *contemporary* global markets, thus gaining high value from it. An example was the successful movie *Crouching Tiger, Hidden Dragon* in 2000 that stormed the theatres world-wide by presenting indigenous legend in the modern language of cinematography. Endogenous knowledge could be described as *crouching intellectual property* because it is unintentional and not fully exploited. It can be locked in someone's brain, isolated as a "knowledge oasis" (Chapter 8) and not used to economic advantage - though it might well benefit individuals and local networks in other non-economic ways e.g. spiritually. Ideally in a knowledge economy, local knowledge is awakened as seed intellectual capital and utilized for competitive advantage in the global knowledge market. If this happens endogenous knowledge can be sustainably developed – including the on-going capacity to preserve, defend and foster the community-level wellsprings of such knowledge.

In addition, technology and knowledge learnt from external sources clearly play a crucial role, for example to help Vietnamese people understand the importance of the environment. Exogenous knowledge has, for example, helped Vietnamese society address the problem of how to reduce environmental devastation by various methods, such as using materials more efficiently, restricting and repairing environmental pollution, or deploying new substitutes for traditional materials or fuels.

An instance was described in Chapter 9 where new knowledge (local or foreign) has helped reduce negative environmental impacts: to earn $10, a Bat Trang craftsman making his historic pots to-day would use ten times less clay and fossil fuels than a Chinese counterpart and thus pollute ten times less than the latter.

16.4 Different metatypes of knowledge promotion in Vietnam

Chapters 8, in considering the phenomenon of "knowledge oases" introduced the notion of metatypes of knowledge promotion. The metaphors suggested by examples cited in Chapters 7 to 13 are: Coconut metatype, Vine metatype, Orchid metatype and Rain Forest metatype. These metatypes are grounded in data collected during the fieldwork among different groups of Vietnamese individuals. In the first two metatypes, the promotion of knowledge and its value is spontaneously conducted by individuals in a bottom-up approach. In contrast, ICT projects in the Orchid metatype are initiated and subsidized by government or companies in a top-down approach. The Rain Forest metatype is commons-based peer production, which is neither bottom-up nor top-down. It is far a more horizontal and distributed network of peer members though some members may be more active than others. Elements in the KP framework in Table 161 will be used to describe the main characteristics of each metatype.

16.4.1 Coconut metatype

In Chapters 8 and 9, the *knowledge oasis* or *Coconut metatype* was identified mainly among non-ICT users. For example, the preservation and trade of indigenous knowledge holders like Quan Ho folk song artisans, the accumulation of new knowledge of S67 who gained knowledge through his tourist service, or S62 who learned the art of designing a golf course.

The purpose of knowledge promotion of this type is *profit making* which is done by *exchanging locally created information* with other people (customers, for instance) to *get some economic value*. S29, S30 and S31 collected Quan Ho songs to sell (by live performance) to customers; S67 learned and developed a business process of eco-tourism service to gain more clients.

The object of promotion is *locally made knowledge* which can be either indigenous knowledge (S29, S30, S31) or new knowledge that the promoter produces based on what she has learnt from experience or external sources (S67).

The scope of a Coconut metatype knowledge promotion is often one individual person or an individual business.

The actors involved consist firstly of the owner of the knowledge and secondly, his/her customers, if there are any.

The key actor or the main builder of the system is often *one person*, the individual owner of the knowledge. Thus a Coconut metatype can be characterized as a *one-person system.*

The main approach of the Coconut metatype is *bottom up* and *spontaneous accumulation* of information and knowledge. Theoretically, we could assume that no horizontal tie is established in a Coconut metatype system. However, in practice, such a relationship is often established, for instance, between the knowledge holder and her peers.

The media used in this type are *non-ICT vehicles*. In the absence of computers and the Internet, individuals use traditional media such as the human brain, paper, or even ceramics to store knowledge. Word of mouth and paper are the main technologies of communication. Recently, electronic media like TV and telephone have become new means of obtaining external information.

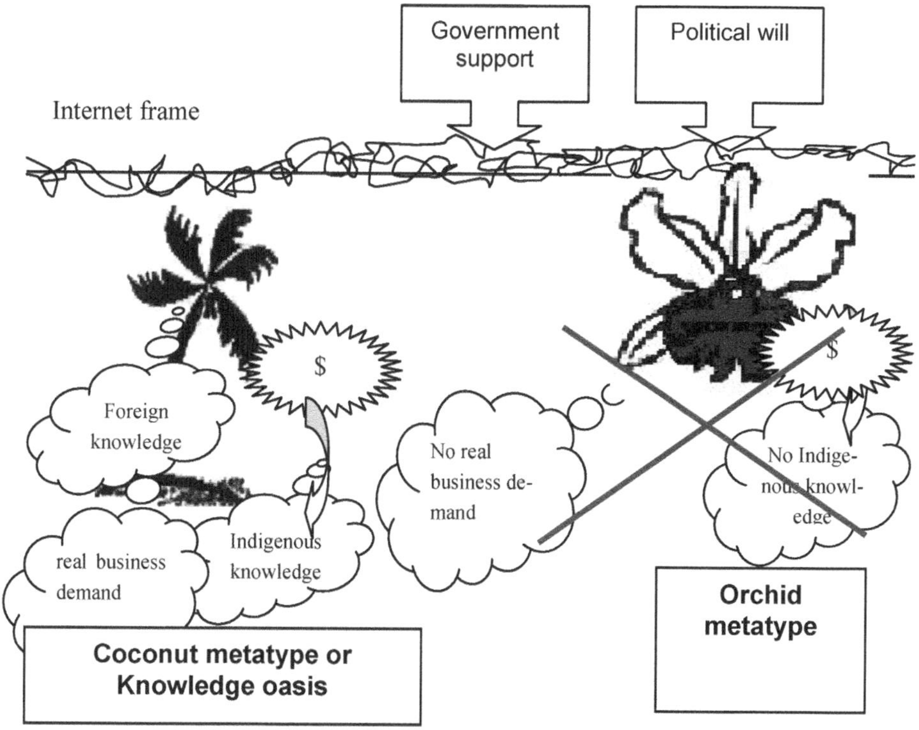

Figure 16.1 The Coconut Metatype of Knowledge Oasis and the Orchid Metatype of Political Will

Strengths: The promotion of knowledge emerges from some real needs of the knowledge holder's business. Old and new knowledge all have some significant value to the livelihood of the holders and in return, bring new earnings to them. With real motive, a Coconut metatype system is livable.

Weaknesses: The accumulated knowledge is locked in the brain of knowledge bearer even if it can be not tacit knowledge. Lack of networking capacity of ICTs, the society can not make access to the knowledge to fully use of the power of individual knowledge multiplication.

For a Coconut metatype system, external obstacles to the networked world include the cost of communication, and the availability of technology infrastructure. Internal obstacles are the poverty, timidity, low education of the holder. In addition, the ambiguity of possible benefits of ICTs stops them as well. The view of knowledge as a business secret prevents its holder to share it with the society (S67, Chapter 8).

Efficiency: From an individual viewpoint, a Coconut metatype system can be cost/effective because the key actor would know the best way to promote knowledge for herself. However, from the social viewpoint, because knowledge is poorly disseminated, the efficiency of a Coconut metatype system is low.

Sustainability: As the practice of knowledge promotion in a Coconut metatype style has existed in villages for ages, it can be considered as highly sustainable. However, the disappearance of the culture that nourishes the knowledge as well as the death of the knowledge holder can terminate the existence of a knowledge oasis (for discussion of the extinction of indigenous knowledge, see Chapter 9).

Suggested government policy interventions: Government interventions for people of the Coconut metatype model could focus on two aims. First, interventions need to improve the effectiveness of existing channels of information to enhance the richness of information and help people to enhance the value of knowledge they create. Secondly, interventions would need to provide more opportunities to people to access ICT facilities by reducing the external and internal obstacles mentioned above.

For that purpose, a consolidation of channels of information for non-ICT users, especially channels of high usefulness and availability like TV could be taken into consideration by responsible agencies and related stakeholders. TV programs would need to be customized and localized to meet the specific needs of each group of people in the Coconut metatype model.

The importance of two-way channels of information, including telephony, professional associations and especially peers are important to villagers and should not be ignored. Local government plays a critical role in promoting those channels as well as potential internet access points, such as CPCP or private internet cafes. As discussed particularly in Chapters 5 and 13, the idea of developing a network of telecentres like CPCP would appear to be in keeping with good ICT policy principles. However, much work would need to be done to improve the effectiveness of such channels in reality (see discussion in the next section).

Since rural villagers, especially farmers, have very-low revenue, they could be targeted for special treatment, such as subsidized telecom tariffs as well as measures to encourage them to use the Internet. There should be special programs to help non-ICT users overcome the timid feeling of the first trial of the Internet. There could be either TV awareness programs or Internet training program at school. Of course for nonusers, many of their reasons for not using a computer may come from the condition of deprivation and lack of motivation (van Dijk, 2005, p. 28) or even some imagination.

In summary, the Coconut metatype is the most simple as well as the most popular method of knowledge promotion in the non-digital world where people use basic, older analogue media to communicate, record and reproduce knowledge. The main communication method is oral and the main medium to store knowledge is human memory. Other secondary facilities are also used to keep data (paper, ceramic) and to receive and transmit information (word of mouth with peers). Knowledge is promoted vertically and singly like a coconut palm growing in an oasis. With the diffusion of mass media and digital media through Vietnamese society, modern technologies (TV, radio, telephone) are gradually infiltrating situations where the Coconut metatype model has long been in place. To a certain degree, the Coconut metatype may change to a new form, the Vine metatype.

16.4.2 Orchid metatype

Examples are VTV2, governmental websites, and CPCPs - Communal Post and Cultural Points (referred to in earlier chapters, especially Chapters 5, 10 and 13).

The purpose of an Orchid metatype KP is non-profit making dissemination of external information to local people to serve a certain political or business objective, or to serve the public interest.

The object of promotion is external knowledge which is introduced by the key actor to inform or educate local people.

The scope of an Orchid metatype KP depends on the design of the key actor. It can operate at a national level (e.g. public broadcaster VTV2 or government websites) or a community level (e.g. CPCP).

Target users are a focused group of local people (VTV2), government officers and business (government website) or community member (CPCP).

Key actors include government, business and NGOs or community-based organizations. Often, an Orchid metatype KP engages little or no participation by target users, such as local people or local authority, in the design and implementation phases.

The main approach of an Orchid metatype KP is *top-down* and politically led.

The media used in this metatype are the Internet, television, radio and other mass media.

Strengths: In the beginning, an Orchid metatype KP grows very quickly because it is supported by political will with good funding and good ICT professional support for the key actors. The project is often designed to serve a certain political and social purpose that profit-making industry wishes not to undertake, for instance, to encourage the use of ICTs in poor areas (CPCP) or to educate farmers about technique of cultivation (VTV2) or to disseminate information about sustainable development and environment protection (www.va21.org). An Orchid metatype KP can be the initial bridge to introduce ICTs to non-ICT users, especially those in poverty or remote areas.

Weaknesses: An Orchid metatype KP often lacks internal motivation from either the key actors or target users. Key actors do not have any direct economic benefit in implementing the project. Target users are not well involved in the design phase, even though this is intended to meet their actual information needs. Such needs are often poorly identified and satisfied because the view of project implementers and local users are not identical in top-down projects. Braun argued that top-down and silo ICT projects often show poor coordination and communication between local people and the state management. Community members generally displayed a low degree of knowledge of and interest in the development of the portal (Braun, 2004). Braun observed that the target users do not actively participate in the project because: i) their information needs are neglected, ii) their local knowledge is ignored, iii) the location of intermediary is not convenient for the target users to access (for instance, VTV2's time schedule conflict with farmer's availability) and equipment is not adequate (for instance, CPCP).

Sustainability: Low, not self-funded. The project can collapse when the top politician or management changes the strategic priority and commitment. When funding to run the project ceases the Orchid metatype dies because an Orchid metatype KP does not get revenue from its operation.

Efficiency: Low. The frequency of use and number of users can be very low (for instance, in the case of CPCP or government websites).

Suggested government policy interventions: The participation of local people and government in the designing and customizing phases would help to define the most appropriate content for them as well as the way that information should be delivered (chapter 9). In addition, pilot projects with the designed prototype need to be conducted before the project is implemented on a large scale to check it that it can work in the field. In the author's experience from the VCIT Project, pilot trials would save a great amount of money, time and effort and substantially improve the quality of the project outcome.

During the implementation phase especially, a feed-back channel may be needed to collect opinions from users so that necessary adjustments can be made in a timely and and appropriate way. The combination of knowledge promotion systems of Orchid and Vine metatypes can enhance the strengths and minimize the weaknesses of the two approaches.

In summary, the Orchid metatype is often found in government and organization knowledge promotion projects. It can be considered as a version of bureaucracy in the networked world. While an Orchid metatype KP can be well equipped and funded, its effectiveness and efficiency may not be high because of low relevance of the content and method of communication to the target users. The combination of a knowledge promotion system of Orchid metatype with Vine metatype is therefore recommended.

16.4.3 Vine metatype

Examples of this metatype are the websites of Bat Trang ceramic potters, or the unsuccessful attempt to build a website of Dong Ho folk printers of S23 (Chapter 9), or personal blogs (Chapters 12).

The purpose of a KP of this type is *profit making* (Bat Trang) or *non-profit making* (personal blog) by gathering and diffusing locally created information about the knowledge promoter's business. A Vine metatype KP is a crucial bridge for local businesses and individuals to communicate their business to the global market and vice versa.

The object of promotion is *local knowledge* which is preserved, collected or created by the promoter.

The scope of knowledge promotion is often one person, one business or a group of peer businesses with the same nature.

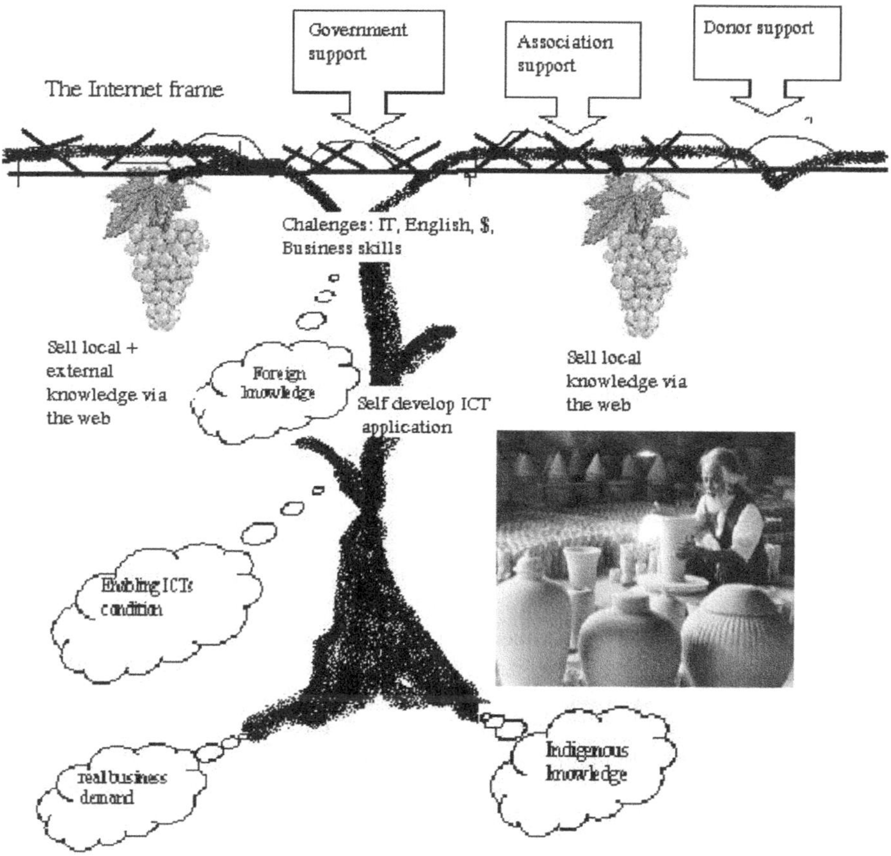

Figure 16.2 The Vine Metatype (bottom up approach)

The target users are *external customers*.

Only *one single person* or *business* is the key actor in this model. The growth of the system often lacks participation of external parties such as government or organizations. A Vine metatype is a single rooted KP system.

The main approach is *bottom up* and *spontaneous*. Some horizontal ties may be built, for instance, hyperlinks from one website to other websites or viewers. The promotion of knowledge is based either on some real business demand (on folk painting or ceramics) or an emotional motive (personal blog). KP is expected to improve the image of the key actor. In the first case, a widely known image is a business advantage. To expand the business (for instance, finding new customers or markets, marketing the product) the key actor takes advantage of ICTs and the Internet to position her business in the global market. At the beginning, simple transactions like online advertisements and information sheets are put in the Internet. Later, subject to the capacity of the key actor and the business outcomes, more sophisticated transactions such as email, or online query and order may be added.

Media of communications: The Vine metatype uses ICT facilities and the Internet to store and diffuse information. The Vine metatype can be regarded as a higher development of the Coconut metatype.

Strengths: Because it originates from real business demand, the Vine metatype KP is practical for its owner and has certain value to her. The use of ICTs and the Internet in a Vine metatype publicizes the key actor or her business to more people thus brings some economic or social value to her.

Weaknesses: At a certain stage in the application of the Vine metatype, the key actor may face a series of challenges, such as the shortage of funds, professional ICT skills, English language, management and business skills that prohibit further growth of the Vine metatype model. The key actor may not sense the real benefit to her of further effort or investment. Without external support to overcome these obstacles, she may lose motivation and neglect the project (for instance, the Dong Ho case or S23, Chapters and 12). The project dies and the Vine metatype downgrades to a Coconut metatype.

Efficiency: A KP of the Vine metatype is often undertaken for a particular business purpose. However, the key actor's weaknesses in ICT skills or other capabilities can severely limit the effectiveness of the KP effort.

Sustainability: If the KP can help the key actor gain some value (getting more customers or viewers), the key actor can maintain her investment in money, time and effort for the project. In other words, the Vine metatype model can be sustained because of its contributions to revenue. Otherwise, the project stops.

Suggested government policy interventions: This metatype holds an important position in the promotion of local knowledge for seeking new value in the networked world. The key actor might need external support from government, organizations or other support agencies to maintain the momentum of the project.

External supporters can play a godmother role to help local business and community to utilize ICTs and exchange their knowledge, at least in the early stages. ICT professional advisors are very important to improve the quality of KP of Vine metatype. An example is the support from Mekong Private Sector Development Facility, the World Bank and Bat Trang Porcelain & Ceramics Association which gave strength to Bat Trang potters' effort in creating a portal for Bat Trang pottery (see Chapter 9).

Unexpected value can arise in a Vine metatype, for instance, marketing Bat Trang pottery for tourist purposes.

In summary, the Vine metatype is a higher development of the Coconut metatype. In the Vine metatype a single key actor uses ICT media to to obtain, keep, transmit and reproduce knowledge. The main communication method is the Internet as a support strategy for growth of the key actor's business. Knowledge in the metatype is promoted vertically from one root then, with the networking power of the Internet, gains access to and communication with other users and counterparts analogous to a vine spreading its branches and joining us with others with the Internet as a frame. To a certain degree, the Vine metatype may either wither away (to a Coconut metatype system) or evolve to a new form, namely the Rain Forest knowledge promotion metatype which in Vietnam is widely dependant on commons-based-peer-knowledge-promotion.

16.4.4 Rain Forest metatype

In the Vietnamese context, the Rain Forest metatype widely includes and depends upon *commons-based-peer-knowledge-promotion.* Examples of a Rain Forest metatype of commons based peer knowledge promotion (CBPKP for short) are given in Chapter 12 and elsewhere. They include Vn99.net, sextvvvvvv.com, http://www.acmilanvn.com, open source projects like Unikey, Vietkey, and the computer game club of Volamtruyenky.

The purposes of a Rain Forest metatype include *non-profit making, profit making* or a *mixture of the two.* It encourages members to share self-created knowledge and collected information with other members in a virtual community for entertainment or professional purposes.

The object of a CBPKP is self-created knowledge, and the collecting and presentation of information that is identified as relevant.

The scope is in a virtual community of people with a common interest. These members may live in different physical places.

The target users are all members of the virtual community. As a virtual community is not limited by a geographical boundary, it can be extended endlessly when other people join the group. The student-run forum of Vn99.net had 143,000 members at its height.

Key actors: CBPP is often claimed to be a distributed entity characterized by peer production, peer governance and a common inclusivity (Benkler, 2002), yet this study identified certain hierarchies in a CBPKP Rain Forest. In a CBPKP, there often exist three main groups of people, namely sponsors, key content contributors and silent profiteers.

Sponsors are one or few persons who set up the system and maintain the regulation and culture of the system. They include administrators and super-moderators. Sponsor members can also include ICT companies, for instance in the Linux project.

Key content contributors are those who are most active in posting threads (web-content) into the system. In some CBPKPs, for instance Vn99.net, there is a mechanism to add a score for each time a person posts a thread. Those with a higher accumulated score get a higher status in the community, for instance VIP member. In general, the number of key content contributors is not many compared with the total number of a virtual community. Sponsors and key content contributors are core members of a CBPKP.

The biggest number of members is *silent profiteers* who exploit the information delivered by key content contributors and others and very seldom contribute any valuable information to the community.

The approach of a CBPKP is peer to peer. Though core members and key content contributors may have more privileges and rights than others (access to many or all categories of content, right to amend or exclude information, right to accept or ban a member), other members have the right to access to most of the content in a CBPKP.

A CBPKP can evolve into a virtual society with various kinds of links among members from weak to strong ties, reflecting bridging to bonding network relationships (Schauder, 2006). These complicated links from member to member or from a group of members to another grow quickly in a CBPKP like the biological growth of rhizomes (key content contributors) and climbing plants (silent profiteers) in Rain Forest whose owners are the sponsors.

Media: The Internet, and particularly the World Wide Web.

Sustainability: A Rain Forest KP is self-funded. However, when core members lose enthusiasm, time or money, the system collapses. Chapter 8 described the early success of Vn99.net in 2004. When the administrators of Vn99.net lost the credit-worthiness to pay the server rent in 2005, they called for financial support from more than 100,000 members to maintain the music forum under a new name, 1vn.net. Fewer than one hundred people (including the author) made some contribution. As a consequence, the students' CBPKP died young (S69 and S70, interview, May 19, 2005, 1vn.com (2006)). This example shows that a very low proportion of members actually contribute any real value in a CBPKP community.

To avoid collapse, a number of CBPKPs experiment with alternative business models to collect money. In Sextvvvvvv.com, VIP members who agree to pay a fee have more rights than a normal member. Vnexpress.net receives income by allowing other companies to place advertisement on the website (S49, interview, Dec 14, 2004). These revenues, besides the spiritual enthusiasm of core members, are the physical basis for sustaining the system.

Efficiency: High.

Strengths: A CBPKP is usually initiated by a person, a group of sponsoring members or a company who provide start-up fund and set rules for the growth of the system, which invites joining and participation from other persons with common interest. The culture of CBPKP encourages all members to share knowledge and appreciates the contribution of active members.

CBPKPs often use Open Source Software, and some are devoted to the development of OSS software. Optimistic interviewees believed that OSS should be encouraged as a good alternative to proprietary software in the context of Rain Forest metatype CBPKPs because of the cheap cost (I12, interview, Jan 25, 2006), the transparency and easy modifiability in collaboration with the developer (I13, interview, Feb 13, 2006). I42 (interview, Mar 23, 2006) argued that OSS had many features that were as good as copyrighted software. With OSS, software developers and implementers could save cost of production, avoid IP violation and more importantly, OSS developers and users can get strong and quick support from the OSS community when they have difficulty. According to I39 (interview, Mar 14, 2006) the resource of OSS was abundant on the web with hundreds of thousands of projects that if properly exploited Vietnamese virtual communities and IT companies could "stand on the shoulders of giants".

Weaknesses: Online activities in most Rain Forest metatype CBPKPs all too often cater mainly to passive consumption. Computers, the Internet and other ICT facilities are used mainly for entertainment purposes and thus contribute little or no economic value to the users other than in their capacity as consumers of entertainment.

Reliability and trust in a CBPKP is not high because most of members use pseudonyms or participate anonymously. As a consequence, the social networking ties among members in a CBPKP are often very loose. To improve these relationships, some CBPKPs organize "offline" (face-to-face) meetings to allow members to know each other in a real environment. The virtual relationship is thus converted into a real relationship and weak ties are converted to strong, bonding relationships.

No compulsory conditions force members to contribute to the community. Like the operation of a communist society such as the former Soviet Union most members of a CBPKP tend to be inactive (silent) in contributing to the commons but keep benefiting from the results of contributions made by others.

As for Open Source software and its potentials for Rain Forest metatype CBPKPs , although OSS is regarded as an alternative solution for Vietnamese users and developers, many interviewees believed that OSS could not compete with commercial software. For both Vietnamese users and OSS developers, the concerns about OSS were considerable, because:

- No clear orientation guides the long term development of the project thus the outcome products are not reliable and there is no guarantee that the coming generation would be compatible with the previous ones;
- Responsibility and accountability of OSS developers are not clear;
- The professional level of OS developers is often lower than that of commercial developers; the quality of products is not high and guaranteed;
- The relation of commercial products is mutually beneficial between the developer and the user while members of OSS community unilaterally exploit fruits of OSS projects (as silent profiteers). The core of large OSS products are often funded by large corporations, e.g. Open Office funded by Sun or Mozilla by Netscape;
- The total ownership cost of an OSS solution (including the cost of training, maintenance, the required capacity to use, the quality of the product, etc.) could be even greater than that of an equivalent commercial solution;
- Software companies only use commercial tools to develop software because it is more convenient for obtaining appropriately trained technical staff to work on the project
- Installation of OSS is more challenging because technical support is often not adequate. Also supporting documents and training program for OSS are limited or not available
- OSS developers are volunteers, few are full-time contributors
- GPL (General Public License) allows a free use of OSS but also does not regulate the accountability of the author, thus does not guarantee any possible loss that end-users may encounter
- OSS is more suitable to professional developers than to normal users, except for very vendor-funded products like OpenOffice or Mozilla
- It is not acceptable to sell OSS based products at high price even though the labor to create may not be low in comparison with vendor-based platforms. The sale value of commercial products is the cost of development while that of an OSS is often after-sale service or value added services.
- OS conditions of participation may cause a developer to lose the right to hold IP rights for herself; commercial companies can steal ideas from OSS and work more efficiently (see Section 13.4.6.3 in Chapter 13).

Suggested government policy interventions: CBPKP is the latest form of KP today. CBPKP attracts a great number of people to join, actively or silently. However, the majority of CBPKP systems in Vietnam favor entertainment uses, such as music, video clips, computer games or pornography. Very few attempt to redirect Rain Forest metatype communities towards knowledge for development or the like.

To improve the quality of CBPKP, government and organizations can support or fund the best CBPKP systems that target members efforts to the goal of economic development. Active contributors can be encouraged by both moral and physical rewards. A hybrid Orchid metatype-CBPKP approach can uphold the strengths of each single type as well as minimize the weaknesses of CBPKP, if growing spontaneously.

In relation to OSS projects, the potential to use OSS as an alternative to commercial software is not obvious due to a series of obstacles, including technical aspects and the ease of use; real market demand and the willingness of customers to go beyond their comfort zones. It appears that unmanaged and non-commercial non-profit making products from OSS movement have difficulty in competing with commercial and profit making packages like Microsoft Windows. However, the emergence of alternatives like OSS is a rallying point for people to fight against corporate monopolies so an even on these grounds alone OSS need to be further encouraged (Chapter 13).

In summary, the CBPKP model can be regarded as a network of many Vine metatype systems with similar interests. The contribution of a CBPKP comes mainly from the root of Vine metatype members (the key content contributors) not from the majority of members who are more likely silent profiteers. Sometimes, a CBPKP can hybridize with an Orchid metatype, for instance, the production of the open source Linux is sponsored by IBM or youtube.com by Google.com. The production and sharing of knowledge in a CBPKP system is similar to the production of wealth in a communist-like society. Because of a lack of good design and physical incentives to encourage the contribution of all members, the system may not be as sustainable and effective as expected.

16.5 Enhancement of domestic value

In the first two metatypes above, Coconut and Vine, ICTs are used in relatively stand-alone modes of promotion of knowledge in Vietnamese communities. The Rain Forest metatype operates at a higher level of complexity and interdependence, but is still ultimately reliant on the initiatives of individuals. Despite the apparent superiority of that metatype, much online activity in the CBPKP context is questionable in value-adding terms – particularly time and effort spent by young people on endless hours of gaming. Although requiring significant ICT skills and literacy, such activity often carries a price in physical self-harm and mental addiction (Chapters 11 and 12). At present, governmental efforts, in the form of Orchid metatype projects, still play a limited role in improving the effectiveness of ICTs for sustainable development as the Rain Forest metatype seems to default all too easily into passive and repetitive consumption of entertainment rather than innovative entrepreneurship.

Having noted these problems in computer gaming, a highly information-intensive economic sector, what can be said about information or knowledge-intensity as a general principle of industry policy aimed at enhancement of domestic value within a framework of sustainable economic development?

There is a range of possible levels of human-made capital intensity in the value of products, acting together with factors of price intervention. According to the theorization that has been presented (Chapter 15) concerning the change of mass for mind, to improve the value of an economy is to improve the H/N ratio in that economy. This could be done by either increasing the human-made intensity in the product or increasing the price intervention to the product. Figure 16.3 is a matrix of those two dimensions as they interact within an economy which attempts to explicate how that process operates. The horizontal axis is human-made intensity (determined by forces of production) and the vertical axis is price intervention. This theorization owes much to Porter and Millar's Information Intensity Matrix (Porter & Millar, 1985, pp. 153-154).

At the left-lower corner is the agricultural economy with commodities containing relatively little knowledge and human-made capital, and also poorly supported by price intervention. Examples of commodities of this group are water that is used as a free gift from nature to irrigate coffee plants (Chapter 8) or clay to make ceramics (Chapter 9). Those products have low economic value though they may have a high utility.

At the right-lower corner is the handicraft economy with commodities that may contain rich knowledge like Dong Ho printing or Bat Trang ceramics (Chapter 9) but also are poorly supported by price intervention, and are thus not of high exchange value. That is the case where high information content nevertheless has low economic value.

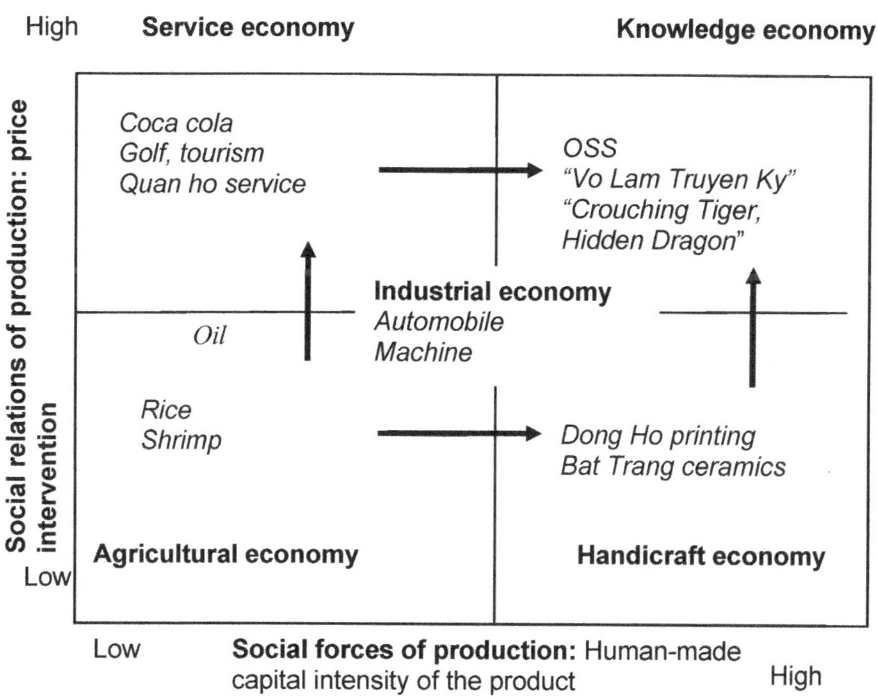

Figure 16.3 Price intervention, human-made input and mode of economy

At the left-upper corner is the service economy. Here price intervention can influence the exchange value of a product of low human-made capital intensity, such as the case of water that is trademarked Coca Cola (Barlow & Clarke, 2002) or crude oil which is influenced by the OPEC cartel (Chapter 15). Entertainment services like golf or tourism (Chapter 9) are other examples of the service economy.

Somewhere in the middle are the cases of the Quanho folk music and Bat Trang ceramics. The Quan Ho folk music served at the Quanho Folk Music and Cuisine in Bacninh province (Chapter 9) can be regarded as an integration of social intervention with cultural goods and services by which more economic value has been added to traditional knowledge. Not everyone approves of enterprises of his kind. Some cultural researchers see the commercialization of indigenous knowledge as a degrading of precious traditions (VietnamNet, 2006a). However commercialization and the application of ICTs as a social intervention bring more economic wealth to the bearers of traditional knowledge and help to maintain the lives of the artists and the sustainability of the art. According to Porter and Millar's model of Information Intensity Matrix, these producers have embedded significant new value-adding activities within the value chain of these traditional industries: activities which support design, production, marketing, delivery and customer support (Porter, 1985; Porter & Millar, 1985, pp. 153-154).

The purpose of a KP of this type is *profit making* (Bat Trang) or *non-profit making* (personal blog) by gathering and diffusing locally created information about the knowledge promoter's business. A Vine metatype KP is a crucial bridge for local businesses and individuals to communicate their business to the global market and vice versa.

The object of promotion is *local knowledge* which is preserved, collected or created by the promoter.

The scope of knowledge promotion is often one person, one business or a group of peer businesses with the same nature.

The target users are *external customers*.

In the right-upper corner is the knowledge economy with commodities that contain high information/human-made intensity and their price is influenced by high price intervention, such as the movie of "Crouching tiger, hidden dragon", Microsoft's Windows Vista operating system, or computer games like "Vo Lam Truyen Ky" (Chapter 12). It is noticed that the distinction of one mode of economy from others depicted in Figure 16.3 is only relative in the global economy today. Different modes of economy can be found in a single company. Similarly, no rigid or clearly defined trajectory is made available for a nation to move from an agricultural economy to a knowledge economy. However, any measure that improves the H/N ratio could lead to that target. Knowledge workers and villagers of Quan Ho villages have enhanced the value of their products and services toward the right-upper corner of the matrix where their crouching knowledge of the Quan Ho songs, with their high information content, gets higher profit. Embedding local knowledge with other services, for instance, tourism, entertainment or cinematography in a creative manner is to infuse fresh value to them. In relation to the author's theorization of the exchange of mass for mind, the effect of such action is *increasing the content of human-made capital intensity in the product.* Together with innovation of this kind comes the necessity for Government and support organizations to assist knowledge workers in protecting and promoting the value of domestic-made knowledge. In relation to the H/N ratio, this means to *increase the price intervention of human-made capital,* for instance, by protecting domestic intellectual property rights or promoting the image and value of knowledge goods and services in the global market.

16.6 Main insights from Chapter 16

Insight 16A: Knowledge promotion is a concept that captures the need to establish a self-generating dynamic for the growth of a national knowledge economy, powered both "top-down" and "bottom-up".

Insight 16B: Knowledge promotion metatypes such as those based on the Coconut, Orchid, Vine and Rain Forest analogies can be explanatory in the review and analysis of situations where knowledge-based solutions may provide effective courses of action.

Insight 16C: In an adaptation of Porter and Millar's Information Intensity Matrix, and updating core Marxist concepts, the desired outcome of knowledge promotion policy at a national level can be theorized as: optimization of value-adding by coordinating decisions on human-made capital intensity (social forces of production), with decisions on price intervention (social relations of production).

Chapter 17
Conclusions and guidelines for policy: Seeking Thachsanh's inexhaustible rice bowl in the 21st century

17.1 Presentation of conclusions and guidelines for policy

This final chapter of the book is divided into two sections.

The first section (17.2) is a consolidation into one sequence of the insights that have concluded each chapter (excluding Chapter 1 Introduction and Methodology and therefore commencing with 2A). The number of each insight indicates the chapter which gave rise to it. Within this section the insights are grouped under the broad headings of foundational analysis, grounded accounts and theorizations, corresponding to the major groupings of chapters within the book. These insights constitute the conclusions of the book.

The second section (17.3) builds on the insights drawn from the grounded accounts and theorizations chapters to derive suggested guidelines for policy development, for consideration by the government of Vietnam. The foundational chapters (chapters 1 to 6) are not used directly for the derivation of proposed policy guidelines as they underlie the empirical and theoretical inquiries developed in the subsequent chapters.

This concluding chapter is subtitled Thachsanh in the 21st century in the belief and hope that the promotion of a knowledge economy can make some contribution to the restoration of the ecosystem as a perpetual source of life and livelihood for humankind, as symbolized in the legend of Thachsanh's rice bowl recounted in Chapter 1. The chapter attempts to synthesize the findings of the book as a guide to future action in the case of Vietnam.

17.2 Consolidated listing of insights from the book

17.2.1 Insights from foundational analyses

Insight 2A: While early Utopian claims for the knowledge economy were exaggerated, ICTs have enabled a shift in what Marx termed the social forces of production, towards much higher knowledge intensity, initially in developed countries but spreading to the developing world.

Insight 2B: With knowledge resources, humankind can minimize the use of physical resources. ICTs enable more efficient monitoring and control of production processes, reducing wastage in the conversion of nature-made to human-made resources. This is one side of the so-called "substituting data resources for physical resources" (Boisot's Principle of Least Action) through changing the social forces of production.

Insight 2C: While knowledge resources are endlessly renewable, many natural resources are finite and exhaustible. Since knowledge production is renewable, it has an advantage over non-renewable natural-resource production. Within the international division of labor, ICTs confer an advantage on countries with more developed knowledge-producing economies compared to countries whose economies depend on the sale of natural resources. That is the other side of "substituting data resources for physical resources" through changing the social relations of production.

Insight 2D: Increasing knowledge-intensity gives an impression of "weightless" production in developed countries, but this is largely achieved by shifting an accelerating burden of environmental resource exhaustion to developing countries under terms of trade and double standards unfavorable to the latter. On a global scale, ecosystem losses from increasing physical resource production in developing countries largely offset ecosystem gains from "weightless" knowledge production in more developed countries.

Insight 3A: The global environmental crisis is an urgent reality. The health of the global ecosystem affects all countries and needs to be dealt with globally through international co-operation. Individual countries are constrained in how far they can deal with ecosystem degradation on their own.

Insight 3B: There is a tension between sustainability and development that must be taken into account. In many ways the term "sustainable development" and thus also "sustainable economic development" is self-contradictory.

Insight 3C: Progress made at the interface between economics and environmental studies have yielded rich concepts and formulations, for example about relationships between human-made and natural capital. These provide a foundation for the present author's attempt, in later chapters, to extend such formal reasoning to include a finer-grained account of information and knowledge as factors in sustainable economic development.

Insight 4A: Vietnam is a society which has experienced and withstood internal and external tensions for centuries, sheltering itself behind the mental construct of the bamboo fence. It demonstrates a great desire for progress in a globalized economy provided that core aspects of traditional culture are maintained.

Insight 4B: Doimoi and other reform measures have helped Vietnam to recover from the US War and become a competitor in the world market, especially in commodities such as coffee and rice, and aquaculture products.

Insight 4C: Increased prosperity has been two-edged. Average literacy rates and incomes have risen, but key social inequities (rich/poor, rural male/female, rural/urban, Vietnamese/ethnic minorities) persist or worsen. Accelerated degradation of the ecosystem has been a major concomitant of economic development.

Insight 4D: With high population pressure and governmental structures still transitioning from a command to a market economy, there is a widely acknowledged need for rapid change, but with risk minimization.

Insight 5A: There is a significant background and momentum in policy development (notably Vietnam Agenda 21 – VA21, and Politburo Directive No. 58/CT –

TW) both for sustainable economic growth and the uptake of ICTs. However the planning horizon for the latter has been relatively short term, and the two areas of policy are in need of closer integration.

Insight 5B: Such policies recognize the severe threats posed to Vietnam by water shortage and pollution; environmental toxicity affecting lands and forests; climate change and similar problems.

Insight 5C: There is recognition that the meeting of human needs, and the development of human capacities to deal appropriately with problems or opportunities, are the prime concerns for both environmental and ICT policy. However current efforts at education, training and on-going support for human capacity development may be insufficient, both in scope and scale, to match the urgency of achieving sustainable economic development.

Insight 6A: There has been a continuity of effort to envision an appropriate framework for KE policy and strategy in Vietnam, starting with reliance on imported models in the 20th century then evolving to more endogenous models in the 21st. An understanding of this trajectory is illuminating for understanding current and emergent KE policy settings.

Insight 6B: At the milestone "2000's Conference on Knowledge Economy and its Issues to Vietnam" education and human capacity building was seen as an issue of vital importance to the welfare of individual citizens, and as a cornerstone of a KE for Vietnam.

Insight 6C: The diverse viewpoints of Vietnamese scholars and experts, both supportive and skeptical, about the potentials of a KE are of value in developing real-world KE policy. The critical voices of KE skeptics are especially valuable in risk assessment and risk management in the formulation and implementation of KE policy.

17.2.2 Insights from grounded accounts

Insight 7A: Pressures in the domestic economy and from globalization are seen as interlinked, driving the nation along courses of action that seek high returns in the short term, but rapidly degrade the environment and deplete non-renewable resources.

Insight 7B: Government is still not fully geared to the conditions of a market economy, and thus is only partially capable of optimizing the opportunities of the free market while managing the risks. Selective but widespread reform is still required of government structures and processes, especially measures that improve capacity for co-ordinated, whole-of-government action.

Insight 7C: Environmental degradation is widespread. Significant human and technological inputs, underpinned by education and research, are needed to reverse the current decline and achieve sustainability and improving living standards. Examples of success need to be further recognized and publicized. The appropriate deployment of ICTs and knowledge strategies is an essential part of this scenario.

Insight 8A: The rural life and economy of Vietnam are under great pressures including population size, spreading urbanization, environmental degradation, slow technological change, and limited capacity for change-management in all levels of government. Youth, women and ethnic minorities are particularly affected by the problem of surplus labor in rural areas.

Insight 8B: There is increasing recognition that educational and knowledge strategies are crucial to addressing problems of sustainable economic development in rural Vietnam, while at the same time preserving core national values and traditions.

Insight 8C: Dramatic knowledge-based advances in areas such as coffee and paper production encourage a view that the nation can "think its way through" rural problems. So do successes in areas such as tourism. However there is a problem of "knowledge oases" where pioneering small businesses remain isolated cases, and small-business know-how spreads only slowly across the rural economy.

Insight 9A: Indigenous, traditional art and craft occupations are endangered by competition and substitution in the open, globalized market economy, and by damage to environmental resources that have nourished these industries for generations. However the new conditions if well managed also open up opportunities, particularly through cultural tourism and Internet marketing of cultural products.

Insight 9B: Practitioners of traditional arts and industries are anxious to see their culturally unique knowledge survive and prosper in the context of new production and information technologies, and are cautiously open to change as long as this does not jeopardize the essentials of their traditional artisanship.

Insight 9C: Government policy frameworks which aim to assist the indigenous artisan sector to develop as part of the knowledge economy need to be subtle and holistic. In this sector particularly introduction of ICTs and other new technologies will succeed only if adopted "bottom up" and not imposed from above.

Insight 10A: Almost all nonuser interviewees were aware of the importance of information and knowledge in the demanding contexts of life and work, but computers and the Internet were not seen as directly relevant to this group. Most of the older nonusers were not interested in using ICTs, but some said that they might encourage their children to use ICTs. Younger people appeared to be more enthusiastic about use of the Internet, although some reported initial fear of using the technology for the first time, and needed the encouragement of peers to assist them in making a start.

Insight 10B: Television and peers were identified by the sample as their main source of information, TV providing various kinds of general information, peers providing specific information such as job availability and knowledge needed for work and learning. Nearly half of those interviewed reported that they could use the telephone for exchanging information with their peers about the market or jobs. There was a small awareness in the study group of radio and communal government and social organizations as possible information sources. Newspapers and books, and CPCPs were not identified by the group of ICT nonusers as channels of information.

Insight 10C: The kinds of factors that were viewed by the nonusers as obstacles to accessing information on the Internet varied with the different occupations, residential localities, ages and economic status of the individuals. The financial costs of ICTs; the lack in rural areas of availability of telecommunication infrastructure and Internet services; time constraints; educational limitations; perceived lack of skills and knowledge; perceptions of the Internet as being non-relevant to work; and a lack of awareness of potential benefits were obstacles identified during the interviews.

Insight 10D: The villagers in the nonuser group viewed schooling beyond a basic level as desirable for a better future, but reported that it was difficult to attain because of factors such as the distance of schools, transportation and terrain difficulties, and the costs of bicycles, clothes and books.

Insight 11A: Two kinds of computer and Internet use were identified in the study – "general users" comprising a group of people under the age of 35 years with diverse occupations and relatively high educational levels, and "gamers" a group of school and university students aged 21 years and under, who played online computer games as their primary form of computer use. The five main places where interviewees used a computer were Internet cafés, homes, workplaces, school, and friends' houses and PTC3s. Internet cafés were the most popular places where users could come to play games, chat, surf the Internet, and make international phone calls at low rates.

Insight 11B: Almost all of the ICT users had at least an email account, all with Yahoo; however activities on which users said they spent most of their time were chatting, playing online multiplayer games, searching for news or information, downloading online music, downloading educational materials, playing solo games, work, other entertainment, creating online content, making friends and relaxing. Bachelor's degree holders spent more time on work, study, news and information, and less time on chat and games. University students were the most active online users who exploited many functionalities of computers and the Internet, with relatively high usage rates for email, chat, and multiplayer games, followed by study, online music and searching news and information. University students also created online content.

Insight 11C: The Internet and computers have radically changed modes of communication, study, work and recreation. Technological advantages of the Internet were perceived by interviewees to include the availability of up to date information, convenient ways of processing and storing information, the availability of content and software at any time and place, the ease of communication, and the availability of different and challenging kinds of entertainment.

Insight 11D: Challenging issues relating to ICT use were seen to include strain injuries and other health problems, inappropriate content on the Internet, the lack of reliability of Internet information and often the costs of retrieving information, a shortfall in the quantity and quality of Vietnamese content, inadequacies of the infrastructure and concerns about Internet security, the role of government measures in controlling Internet use, the need for English language competency, IT literacy including shortfalls in Internet searching techniques and the role of educational institutions, and the disadvantages experienced by rural ICT users.

Insight 12A: Positive aspects of online gaming include its role in bringing people together and in attracting people to enter the Internet for broader purposes; the capacity of games to provide useful learning experiences; the opportunity to improve English language skills when playing games in the English language; and the use of games forums as a vehicle for meeting other gamers face to face. Negative aspects of online gaming include the social impacts of the common theme of violence, and also the possibility of addiction.

Insight 12B: While online gaming provides different kinds of business opportunities in Vietnam large Vietnamese gaming software companies have been faced with the difficulty of competing with overseas suppliers, and with each other.

Insight 12C: Key benefits of virtual communities are that they allow individuals to present their knowledge and experience on the web, and to learn from, or enjoy the contributions of others; connection between a community's participants is not limited by physical space limitations, time, personal background, race or geography; and a diverse range of services, activities and topics are available to meet different needs.

Insight 12D: Challenges in relation to virtual communities include the issue of trustworthiness and integrity in the creation of virtual relationships; the dangers posed by illegal online activities and extremists; the real costs of virtual communities to those who act as managers and hosts; and the extent to which virtual community participants can be relied upon as sources of assistance to individuals.

Insight 13A: The ICT industry ranges from Internet café micro-businesses to large foreign and state-owned corporations. In addition to Internet cafes, strategy for rural service provision is largely based on CPCPs (contracted out to micro-business private contractors) and PTC3s (directly operated by government). There remains scope for government policy to facilitate the closing of gaps through public-private sector initiatives.

Insight 13B: Local ICT companies have greatly improved their capacity to provide services to customers, to the point of dominating the market – all small and medium IT projects are undertaken by Vietnamese companies. Local companies are also providing some services to overseas markets. Nevertheless there are still issues concerning state monopolies, project development expertise and due diligence. Also access to capital for startup and expansion of enterprises lacks clear guidelines.

Insight 13C: There is a major move towards content based products and services (including e-news media and e-forums), although the original focus of the Vietnamese ICT industry on software development, hardware sales and infrastructure services remains strong. Major issues affecting the progress of software and content development especially relate to open source versus proprietary business models and related legislative frameworks. Outsourcing is becoming an ever-more prominent feature of ICT service provision.

Insight 13D: The deregulation in the telecommunications sector has facilitated growth in infrastructure and brought down prices. However there is a wide difference between the availability and reliability of high bandwidth communication in the major cities compared to rural Vietnam.

Insight 13E: Different views were reported in regard to levels of human resources available in the IT sector, the quality of formal education, and existing projects which provide supplementary training for new graduates.

17.2.3 Insights from theorizations

Insight 14A: If declining sustainability is conceptualized as increasing entropy (i.e. increasing disorder or randomness in a system), it is possible demonstrate through logical argument that in organic beings and communities information and knowledge acts to reduce the rate of entropy and increase sustainability.

Insight 14B: From the perspective of developed countries it may seem that a knowledge economy relieves exploitation of the physical environment, but in fact environmental damage is shifted to less powerful developing economies through WTO-style price interventions. Nevertheless switching to a knowledge economy is the best way to expand the knowledge accumulation in all capitalist economies to buy time and options for the future (retard entropy in the economic sphere).

Insight 14C: Substitution – whether of knowledge resources for non-renewable physical resources (e.g. use feedback monitoring to vastly increase efficiency of use), or of renewable resources for non-renewable resources – is the core challenge in sustainable economic development. The pessimistic scenario is that substitution fails and the world descends into turbulence with violence, war, and massive environment-induced emigration at the global scale. The positive scenario is characterized by the triumph of substitution, captured in Georgescu-Roegen words "the entire stock of natural resources is not worth more than a few days of sunlight" (Georgescu-Roegen, 1971, p. 21).

Insight 14D: While the problem of environmental sustainability is global, each nation needs both collaborative and competitive strategies to achieve an optimal national trajectory on sustainable economic development. This is because powerful nations can more easily influence the price-intervention settings in the world trade order governing the exchange of renewable knowledge resources for non-renewable physical resources.

Insight 15A: The long-respected doctrines of Adam Smith and David Ricardo which hold that comparative advantage is best achieved if all countries specialize in exports that they can produce at the lowest cost, appear to be inadequate in relation to the global knowledge economy and sustainable development. The present author argues that human-made, knowledge intensive products in fact appear to have a structural advantage over natural resource based products in the world trade order.

Insight 15B: The present author proposes a new mathematical formula (the Natural-Human ratio - N/H ratio) for terms of trade. According to this formula, terms of trade are composed of the trade protection not just of one country but both parties, together with the real values of human-made and natural components.

Insight 15C: Scientific studies show that terms of trade for primary products have continued to deteriorate since 1870, except in rare periods of history (during the world wars or oil crises) when primary products reveal their real value with higher value of the N/H ratio. When oil demand of the world, especially from the U.S. and the growing Chinese economy, sharply surged in the chaos of the Iraq war, oil suddenly broke the artificially low human-made H/N ratio to partly reveal its real economic value e.g. rising from USD20 per barrel to USD54 on October 12, 2004 (Barbajosa, 2004) and up to about USD100 in 2007.

Insight 15D: The systemic, structural advantages of human-made resources over natural resources in the world trade order pose particular difficulties for developing countries such as Vietnam in achieving sustainable economic development, and lend credence to accusations that the global knowledge economy is a new form of imperialism.

Insight 16A: Knowledge promotion is a concept that captures the need to establish a self-generating dynamic for the growth of a national knowledge economy, powered both "top-down" and "bottom-up".

Insight 16B: Knowledge promotion metatypes such as those based on the Coconut, Orchid, Vine and Rain Forest analogies can be explanatory in the review and analysis of situations where knowledge-based solutions may provide effective courses of action.

Insight 16C: In an adaptation of Porter and Millar's Information Intensity Matrix, and updating core Marxist concepts, the desired outcome of knowledge promotion policy at a national level can be theorized as: optimization of value-adding by co-ordinating decisions on human-made capital intensity (social forces of production), with decisions on price intervention (social relations of production).

17.3 Suggested guidelines for future policy development on ICTs knowledge and sustainable economic development

17.3.1 Policy guidelines from grounded accounts

Making explicit the interconnection of ICTs, ecosystem protection and sustainable economic development

Chapter 7 presents a picture of a nation in transition from an economy focused locally to integration with global trade; from a government reorienting from a command to a market economy and an ecosystem rapidly degrading under short term pressures for high commercial returns. Insights 7A to 7C endorse the current directions of Vietnamese government policies as they relate to market reform locally and globally, and its current measures and aspirations on environmental protection – while urging that action in both areas be taken forward vigorously, albeit with systematic management of risk. However, the chapter 7 insights critique the loose integration of strategies on markets and sustainability with those on information, knowledge, education and training. Understanding of examples of economic success are too little, and too slowly shared through knowledge networks, and it is not clear that the vital nexus between knowledge, protection of the ecosystem and economic development are sufficiently supported through education and training at all levels.

Guideline 1: The Vietnamese government should seek to make explicit the interconnectedness of information, education and training strategies with sustainable economic development. High competencies in ICTs should increasingly be part of everyone's education and training, both initial and ongoing, from childhood through adulthood, with a view to optimizing the contribution that information and ICTs can make to sustainable economic development in all aspects of working and living.

Rural Vietnam and knowledge oases

Insights 8A to 8C deal with the extreme pressures on economic sustainability in rural Vietnam flowing from population size, spreading urbanization, environmental degradation and a too-slow pace of technological change. They draw attention to the particular problems affecting young people, women and ethnic minorities, and the urgent need to offer alternative opportunities to those who are unemployed or underemployed. They also highlight that change must be accompanied by preservation of core national values and traditions that are deeply held in rural Vietnam. Chapter 8 provides examples of dramatic economic achievement in rural Vietnam which are both built upon, and which generate highly valuable knowledge – pointing the way to a more prosperous and more sustainable economic future for rural Vietnam. However the chapter notes that such knowledge is all too often an "oasis", with know-how being too slowly shared through knowledge networks.

Guideline 2: The Vietnamese government needs to continuously review all policies affecting rural Vietnam to ensure that the nexus of information and knowledge strategies with training, enterprise and environmental strategies are optimized. A focal area for such an approach is the continuous improvement of farm management and technology for profitability with sustainability. Special attention should also be given to the livelihood needs of young people, women and ethnic minorities.

Indigenous knowledge

Chapter 9 focuses on the issue of indigenous, traditional art and craft occupations which have been nourished for centuries by resources from the natural environment of Vietnam. These are threatened, both by pressures on the natural resources that support them and by immensely increased competition for the consumer dong (or dollar) with the advent of the market economy. Insights 9A to 9C reveal that the practitioners of traditional arts and industries are cautiously open to the adoption of new production and information technologies, as long as these do not jeopardize the essentials of their traditional artisanship, and provided that the ways such technologies are deployed are initiated and guided by them, not enforced from above.

Guideline 3: The survival of traditional indigenous artisanship and industries, and the long, living tradition of knowledge transmission that they embody are a touchstone for the appropriateness and sensitivity of government policy initiatives relating to knowledge and sustainable economic development.

ICT nonusers

Insights from chapter 10, which focuses on the approximately sixty million Vietnamese people who are living without computers and the Internet, reveal that even when people are not personally users of digital technology, they nevertheless have an appreciation of its far-reaching importance in national development and daily life. Indirectly, the availability of ICTs to sources that nonusers turn to for information means that the quantity, currency and variety of information available to them is being affected by the advent of digital technology. Such sources typically include television and peers as frequent sources of information, and extend to radio, government agencies at commune level, but only seldom newspapers and books and the postal service points called CPCPs. While not seeing the utility of ICTs for themselves personally, nonusers are generally supportive of removing barriers to ICT access and competencies, especially for the young. These include obstacles to schooling at grades 10 to 12, for example the cost of textbooks, clothing and transportation.

Guideline 4: Development and implementation of policy integrating information and knowledge strategies with sustainable economic development should include and involve non-ICT users, both for reasons of social inclusion and to ensure that they can benefit indirectly from the enhancement of ICT based knowledge sharing networks and media. Also nonusers young and old should continuously be offered practical opportunities to become users, and thus direct participants in the knowledge economy.

ICT users

Chapter 11 and the insights derived from it present a picture of vigorous and diverse Internet usage in Vietnam, especially amongst younger people who form the majority of the population. Most notably Internet cafes, but also homes, workplaces, schools, friends' houses and post and telecommunication centers are places where computers are used. Game playing, especially online multiuser gaming is a major ICT based activity, and other entertainment uses, such as music downloading also figure prominently. Advantages seen in the use of ICTs include access to current information, convenience in processing and storing information, and the availability of content and software at all hours and anywhere. Disadvantages included strain injuries and other health problems, costs, unreliability of network connections, security, inappropriate content, shortage of Vietnamese content, inadequate English language competency, and shortfalls in ICT skills, for example information searching skills. Finally there were concerns about the extent of government measures for control of Internet use.

Guideline 5: Current government policy aimed at strengthening infrastructure throughout Vietnam should be sustained. In addition, a range of policy measures are required to:

- expand the amount of good quality Vietnamese content;
- develop Internet-related skills, ranging from information searching to English language competencies;
- develop discernment about trustworthiness and quality of content;
- raise awareness of ICT-related health risks and their counter-measures;
- expand personal scope of action to use the resources of the Internet freely as quickly as the social, political and historical context of Vietnam can allow.

The virtual worlds of gaming and Internet communities

Chapter 12 provides an overview of computer gaming and virtual communities. With the advent of multiplayer online games, these two vigorous phenomena in the Vietnamese ICT landscape have a considerable degree of overlap. Insights 12A to 12D highlight the strengths and potentialities of the gaming and virtual community movements for the development of personal and national ICT related capacity. Knowledge-sharing habits and skills developed in virtual communities are highly relevant to grassroots engagement with issues such as sustainable economic development. There are also serious dangers relating to both gaming and online community activity. Gaming addiction is a major concern. In virtual communities there are challenges of trustworthiness and integrity in the creation of virtual relationships and dangers posed by illegal online activities and extremists.

Guideline 6: Government strategy on ICTs should seek to leverage the confidence, skills and knowledge sharing habits cultivated through mulituser gaming and online community participation in the cause of sustainable economic development.

ICT industry

Insights 13A to 13E profile the extent and energy of the Vietnamese ICT industry, as well as issues and challenges facing its further development. A particular challenge is that access to capital for startup and expansion of enterprises lacks clear guidelines. Enterprises large and small are providing a wide range of ICT goods and services in the domestic market, and are even making inroads internationally. A feature of the current scene is an increasing focus by the industry on content production. The state of the industry demonstrates an accelerating readiness for Vietnam's development as a knowledge economy.

Guideline 7: Policy frameworks for the ICT industry should encourage initiatives that address issues of ecosystem protection and sustainable economic development. These might range from engineering dimensions of the industry, for example environmental sensing and monitoring, to facilitation of virtual forums and action groups. Guidelines for accessing industry support schemes provided by government should be clear, practical and well publicized.

17.3.2 Policy guidelines from theorizations

Mind and mass: fundamental rationale for integrated policy linking information, knowledge and sustainable economic development

The core theoretical insight that justifies the creation and maintenance of policy that simultaneously addresses the three key areas of knowledge and ICT strategy, environmental sustainability and economic development is provided by Insight 14A, namely that declining sustainability is cognate to the displacement of order by randomness as explained by the concept of entropy. Economic decline resulting from environmental degradation is comparable to the notion of "heat death" in thermodynamics. The argument made in chapter 14 is that information acts to slow the entropy of declining sustainability because greater knowledge enables tighter monitoring and control of resource use and supports innovation towards greater efficiency in resource use. In these ways information can "stretch" the period before key resources are exhausted.

Insight 14B (extending Insight 2D) conveys that the apparent displacement of "dirty" physical production by "clean" knowledge production is a false dawn. In something like a zero sum shift, production in developed countries has become less polluting but the environmental damage has merely shifted to the developing world in the context of globalization. Thus Insight 14C explains why the stretching of time through more effective information and knowledge deployment is so important. Humankind needs to "buy time" for genuine substitution technologies to be devised and applied, as captured in Georgescu-Roegen's dictum "the entire stock of natural resources is not worth more than a few days of sunlight". In other words, if truly effective technology could be introduced for substituting solar energy for fossil fuels sustainable economic development could continue without the self-defeating destruction of the world's ecosystem.

Guideline 8: The integrated development of policy for information and knowledge on one hand and sustainable economic development on the other should be premised on an understanding that more effective deployment of information and knowledge makes more likely survival until authentic substitute resources can be brought into play.

Sustainable development, permanent scarcity and the centrality of price intervention

It is a truism that there is no such thing as a perfect market. This is especially so at the dawning of the information age where national power deriving from the past, not least the age of empire and the western industrial revolution is used to set up terms of trade that favor nations that are currently wealthy and powerful, and where price intervention is subtly concealed within the Trojan horse of the world trade order. Insights 15A to 15D demonstrate that in an ideal world all would benefit from a migration of labor capacity towards production of knowledge-intensive goods and services, rather than production that depletes the physical ecosystem. In a perfect market this migration would be driven by an ever accelerating rise in the price of exhaustible natural resources compared to inexhaustible knowledge resources. However as the discussion in chapter 15 of the natural/human N/H ratio demonstrates, using the example of crude oil, nations with significant economic and military power can and do intervene to suppress the price of vital exhaustible resources. In contrast, the intellectual property regime often serves to inflate the price of knowledge goods, e.g. patent licenses, beyond what a "perfect" market would justify.

Guideline 9: The reality of price intervention in the exchange of natural resources for knowledge resources needs to be recognized and accommodated even by governments fundamentally committed to an open world trade order. However it should be a priority for such governments to mutually demonstrate and act towards the reduction of such interventions in international trade relationships.

Metatypes for knowledge promotion: the importance of human information networking

Insights 16A to 16C use problems presented by the phenomenon of isolated concentrations of know-how in Vietnam to present a meta-typology of human networking. They relate these metatypes to Marx's concepts of the social forces and relations of production on one hand, and to Porter and Millar's Information Intensity Matrix on the other. Chapter 16 seeks to demonstrate how the promotion of knowledge sharing networks positively affects the capacity to add value in a national economy. This reinforces Guideline 1 concerning the benefits of knowledge creation and sharing in stretching the utility of diminishing natural resources. While the common advantages of knowledge sharing are urged, the Chapter 16 and the prior chapters it draws on recognize and uphold the need for trade secrets and proprietary knowledge, and for risk management in information-sharing regimes. Also recognized is the fundamental importance of continuing development of information and knowledge-related skills among all members of a society aspiring towards effective knowledge transfer relationships.

Guideline 10: Governments should adopt knowledge promotion strategies that encourage the sharing of information and know-how while protecting trade secrets, proprietary information, and indigenous cultural knowledge.

17.4 Future research

It is hoped that this book can assist in identifying topics for future research endeavors, and provide a starting point for them.

At a practical level, the suggested policy guidelines, if used as a basis for operational initiatives, require follow-up research to develop detailed program specifications.

At a theoretical level, it is hoped that the theorizing presented will contribute to an ongoing conceptual dialogue seeking greater power of explanation and prediction on the interplay of information, knowledge and sustainable economic development.

The issues are urgent and much work remains to be done. May the spirit of Thachsanh inspire others in their research quests for a sustainable world.

Appendices

Appendix 1: Glossary of acronyms, abbreviations, notations and terms used in the book

	Full name
AC	Annual consumption
ADB	Asian Development Bank
AFTA	Free Trade Area of the Association of Southeast Asian Nations
AR	Accessible resource
ARPA	Advanced Research Projects Agency
ASEAN	Association of Southeast Asian Nations
AUD	Australian dollar
AV	Actual value
BSA	Business Software Alliance
BTEP	Bat Trang Ceramics Export Promotion Center
Btu	British Thermal Unit
C	Level of consumption
CBPKP	Commons-based peer knowledge promotion
CBPP	Commons-based peer production
CEA	Canadian Executing Agency
CEO	Chief executive officer
CER	Conventional earth resources

CERN	Centre Europeen pour Recherche Nucleaire
CF	Capital formation
CFC	Chloro-fluoro-carbon
CIA	Central Intelligence Agency
CIDA	Canadian International Development Agency
CIO	Chief information officer, or head of the information technology or information management section
CPCP	Communal posts and culture point
CPNET	Chinh Phu Net - an intranet of the Vietnamese Government
CS	Consumption
DP	Domestic production
DPT	Department of Posts and Telematics
DWDM	Dense wave length division multiplexing technology
ec	Efficient coefficient of resource used in production
EC	European Community
EDP	Eco domestic product
EIU	Economist Intelligence Unit
ER	Earth resources
EQ	Environmental quality
E-readiness	Electronic readiness
ET	Existent threshold
EU	European Union (formerly known as European Community - EC)
EV	Existence value
EVN	Viet Nam Electricity Group
EX	Export

FDI	Foreign direct investment
FPT	Finance and Technology Promotion Group
G7	Group of Seven (developed countries)
G8	Group of Eight (developed countries)
GAIA	Global Alliance for Infostructure Advancement
GDP	Gross domestic product
GoV	Government of Vietnam
GPL	General Public License
H	Human-made input
H/N ratio	Human / nature exchange ratio
HC	Human-made capital
HR	Human resources
ICT	Information and communication technology
ICT4D	Information and telecommunication technologies for development
IK	Indigenous knowledge
IM	Import
IMF	International Monetary Fund
INEGI	Mexican Instituto Nacional de Estadistica, Geografia e Informatica
IP	Intellectual property
IPR	Intellectual property right
IT	Information technology
IT/PIAP	Information Technology Policy Implementation Assistance Project
ITM	Intangible natural input
ITU	International Telecommunication Union
IXP	Internet Exchange Provider

KE or KBE	Knowledge economy or knowledge-based economy
KM	Knowledge management
LAN	Local area network
M	Accountable natural input
MARD	Ministry of Agriculture and Rural Development
Mbd	Million barrels per day
Mb/s	Megabytes per second
MDF	Mekong Private Sector Development Facility
MDGs	Millennium Development Goals (set by the United Nations)
MMRORPG	Multiplayer online role playing game
MNC	Multinational corporation
MONRE	Ministry of Natural Resources and Environment
m.p.h	Miles per hour
MPI	Ministry of Planning and Investment
MPT	Ministry of Posts and Telematics (former), Ministry of Information and Communications (present)
N	Life length of the deposit of a natural resource
N	Natural input
NC	Natural capital
ND	Newly discovered amount of a natural resource
NGO	Non governmental organization
NPIT2000	National Program on Information Technology 1996-2000
NPV	Net present value
ODA	Official development assistance
OECD	Organization for Economic Cooperation and Development

OSC	Office of the Steering Committee
OSS	Open source software
P	Market price
P&T	Posts and telecommunications
PC	Planet capital
PDA	Personal digital assistant
PI	Price intervention
PId	Price intervention of demand
PIh	Price intervention of human-made input
PIn	Price intervention of natural input
PIs	Price intervention of supply
Pn	Population
Politburo	Political Bureau
ppm	Parts per million
PPP	Purchasing power parity
PTC1	Posts and telecommunications company level 1 or provincial level
PTC2	Posts and telecommunications company level 2 or district level
PTC3	Posts and telecommunications company level 2 or communal level
R	Discount rate
Rc	Rate of consumption of a natural resource upon the total stock
RER	Renewable/reusable earth resources
RPG	Role playing game
rpmu	Reach per million users
RR	Resource rent
RSI	Repetitive strain injury

RV	Resource value
SC58	Steering Committee on Implementation of the Directive number 58
SC-NPIT	Steering Committee of the National Program on Information Technology 1996-2000
SEC	Stock earth resources
SEEA	System of Integrated Environmental and Economic Accounting
SER	Stock earth resources
SHTP	Saigon Hi-Tech Park
SNA	Systems of National Account
SR	Spatial resources
T	Technological development level
TAC	Total actual costs of a commodity in its entire life
TD	Total demand
TEV	Total economic value
Tp. HCM	Hochiminh city
TRIPS	Trade Related Aspects of International Property Rights of the World Trade Organisation
TS	Total stock (of a non-renewable resource on the earth)
TSP	Total supply
UCER	Unconventional earth resources
UN	United Nations
UNDP	United Nations Development Program
UNEP	United Nations Environment Program
UNESCO	United Nations Educational, Scientific and Cultural Organisation
UNSTAT	United Nations Statistical Division
US or USA	United States of America

USD	United States dollar
USSR	Union of Soviet Socialist Republics
VA21	Vietnam Agenda 21
VAIP	Vietnam Information Processing Association
VCIT	Vietnam-Canada Information Technology Project
VEA	Vietnamese Executing Agency
VEIA	Vietnam Electronic Industries Association
VEPA	Vietnam Environment Protection Agency
VINASA	Vietnam Software Association
VLTK	Vo Lam Truyen Ky
VND	Vietnamese currency Dong
VNIX	Domestic Vietnam National Internet eXchange Company
VNPT	Vietnam Posts and Telecom Corporation (former) or Vietnam Posts and Telecommunications Group (present)
VTV4	Vietnam Television Channel 4
WCED	World Commission on Environment and Development
WSIS	World Summit on the Information Society
WTO	World Trade Organisation
Δ	Difference
$\lambda(t)$	Price index of a commodity at t relative to t0

Appendix 2: Provisional definitions of terms, notations and concepts used in the book

Term	Definition
Accessible resource (AR)	Level of the planet capital that human kind can exploit at a given level of human-made accumulation.
Accountable natural input (M)	Costs of production that are accountable by the SNA before 1993.
Capital formation (CF)	The net additions to the (physical) capital stock in an accounting period.
Coconut typology or knowledge oasis	Knowledge promotion identified among non-ICT users where only one person or a single company accumulates knowledge for herself without sharing that knowledge with others.
Commons-based peer promotion	A production whose "central characteristic is that groups of individuals successfully collaborate on large-scale projects following a diverse cluster of motivational drives and social signals, rather than either market prices or managerial commands (…) relies on decentralized information gathering and exchange (…) and human creativity available to work on information and cultural resources." (Dowlah, 2004, pp. 58-59).
Commons-based peer knowledge promotion (CBPKP)	A commons-based peer production to create knowledge or exchange information.
Communal posts and culture point (CPCP)	A type of tele-center or Internet kiosk initiated by VNPT to provide basic Internet, postal, communication and librarian services to communal residents in rural areas of Vietnam.
Directive #58	Vietnam Communist Party Political Bureau's Directive number 58-CT/TW on Accelerating the Use and Development of Information Technology for the Course of Industrialization and Modernization.
The Doha Ministerial conference	The Fourth Ministerial Conference of the WTO that was held in Doha of Qatar in November 2001 to negotiate on a range of subjects and other work. The negotiations included those on agriculture and services, which began in early 2000 (James, 2000)

Doimoi	The socio-economic reform program of Vietnam started from 1986.
Economically recoverable oil reserve	Oil and gas reasonably certain to be producible using current technology at current prices, with current commercial terms and government consent (United Nations et al., 2003, p. 5).
Environment quality (EQ)	Quality of the sink functions and service functions of the nature.
Environmental quality loss (EQL)	Loss of environmental quality due to the existence of a commodity in its entire life
Environmental sink functions	Functions of the eco system that "absorb the unwanted by-products of production and consumption; exhaust gases from combustion or chemical processing, water used to clean products or people, discarded packaging and goods no longer wanted. These waste products are vented into the air, water (including sea water) or are buried in landfill sites." (Beckman, 1999, p. 5)
Existent threshold (ET)	Minimum level of exploitable natural resources required for a positive growth of capital production and consumption per capita.
Explicit knowledge	"knowledge that can be expressed, formalized or made explicit to have some significant value" (wowwiki.com, n.d.).
Game player	A person who played online or offline computer games as their primary purpose for computer use.
GNU	A recursive acronym for GNU's Not Unix launched in 1984 by Richard Stallman.
Gold farmer	Game player who seizes in-game currency in a MMORPG by collecting virtual properties, such as in-game money or armor through defeating rivals within the game. She then sells the collected virtual items to other for real money (1989)
Green economy	Environmental accounting theories initiated by Pearce, Markandya and Barbier (Heeks, 2005) in 1989 to measure the effect of environmental changes on welfare.
Gyandoot	An initiative of computer kiosks in rural India (Wilson, 2002).
H/N ratio	Human / Nature exchange Ratio, measured by $\frac{H}{N}\frac{PI_h}{PI_n}$

Human-made capital (HC)	All kind of natural capitals that humankind have exploited, appropriated and used up until a given time.
Human-made Input or human-made cost	Total human labor costs that are necessary to make natural inputs into a commodity during its entire life.
Human resources	Labor force in an organization or company
Implicit knowledge	Knowledge in the human brain that is unexpressed but can be expressible (UNESCO, n.d.).
Indigenous Knowledge (IK) or traditional or local knowledge	"the large body of knowledge and skills that has been developed outside the formal educational system. IK is embedded in culture and is unique to a given location or society. IK is an important part of the lives of the poor. It is the basis for decision-making of communities in food security, human and animal health, education and natural resource management."(OECD, cited in Debowski, 2006, p. 310).
Information and communication technology (ICT)	Devices or features that facilitate humankind in processing and communicating information. ICTs can vary on different plattforms, for instance, manual, electronic, optical or organic basis.
Intangible natural inputs (ITM)	Costs of production that are not accounted as cost in SNA before 1993.
Knowledge	What is known by a person, consciously (explicit) or unconsciously (tacit or implicit) by a person. Knowledge does not always mean correct.
Knowledge economy, or knowledge-based economy	Economy where "knowledge is the main source of wealth, growth and employment, with a strong reliance on information technologies" (United Nations et al., 2003, p. 5).
Knowledge promotion	Process in an entity (a single person, a group of people, a community, a company or a nation) to promote favorable conditions and culture in which old and new knowledge can be expressed, shared, exchanged, accessed, learned, used, created or re-created by each member of that entity at a relevant, precise, sufficient and timely manner. The target of knowledge promotion is not knowledge, but different ways of

	knowing to gain value.
Listserv	Email mailing lists through which members can send mails to each other
Natural capital (NC)	Natural resource stocks, land and ecosystems (Aplin, 2002, p. 16)
Natural input (N) or natural cost	Any natural resources and environment losses due to the production, consumption and demolition of a commodity during its entire life.
Natural-Human ratio (N/H ratio)	Measured by $\frac{N}{H}\frac{PI_n}{PI_h}$
Negative price intervention zone	Zone in between the market price and the absolute value of a physical commodity
Negative surplus value	Difference between the absolute value and the market price of a physical commodity
News-online, or cyber-news	Electronic form of newspaper
Orchid typology	Knowledge promotion of top-down or political lead. Media used in this typology are often the Internet, television, radio and other mass media
Permanent scarcity era	Era when essential natural capitals are left below a minimum threshold adequate for socio-economic activities of humankind but no sufficient discoveries of new deposits or substitutes to replace them.
Planet capital (PC)	Total of mass and energy in any form on the earth planet
Positive surplus value	Difference between the absolute value and the market price of a human-made commodity
Price intervention (PI)	Socio-economic interventions that deviate market price of a commodity from its total actual costs
Price intervention of demand (PId)	Degree of responsiveness of quantity of a commodity demanded by humankind to a change in price of the commodity
Price intervention of human made input (PIh)	Price interventions which affect the exchange value of the human-made component of a commodity

Price Intervention of natural inputs (PIn)	Price interventions which affect the exchange value of the natural input component of a commodity
Price intervention of supply (PIs), or elasticity of supply	Degree of responsiveness of the level of total stock associating with the natural input ingredients of a commodity to a change in price of that commodity.
Project 112, or SAMCom Project	Project number 112/2001/QD-TTg on Computerization in State Administrative Management Project
quad	One quadrillion (1015) Btu's – British Thermal Units
Rain Forest typology, or commons based peer knowledge promotion	Knowledge promotion that is based on ICTs, especially the Internet to create and share knowledge where every member presumably has same right and can link or contribute to any other member or a group of members in the entity.
Renewable/reusabl e Earth Resources (RER)	Resources "whose stocks are clearly depleted with use, but which can be replenished, perhaps by nature but often through human intervention" (United Nations et al., 2003, p. 5)
SC58	National ICT Steering Committee on Implementation of the Directive Number 58-CT/TW
Service functions	The habitat function of nature for all living beings, such as land, air, water, and landscape (United Nations et al., 2003, p. 5)
Sink functions	The functions of nature that absorb the unwanted by-products of productions and consumption and the service functions that provide the habitat for all living beings, including mankind (United Nations et al., 2003, p. 6).
Social output	Examples of social outputs are factors relating to sustainable development, standard of living and return on social investments
Spatial Resources	Resources or energy originated from the space, such as solar beam or wind powered by the sun or tide powered by the moon
Stock earth resources (SEC), or non-renewable resources	Natural resources on the earth which only exist as a finite stock, which once used will no longer be available for future economic activity, e.g. fossil fuels.

Strong sustainability	Development that requires "all forms of capital be maintained intact independent of one another. The assumption implicit in this interpretation is that different forms of capital are mainly complementary; that is, all forms are generally necessary for any form to be of value. Produced capital used in harvesting and processing timber, for example, is of no value in the absence of stocks of timber to harvest. Only by maintaining both natural and produced capital stocks intact, the proponents of strong sustainability argue, can non-declining income be assured" (Shapiro & Varian, 1999).
Sunk costs	Costs that have already been incurred and which cannot be recovered to any significant degree (World Commission on Environment and Development, 1987, p. 43).
Sustainable Development, or Sustainability of development	"development that meets the needs of the present without compromising the ability of future generations to meet their own needs" (Pearce et al., 1989, p. 33)
Sustainable Economic Development	Economic growth in which "real GNP per capita is increasing over time and the increase is not threatened by "feedback" from either biophysical impacts (pollution, resource problems) or from social impacts (social disruption)" (Polanyi, 1958).
Tacit knowledge	"the idea that certain cognitive processes and/or behaviors are undergirded by operations inaccessible to consciousness" (International Telecommunication Union, 1998)
Teledensity	The number of landline telephones in use for every 100 individuals living within an area. (Singer, 1950)
Terms of trade	Relative index of prices of a country's exports and imports. It is calculated by dividing the index of export prices by the index of import prices. (United Nations et al., 2003, p. 6)
Thach Sanh Rice Bowl Model	A model developed by the author that identifies and integrates key factors and relationships affecting the impacts of knowledge economy, ICTs on sustainable economic development.
Total actual costs of production (TAC)	Total natural costs and human-made costs due to the existence of a commodity during its whole life cycle from initialization to dispose.

Total supply (TSP)	Total supply of goods and services
Unconventional Earth Resources	Resources which are or will be made available for human use by either new technologies or by urgent demand of the post-scarcity period
User (of ICT/Internet)	a person who had at the time been using a computer or the Internet for any purpose
Viet Kieu	a Vietnamese term to refer to Vietnamese people living overseas
Vine typology	Knowledge promotion based mainly on computer and the Internet where only one single person or business is the key actor of collecting and providing information to others. The growth of the system often lacks participation of external parties such as other individuals, government or organizations. A vine knowledge promotion thus is a single rooted system.
Vo Lam Truyen Ky (VLTK)	A Vietnamese version of "the First Myth", a multiplayer online role playing game developed by Kingsoft in China
Weak sustainability	Development that seeks to maintain from year-to-year the per capita income generated from the total capital stock available to a nation (measured in monetary terms). No regard is given to the composition of this stock, as it is assumed that all forms of capital are substitutes for one another. Weak sustainability clearly allows for the depletion or degradation of natural resources, so long as such depletion is offset by increases in the stocks of other forms of capital (for example, by investing royalties from depleting mineral reserves in factories) (National Assembly of Vietnam, 2005a).

Appendix 3: Fieldwork - First round from 13 October 2004 to 19 May 2005

Researcher: Nguyen Thanh Tuyen
PhD candidate, School of Information Management and Systems, Faculty of Information Technology, Monash University

Appendix 3.1: List of Interviewees

The following persons have been interviewed by the researcher

No	Code	Title/ Institution	Date of interview
1	S01	Chairman, Party's IT Steering Committee	28/10/2004
2	S02	Senior Advisor Hoalac Hi-Tec Park	27/10/2004
3	S03	Former Vice Minister of Industry	19/10/2004
4	S04	Researcher Center for Analysis and Forecast of Socio-economics Information, MPI	28/10/2004 & 10/11/2004
5	S05	Former deputy Chairman of the National IT Committee	21/10/2004
6	S06	Director, Computer Center, Hanoi University of Technology	26/10/2004
7	S07	Congressman, Science-Technology & Environment Com.tee. National Assembly	19/10/2004
8	S08	Chairman, Hochiminh Computer Association	15/10/2004
9	S09	CIO, Ministry of Health	29/10/2004
10	S10	General Director Office of the National IT Steering	29/10/2004

		Committee	
11	S11	Reporter, Vietnam Posts	02/11/2004
12	S12	Vietnam University of Hanoi	04/11/2004
13	S13	General Secretary Vietnam Chamber of Commerce and Industry	05/11/2004
14	S14	Web Moderator, UNDP	03/11/2004
15	S15	Sr. Sus. Dev.Ad. UNDP	03/11/2004
16	S16	Director, IT Center Ministry of Training and Education	04/11/2004 & 09/11/2004
17	S17	Chair of Science Committee, Institute of Information Technology	04/11/2004
18	S18	Director, DECENT	03/11/2004
19	S19	Partner, Vietnam Electronics & Informatics Group	15/11/2004
20	S20	Vice Director Center for Analysis and Forecast of Socio-economics Information, MPI	15/11/2004
21	S21	Health national officer UNICEF Vietnam	09/11/2004
22	S22	Senior officer, Department of Europe Market, Ministry of Trade	16/11/2004
23	S23	Vice Director, Institute of Sociology	16/11/2004
24	S24	Senior researcher, Institute of Sociology	16/11/2004
25	S25	Former CIO, Ministry of Home Affairs	17/11/2004
26	S26	Reporter, PC World Vietnam	10/11/2004
27	S27	Director of the Training and Consulting Center, Central Institute for Economic Management, Ministry	18/11/2004

		of Planning and Investment.	
28	S28	Vice General Dir. Office of ICT Committee	18/11/2004
29	S29	CIO, Vietnam National Administration of Tourism	18/11/2004
30	S30	General Director E-Commerce Dept. Ministry of Trade	19/11/2004
31	S31	Manager, SIDA S&T Cooperation Program	22/11/2004
32	S32	Senior Advisor, MPI	23/11/2004
33	S33	Institute of S&T Strategy	25/11/2004
34	S34	Import/Export, Petrolimex	25/11/2004
35	S36	CIO, MARD	30/11/2004
36	S37	Accountant, Vinhthang Ceramic Co. Ltd	30/11/04
37	S38	Chief Officer Battrang Porcelain Ceramic Association	30/11/2004 A group discussion of two persons
38		Manager, Battrang Ceramics Export Promotion Center	
39	S39	Vice Gen. Dir. of Legal Dept.	01/12/2004
40	S40	CIO, National Academy of Science and Technology	30/11/04
41	S41	Dean, Science Management Faculty, National University of Hanoi	02/12/04
42	S42	Project manager, National Office on Sustainable Development, MPI	02/12/2004
43	S43	Director, Institute of Development Studies	07/12/2004

44	S44	Dept. of American Market, Ministry of Trade	07/12/2004 www.vninvest.com
45	S45	Sr. Official, Mechanical, Metallurgical & Chemical Department, Ministry of Industry	07/12/2004
46	S46	Vice CIO, Ministry of Finance	08/12/2004
47	S47	Vice Dir. Planning and Finance, Ministry of Health	08/12/2004
48	S48	Reporter, Vnexpress.net	09/12/2004
49	S49	Editor Secretary, Vnexpress.net	14/12/2004
50	S50	Officer, S&T Division, PetroVietnam	14/12/2004
51	S51	Vice Director, NISTPASS	14/12/2004
52	S52	Vietnam Academy of Science and Technology	17 & 23/11/2004
53	S53	former director National Institute for Agriculture Planning and Projection	17/12/2004
54	S54	Institute for Industrial Strategy and Policy	21/12/2004
55	S55	National University of Hanoi	21/12/2004
56	S56	Chairman of genetics and plant breeding Vietnam Agricultural Science Institute	21/12/2004
57	S57	Former minister of Science, Technology and Environment Chair of the S&T Council	23/12/2004
58	S58	Senior advisor, MPI	22/12/2004
59	S59	Director of Aquaculture Planning Institute, Ministry of Fisheries	5& 6/01/2005
60	S60	Head of researching environment	05/1/2005

		costs Central Institute of Economic Management	
61	S61	Marketing & Sales Director Southern Rubber Industry Company (CASUMINA)	24/01/2005
62	S62	Chief Executive Longthanh Golf Club & Residential Estate Longthanh, Dongnai	25/01/2005
63	S63	Sales & Marketing Manager NamHoa Trading & Production Corp.	25/01/2005
64	S64	Northern Food Company Cuulong Food Branch Vinhlong Province	26/01/2005
65	S65	Director, Vinhlong Department of Industry Vinhlong Province	26/01/05
66	S66	Vice Director Vinhlong Food Company	27/01/2005
67	S67	Owner of an eco-Tourism	27/01/2005
68	S68	Head, Intl. Dept. Vietnam Development and Investment Bank, DBA student of Monash	29/08/2004
69	S69	vn99.net web moderator	19/05/2005
70	S70	vn99.net Admin	19/05/2005

Appendix 3.2: List of Interviewees in pilot phase from September 12 to October 13, 2004

No	Code	Occupation	Sex	Date	Description
1	PS1	Postgraduate student, Banking and Finance, Monash University	M	12-9-2004	Interview in Vietnamese
2	PS2	Postgraduate student, IT Monash University	M	19-9-2004	Interview in Vietnamese
3	PS3	Postgraduate student, Education, Monash University	F	18-9-2004	Interview in Vietnamese
4	PS4	Postgraduate student, Public Policy Monash University	M	26-9-2004	Interview in Vietnamese
5	PS5	DBA student, Monash University	M	27-9-2004	Interview in English
6	PS6	Postgraduate student, Education, Melbourne University	F	30-9-2004	Interview in Vietnamese
7	PS7	Researcher, Vietnam Economics Institute	F	13-10-2004	Final trial to check the questions and settings at field in Vietnam

Appendix 4: Storyboard of questions for the first round of field-work 2004-2005

MONASH University **Ministry of Posts and Telematics of Vietnam**

Faculty of Information Technology
School of Information Management and Systems

Appendix 4.1: Explanatory Statement

Project Title: Policies on ICTs for a sustainable economic development in Vietnam

Date

My name is Nguyen Thanh Tuyen, and I am a government officer of the Ministry of Posts and Telematics of Vietnam who is undertaking a PhD program in the Monash University, Australia. This research project is conducted under the supervision of Professor Don Schauder, Dr. Graeme Johanson and Dr. Steven Wright of the Faculty of Information Technology at Monash University.

The aim of this project (Monash University application number 2004/605) is to collect data on Vietnam's economic development in general and information and telecommunications (ICTs) development in particular. These data are used to ground the conceptual argument: Vietnam as a developing economy and society, and 'commons-based peer production' as an emergent mode of ICT praxis (e.g. as exemplified by the open source and community networking movements). The intended outcome of the book is a methodology for generating and evaluating ICT policy in developing countries, particularly in Vietnam.

We seek your participation in a number of ways. Some of you may be delivered a self-administered questionnaire to seek your views and opinions about economic development and the use of ICTs. We will be circulating a survey to elicit your views and we may talk to you by phone or email. We especially seek your participation for in depth interviews of about 30 minutes (date to be determined) in the 1st quarter of 2006.

The information you provide will remain confidential. You can select to be

☐ disclosed,

☐ pseudonym or anonymous

in the research reports.

Only with your consent, when the project is written up for the Ministry of Posts and Telematics and Monash University, interviewed individuals and organizations will be acknowledged by name as having made a contribution to the process, and copies of the results will be distributed to you by email or via website. You are welcome to contact me about any other aggregated materials and reports which may be published in the future. Of course, in any report writing for academic purposes, anonymity of all participants, if requested, will be preserved.

If there are any issues you wish to discuss that you would prefer to discuss anonymously, you are free to do so in a one-to-one interview.

For further contact, could you allowed me to have your contact details: Name, email or telephone number.

If you have any queries, please contact

Prof. Don Schauder, School of Information Management and Systems, Faculty of Information Technology, P.O. Box 197, Caulfield East 3145. Phone: 61-3-9903 2626 Fax: 61-3-9903 2005 E-mail: don.schauder@infotech.monash.edu.au .

Nguyen Thanh Tuyen, In Vietnam: Department of Information Technology Industry, Ministry of Posts and Telematics of Vietnam, 18 Nguyen Du, Hanoi. Phone: 84-4-9436733, 61-3-9903 2445, Fax: 84-4-9436734. Email: nttuyen@mpt.gov.vn. In Australia: School of Information Management and Systems, Faculty of Information Technology, P.O. Box 197, Caulfield East 3145. Phone: 61-3-9903 2445 Fax: 61-3-9903 2005 E-mail: ttngu17@student.monash.edu.au.

Should you have any complaint concerning the manner in which this research (project number 2004/605) is conducted, please do not hesitate to contact the Monash University Standing Committee on Ethics in Research Involving Humans at the following address:

The Secretary
The Standing Committee on Ethics in Research Involving Humans (SCERH)
Building 3D
Research Grants & Ethics Branch
Monash University VIC 3800
Tel: +61 3 9905 2052 Fax: +61 3 9905 1420 Email: scerh@adm.monash.edu.au

Thank you
Nguyen Thanh Tuyen

Appendix 4.2: Informed Consent Letter

Policies on ICTs for a sustainable economic development in Vietnam

I agree to take part in the above Monash University research project. I have had the project explained to me, and I have read the Explanatory Statement, which I keep for my records. I understand that agreeing to take part means that I am willing to:

- ☐ participate in an interview with the researcher upon an agreed date.
- ☐ participate in follow up from the interview (eg comments or contribute to the report and recommendations)
- ☐ I understand that any information I provide for the interview can only be personally attributed with my permission. This consent form provides that permission.
- ☐ If there is any information I wish to contribute anonymously on a one-to-one basis, I can.
- ☐ I understand that the research results will be communicated in a written form and recommendations used in making policies on ICTs for a sustainable economic development in Vietnam.
- ☐ I also understand that participation is voluntary, that I can choose not to participate in part or all of the project, and that I can withdraw at any stage of the project without being penalised or disadvantaged in any way.

Please tick the appropriate box:

- ☐ The information I provide can be used in further research projects which have ethics approval as long as my name and contact information is removed before it is given to them
- ☐ I am agreed to be named as a contributor to the study
- ☐ I wish to check the notes of the interview or quotations used in the report drafts for accuracy

Name:

Signature:

Address (mail
or email)

Date:

Appendix 4.3: Interviewee's storyboard on ICTs for a sustainable economic development in the country

The purpose of this discussion is to gather data about sustainable economic development of the country, and the possibility to leverage knowledge and ICTs to improve the sustainability in the context of more deterioration of physical resources and technology/knowledge dependent. Sustainability is defined at the Earth Summit 1992 as that "meets the needs of the present without compromising the ability of future generations to meet their own needs". *We would like to discuss with you, as a senior expert in your field, some critical issues in the next 60 minutes.*

The interview, if kindly allowed by you, may be recorded for the purpose of researching. This recording will be kept confidentially.

General information

Date and time of interview: ....

Location: ....

Name of the interviewed institution:...

What is your main duty?

How long have you been working in this sector ....Years

What is the major mission of the institution?

Economic Development and Sustainability

What were the major achievements of your enterprises/ sector in the last 5 years?

How does the context of more deterioration of physical resources and technology/knowledge dependence affect your enterprise(s)/sector?

Are any major physical resources (e.g. energy, raw material, land, water) needed by your enterprises/ sector/(for general agencies - The country) becoming too scare or expensive?

What could be the solution?

What would be the most critical factors (e.g. legal framework, market, labour, technology, money, energy, raw material, land, water) to your enterprises/sector now and for the sustainability of your enterprises in the next 10 years?

What would be other main challenges to your enterprises/sector in the next 10 years?

Could there be any new cooperation arrangements (e.g. in finance, market, client, technology, knowledge and information as the way OPEC members do) that could assist the success your enterprises?

Are there important ways in which the international market framework (such as WTO, EU, BTA, AFTA) is unfair against your enterprises?

Could you please give your estimation (by ticking in an relevant box) of the contribution of the two key factors, i.e. physical resources vs. knowledge and other immaterial components to the value of your business

Now

	The level of contribution				
	Least valuable	Low	Medium	High	Most valuable
Physical Resources					
Raw materials, energy					
Fixed assets					
Current assets					
Salary					
Environmental Impact					
Immaterial and artificial resources					
Money & Financial factor					
Knowledge & information					
Technology					
R&D					
Management					
Market factor					
Time					
Others					

In the next 10 years

	The level of contribution				
	Least valuable	Low	Medium	High	Most valuable
Physical Resources					
Raw materials, energy					
Fixed assets					
Current assets					
Salary					
Environmental Impact					
Immaterial and artificial resources					
Money & Financial factor					
Knowledge & information					
Technology					
R&D					
Management					
Market factor					
Time					
Others					

Knowledge and ICTs

What are the roles of knowledge to support the above most critical factors for your enterprise /sector?

What are the roles of ICTs to support the above most critical factors for your enterprise /sector?

(for general agencies) The country

Is there any mission critical knowledge your enterprises/sector need protected by patents or copyright? Is this a problem and how could it be solved?

What are the main factors preventing your enterprises/sector from ICTs and KNOWLEDGE ECONOMY ?

(For general government agencies) What are the main factors preventing The country from penetrating ICTs and KNOWLEDGE ECONOMY?

What changes in organizational/national ICT policy could assist your enterprises/sector?

What changes in International ICT policy could assist your enterprises/sector?

What are the roles of your institution in improving the sustainability (to physical and immaterial factors, life style) of the country?

How your information will be used

Thank you very much for your time and support. The information you provide will be remained confidential. Would you:

- ☐ wish to be anonymous under all circumstances
- ☐ be willing to be named as a contributor to the study
- ☐ wish to check the notes of the interview for accuracy

Name of Interviewee ....
Contact address ....
....

Name of Interviewer.....................................
Contact address...
Completed time: ....

Appendix 5: List of Interviewees of the second round of field-work from January to December 2006

Appendix 5.1: List of interviewed Non-ICT Users

No	Code	Age	Date of Interview	Job	Sex	Province	Note
1	I21	71	20/2/06	Folk Artisan	M	Bacninh	Artisan of Dongho folk paintings in Bacninh province
2	I26	21	24/2/06	Student	F	HaTay	Student of a University in Hanoi
3	I27	50	25/2/06	Retired	M	Bacninh	A pensioner in Bacninh province
4	I28	48	25/2/06	Part time worker	F	Bacninh	A garment worker in Bacninh
5	I29	46	25/2/06	Folk Singing manager	F	Bacninh	Managing director & Quanho folk-song singer
6	I30	25	25/2/06	Folk singer	F	Bacninh	Quanho folksong singer
7	I31	34	25/2/06	Folk singer	M	Bacninh	Quanho folksong singer
8	I35	34	7/3/06	Farmer	F	Haiphong	Farmer in Doson town
9	I36	34	7/3/06	Head of Farm	M	Haiphong	Head of a seafood breeding production in Doson town
10	I37	49	7/3/06	Farmer	M	Haiphong	Farmer in Doson town
11	I38	50	7/3/06	Fisher	F	Haiphong	Fish woman in Doson town
12 13 14	I51 (03 persons)	75 71 51	7/4/06	Farmer Farmer Farmer	M	Long An	3 farmers in a join discussion in the South of Vietnam
15	I52	51	7/4/06	Business man	M	HCM city	Owner of a seafood restaurant
16	I53	25	7/4/06	Worker	F	Quang Tri	A ex-farmer now works in Hochiminh city

Appendix 5.2: List of ICT Users

No	Code	Age	Date	Job	Sex	Location
1	I01	26	19/01/06	sociology researcher	F	Hanoi
2	I03	23	20/01/06	hair dresser	F	Hanoi
3	I04	22	20/01/06	university student	F	Hanoi
4	I05	29	24/01/06	university cark	F	Hanoi
5	I06	25	24/01/06	university cark	F	Hanoi
6	I10	28	24/01/06	Public relation	F	Hanoi
7	I12	29	25/01/06	government officer of Ministry of Posts and Telematics	F	Hanoi
8	I22	18	25/1/06	high school student at a rural Internet cafe in Ho Town	M	Bacninh
9	I23	35	20/2/06	manager of Dong Ho, Showroom, 3rd son of I21	M	Bachninh
10	I34	30	6/3/06	medical doctor	M	Haiphong
11	I44	16	5/4/06	guard keeper cum operator of an Internet cafe	M	HCM
12	I45	22	5/4/06	super moderator of ACVN forum at www.acmilanvn.com	M	HCM
13	I48	26	5/4/06	A government officer in Ministry of Home Affairs	M	HCM
14	CP05 I65	15	18/12/06	Users of PTC3	F	Haiphong
15	CP06 I66	20	12/12/06	Users of PTC3	F	Hatay

Appendix 5.3: List of Game Players

No	Code	Age	Date	Sex	Province	Note
1	I07	21	24/1/06	M	Hanoi	A university student
2	I08	18	24/01/06	M	Hanoi	A high school student
3	I09	18	24/01/06	M	Hanoi	A high school student
4	I15	21	15/2/06	M	Hanoi	A jobless
5	I18	21	16/2/06	M	Hanoi	A university student
6	I19	19	16/2/06	M	Hanoi	A university student
7	I33	12	6/03/06	M	Haiphong	A secondary school student

Appendix 5.4: List of ICT Business Interviewees

No	Co de	Type of business	Date	Sex	Exper ience (in year)	Location	Position
1	I02	Net shop	20/01/06	M	5	Hanoi	Manager
2	I11	Foreign consulting	25/1/06	M	9	Hanoi	General Director, Link Development Interview in English
3	I13	Software	13/02/06	M	10	Hanoi	Director, MISA Soft Co, spec
4	I14	Netshop & shopware	15/02/06	M	5	Hanoi	Managing Director, of a company specializing in software for small shops & companies.
5	I16	E-commerce	16/02/06	F	1	Hanoi	Chief of Ecommerce Dept. Vietnam In charge of an e-commerce pilot project
6	I17	Network & cable	16/2/06	M	12	Hanoi	Director, Connectic Systems Co., Ltd – specializing in network, cable installment
7	I20	Online magazine & service	18/2/06	M	16	Hanoi	Acting Director of Planning, VASC Co.,

							specializing in software, communication and value added products on mobile phone.
8	I24	Foreign game for telephone	22/2/06	M	10	Hanoi	Deployment Manager An American company of game for mobile phone
9	I32	Netshop	6/3/06	M	1	Haiphong	Owner of an Internet shop in Haiphong city
10	I39	Open source	14/3/06	M	16	Hanoi	Director, an open source software company
11	I40	Telecom	14/3/06	F	1	Hanoi	Salesman, FPT Telecom
12	I41	Open source	21/3/06	M	9	Hanoi	Manager of Vietsoftware Training Center, Vietsoftware JS Co.
13	I42	Open source	23/3/06	M	12	Hanoi	General Manager, Tinhvan Group
14	I43	Net shop	5/4/06	M	3	HCMC	Move from a guard keeper. In HCMC
15	I46	Software solution	5/4/06	M	14	HCMC	Gen Dir of IES Co. Ltd. Helsinsky Branch

16	I47	Integrated services	5/4/06	M	13	HCMC	Chief Operating Officer, Informatics Service Provider Inc.
17	I49	Hardware retailer	6/4/06	M	13	HCMC	Vice Dir of an IT Trade & Service Ltd.
18	I50	IT Dept.	5/4/06	M	13	HCMC	Regional Project Portfolio Manager of British Petroleum Inc.
19	I55	IT Consulting	11/4/06	M	11	Hanoi & HCMC	Director ISTT Co.
20	I61 (CP 01)	CPCP dealer	12/12/06	F	7	Hatay	CPCP Dealer Phuluong, Hadong, Hatay
21	I62 (CP 02)	CPCP dealer	18/12/06	M	7	Haiphong	Doanxa, Kienthuy, Haiphong
22	I63 (CP 03)	PTC3 manager	12/12/06	F	4 months	Hatay	Phulam, Hadong, Hatay
23	I64 (CP 04)	PTC3 employee	18/12/06	F	2	Haiphong	Bangla, Doson, Haiphong

Appendix 5.5: List of ICT Government Manager Interviewees

No	Code	Date	Sex	Experienc e (in year)	Note
1	I20	17/2/06	M	10	Vice-general director of IT Office of MOST & Director of Open source project
2	I54	8/10/06	M	26	Chairman, HCMC IT Association
3	I56	12/4/06	M	12	Vice Director, National ICT SC Office
4	I57	12/4/06	F	6	Vice Gen. Dir. Telecom dept. MPT
5	I58	14/4/06	M	14	Sr. Officer, IT Industry Dept. MPT
6	I59	18/4/06	M	34	JICA senior advisor to MPT Interview in English
7	I60	19/4/06	M	27	General Director, IT Industry Dept. MPT

Appendix 5.6: Interviewees in pilot phase of the second round

No	Code	Sex	Date	Occupation	Description
1	PI1	F	05/11/2005	Master Student of IT, Monash University	Initial pilot interview as with an IT user, in English
2	PI2	F	10/11/2005	PhD Student of IT, Monash University	Pilot interview as with an IT provider, in English
3	PI3	F	14/11/2005	Pensioner	Pilot interview as with a non IT provider, in English
4	PI4	F	15/11/2005	Master Student of Finance, Monash University	Pilot interview as with an IT user, in English
5	PI5	F	21/11/2005	Master of Banking Student, Monash University	Pilot interview as with an IT user, in Vietnamese
6	PI6	M	27/11/2005	PhD, Senior Lecturer of the Nguyen Ai Quoc National Politics Institute	Pilot interview as with an IT user in public sector, in Vietnamese
7	PI7	F	22/11/2005	High school Student, Hochiminh city	Interview via Yahoo Messenger as IT user in Vietnamese

Appendix 6: Storyboard for the second round of fieldwork 2006

Appendix 6.1: Interviewee - Government Managers

The purpose of this discussion is to gather data about policy to leverage knowledge and ICTs to improve the sustainability of Vietnam. We would like to discuss with you, as a senior expert in your field, some critical issues in about 30 to 60 minutes.

The interview, if kindly allowed by you, may be recorded for the purpose of researching.

I. General information

Date and time of interview: ....

Location: ....

Name of the interviewed institution:...

What is your main duty?

Sex □ M □ F

How long have you been working in this sector ....Years

What is the major mission of the institution?

II. ICT Issues

How and in what way ICT can help villagers to harness a sustainable and prosperous life?

How and in what way ICT can help villagers to maintain and promote endogenous knowledge?

Which ICT policy that government need to issue to enable farmer using ICT applications for their life and work?

Which ICT policy or program that government needs to carry out to increase the effectiveness and usefulness of the Internet and ICT applications?

Online games are profitable to game companies and Net shops but seems not have appropriate usefulness to game players, and even create a negative image of the Internet, thus, which ICT policy that government need to issue to reshape this phenomenon in a more positive way?

What policy or action should be taken to improve the telecom and Internet infrastructure?

Which strategy, policy or action should be taken in relation to Intellectual property?

How the management of ICT should be improved?

Which policy or action the government should take to about ICT procurement in government sector?

Which policy or action the government should take to improve the quality of IT high-end human?

Which action the government should take to quickly concretize and enforce the law on e-transactions?

Which ICT policy or action that government needs to take to in relation to open source?

Which ICT policy or program that government needs to carry out to increase the effectiveness and usefulness the virtual community?

Which ICT policy or action that government needs to carry out to increase the effectiveness and usefulness of the legal environment for ICT?

Thank you very much for your time and contribution.

Appendix 6.2: Interviewee - ICT Business

The purpose of this discussion is to gather data about policy to leverage knowledge and ICTs to improve the sustainability of Vietnam. We would like to discuss with you, as a manager in ICT business, some critical issues in about 30 to 60 minutes.

The interview, if kindly allowed by you, may be recorded for the purpose of researching.

I. General information

Date and time of interview: ....

Location: ....

Name of the interviewed business:...

Sex □ M □ F

What is your main duty?

How long have you been working in this sector ....Years

What is the major mission of the business?

II. ICT Issues

What are your priorities in ICT policies and plans? (Infrastructure, Financial/tax policy, Competition, Intellectual property issues, Social and culture, Network environment, Security, Technology, market, human, etc.)

What are the critical difficulties facing your Net/ ICT business (e.g. legal environment, Human resources, Financial challenges, Technology challenges, Market promotion, Network communication, Culture, Security, Capacity to provide ICT services to client, Capacity to expand, foreign factor, others)

What is your view on intellectual property rights, such as copyright, patent, business secret, copying or downloading a music MP3 or software without paying the copyright?

What are the main issues and policies for your enterprise in relation to Open source development and/or application, such as Linux or Mozilla?

12.a) Advantages

12.b) Disadvantages

Do you have any other comments that you believe are relevant to this discussion?

Thank you very much for your time and contribution.

Appendix 6.3: Interview - Internet/ICT users

The purpose of this discussion is to gather data about policy to leverage knowledge and ICTs to improve the sustainability of Vietnam. We would like to discuss with you, as a ICT user, some critical issues in about 30 to 60 minutes. The interview, if kindly allowed by you, may be recorded for the purpose of researching.

I. General information

Date and time of interview: ....

Location: .... (□ Urban/ □ Rural)

What is your occupation ....

Your age ....Year

Sex □ M □ F

Your average monthly income ....thousands VND

Education background: □Primary □ Secondary

□ High-school □Tertiary student

□ Higher □ Other (specify ....)

How long have you been using the Internet ....year

II. ICT Issues

What are the primary purposes for which you use the Internet/ ICTs (game online, chat, entertainment, study, business, other)?

What makes you use the Internet/ICT rather than other alternatives (e.g. visual rather physical)?

How do you think about the capacity that the Internet helps to build networked-peer-community?

10.a) Advantages

10.b) Disadvantages

Do you access to the Internet from: □ Netcafe □ Home □ Business □ School

□ Other (specify ....)

What difficulties do you face in using Net/ ICT business (e.g. legal environment, skills & Technology challenges, facility availability, Financial difficulties, Culture, Security, English language, others)

What is your view in Intellectual property rights, such as copyright, patent, business secret, copying or downloading a music MP3 or software, photocopying a book without paying the copyright?

Have you ever heard about open source code or free software, such as Linux or Mozilla? If yes, how do you think about using this?

7.a) Advantages

7.b) Disadvantages

Do you have any other comments that you believe are relevant to this discussion?

Appendix 6.4: Interview - Non-ICT users

The purpose of this discussion is to gather data about policy to leverage information and computer. We would like to discuss with you the issue in about 30 minutes.

The interview, if kindly allowed by you, may be recorded for the purpose of researching. This recording will be kept confidentially.

I. General information

Date and time of interview: ....

Location: .... (□ Urban/ □ Rural)

What is your occupation ....

Your age ....Year

Sex □ M □ F

Your average monthly income ....thousands VND

Education background: □ Primary □ Secondary

□ High-school □Tertiary student

□ Higher □ Other (specify ....)

II. ICT Issues

What are the primary information or knowledge that are necessary for you/your life/your career/entertainment)?

What are the primary channels through which you obtain the needed information? (TV, radio, telephone, peer, association, other – please, specify)

Do you intend to use telephone, computer or the internet?

Why don't you use?

If yes, what could be the major obstacles (e.g. skills, facility availability, Financial difficulty, Culture, Security, English, language, others)

Do you have any other comments that you believe are relevant to this discussion?

Thank you very much for your time and contribution.

Appendix 7: Storyboard on ICT policy for sustainable economic development in Vietnam

Appendix 7.1: Interview - Communal Posts and Cultural Point and Posts and Telecom 3 manager

The purpose of this discussion is to gather data about policy to leverage knowledge and ICTs to improve the sustainability of Vietnam. We would like to discuss with you, as a manager in ICT business, some critical issues in about 30 to 60 minutes.

The interview, if kindly allowed by you, may be recorded for the purpose of researching. This recording will be kept confidentially.

I. General information
Date and time of interview: ....
Location: ....
Name of the interviewed business:...
What is your main duty?....................................
II. Management Issue
How long has your CPCP been providing business ....years?
Has your CPCP been a postal office before or newly built?
Do your CPCP have separated building or share the space with other
Do the staff work full-time or part-time for CPCP?
Are the staff paid by the company/government or are they volunteers?
How much did the CPCP earn last month?

III. ICT Issues
What are the major services of CPCP defined by the government regulation?
Do you think that you are able to fulfill the expected task?
Do you do any extra thing than the expected task?
Do you provide Internet service?
Do customers have to pay for Internet access?
Does your CPCP provide any form of e-government services?
Who are major customers?
How many customers you served yesterday? averagely in the last month ....
Do customers pay for Internet and computer user? If yes, how much per hour?
What are major challenges to your CPCP? (e.g. legal environment, Human resources, Financial challenges, Technology challenges, Market promotion, Network communication, Culture, Security, Capacity to provide ICT services to client, Capacity to expand, foreign factor, others)
In your opinion, how to improve the CPCP?
Is there net café in the commune? What are advantages of CPCP?
Do you think that CPCP can cooperate with net-café in delivery of services to the community?
Do you have other opinions about the topic that we have not covered?
Thank you for your time and support.

Appendix 7.2: Interview - Communal Posts and cultural point user

The purpose of this discussion is to gather data about policy to leverage knowledge and ICTs to improve the sustainability of Vietnam. We would like to discuss with you, as a manager in ICT business, some critical issues in about 30 to 60 minutes.

The interview, if kindly allowed by you, may be recorded for the purpose of researching. This recording will be kept confidentially.

I. General information

Date and time of interview: ....

Location: ....

What is your occupation ....

Your age ....Year

Education background: □ Primary □ Secondary □ High-school

Tertiary student □ Higher □ Other (specify ....)

Average monthly income

II. ICT Issues

How long have you been using computer?

How long have you been using the Internet?

How many hour you use the Internet per week?

Do you access to the Internet from:

□ Netcafe □ Home □ Business □ School

□ CPCP □ Other (specify ....)

What are the primary purposes for which you use the CPCP here?

Do you find what you need here? What are the benefits you get from CPCP?

Do you have to pay for Internet/computer use at CPCP? If yes, how much per hour?

Is there net café in the commune?

Why do you go to CPCP? What are the advantages/disadvantages of CPCP compared with net-café?

What are difficulties you get?

In your opinion, how to improve the CPCP?

Do you have other opinions about the topic that we have not covered?

Appendix 8: Major Vietnamese ICT related policies and legal documents since 2000

Source: tabulated by the author from various sources indicated in the table

Document Type & Number	Issuer, date, source	Title	Core Issues
Law and Ordinances of the National Assembly			
Ordinance 43/2002/PL-UBTVQH10	National Assembly 25/5/2002 (National Assembly of Vietnam, 2005d)	On Posts and Telecommunication	Legal document on posts and telecommunication
Civil Code 2005 33/2005/QH11	National Assembly 14/06/2005 (National Assembly of Vietnam, 2005a)	Civil Code 2005	Articles 738: Authorship of computer program Article 747.o: Inclusion of computer program Article 766.2: Protection duration:life+50 years
Law No.51/2005/QH11	National Assembly 29/11/2005 (National Assembly of Vietnam, 2005b)	On Electronic Transaction	E-business: electronic data, message, signature, certification, contract in general and e-transaction in government sector
Law 50/2005/QH11	National Assembly 29/11/2005 (National Assembly of Vietnam,	On Intellectual Property Right	Protection of authorship of art, science and literature works, industrial invention or design, gene and breed. IPR related laws are enforcement of international treaties rather than domestic

	2005c)		demands
Law No. 67/2006/QH11	National Assembly 12/7/2006 (National Assembly of Vietnam, 2006)	On Information Technology	Legal document on use and development of IT
Strategies and Plans			
Directive 58-CT/TW	CPV Politburo 17/10/2000 (Vietnam Communist Party Poliburo, 2000)	On accelerating the use and development of information technology for the cause of industrializatio n and modernization	General ICT vision to 2020
Decree 07 /2000/NQ-CP	Government 5/6/2000 (Government of Vietnam, 2005b)	Government's Decree on development of the software industry in 2000-2005	5 years software industry plan 2001-2005
Decision 95/2002/QĐ-Ttg	Prime Minister 17/7/2002 (Government of Vietnam, 2002)	Approval of the Master Plan on Use and Development of IT to 2005	5 year national IT plan 2002-2005
Decision 331/QD-TTg	Prime Minister 2004	to approve the program on IT human resource development to 2010	IT Human resource, training and education
Decision	Prime	Use and development	Open source Plan. Key themes: Awareness on

235/QĐ-CP	Minister 2/3/2004 (Government of Vietnam, 2004c)	of Open source software in Vietnam in 2004-2008	copyright and OSS; research and train on OSS, develop OSS business model 2004-2008
Decision 191/2005/QĐ-TTg	Prime Minister 29/7/2005 (Government of Vietnam, 2005c)	Decision of Prime Minister on Proposal supporting IT application in businesses for global integration and development in 2005-2010	Government support to IT utilization in businesses
Decision 246/2005/QĐ-TTg	Prime Minister 6/10/2005 (Government of Vietnam, 2005d)	Decision of Prime Minister on approval of the strategy of information and telecommunica tion to 2010 and vision to 2020	ICT Strategy 2010 and vision 2020
Decision 32/2006/QĐ-CP	Prime Minister 7/2/2006 (Government of Vietnam, 2006a)	Approval of the Plan of Telecommunic ation and Internet Development in Vietnam to 2010	Telecom and Internet: develop a fair market, reliable, ubiquitous and advanced network and diversified services 2006-2010
Decision 51/2007/QĐ-TTg	Prime Minister 12/4/2007 (Government of Vietnam, 2007e)	Decision of Prime Minister on approval of the Program of software industry in Vietnam to 2010	Software industry to 2010
Government Regulations			

Decree 55/2001/NĐ-CP	Government 23/8/2001 (Government of Vietnam, 2001)	Management, supply and utilization of Internet Service	Regulation on Internet service
Decision 74/2006/QD-TTg	Prime Minister 27/07/2001 (Government of Vietnam, 2006b)	Establishment, structure and operation of the Vietnam Public Telecom Service Fund	Public telecommunication services: provision of public telecom services to needed communes by state funded project with the partnership of SOEs and private businesses
Decision 169/2006/QD-TTg	Prime Minister 17/07/2006 (Government of Vietnam, 2006c)	on the Regulation of IT procurement by government budget in agencies and organizations	IT procurement in public sector
Decree 64/2007/NĐ-CP	Government 10/4/2007 (Government of Vietnam, 2007b)	Use of Information technology in government agencies	ICT use in public sector
Directive 04/2007/CT-TTg	Prime Minister 22/02/2007 (Government of Vietnam, 2007a)	Prime Minister's Directive on strengthening the protection of software copyright	Protection of Software copyright
Decree 26/2007/NĐ-CP	Government 15/2/2007 (Government of Vietnam, 2007c)	Detailed regulation of digital signature for enforcement of Law on electronic transaction	Guideline on digital signature for enforcement of Law on electronic transaction

Appendix 9: Summary of challenges to and proposed solutions for the Vietnamese ICT industry

Services	Finance	Infrastructure	Client/market	HR
Net café		Telecom not reliable Infrastructure supply doe not meet faster demand	High competition brings down revenue Clients more demanding	
Copyrighted software	Few investors	Telecom not developed in rural and remote areas Online help is rarely provided Difficult to extranet with branches.	Unfair competition and direct bid, Bribe and collusion, business managers dot not understand IT needs. Market, apart from government, is small No standardization of business process	Difficult to recruit good HR
Open source software	Lack fund Difficult to get back investment			Difficult to recruit good HR Few full-time OSS contributors
IT consulting	Client can pay foreign but not national consulting service	Not much improved	Clients do not re-organize business process, corrupt	Hard to find high-level personnel
Hardware				

retailer				
E-content	Online payment is not applicable		Business not familiar with e-commerce Low trust and commitment on virtual trade Competition harder	Lack staff with e-commerce, technology and business skills
IT Training	Joint-fund & cooperation with university needed	Telecom not satisfactory		New graduates need 6-12 month retraining
Game Production		Telecom not satisfactory		Soft skills are low IT community in Vietnam is small Training system not qualified
Network & cable		DSL not available in provinces	High monopoly in telecom High rate of subscription abandonment	
CPCP	Low income	No Internet and computer Location is not convenient to clients	Local people not well informed Few customer	Low salary Employee must work every hour of the week 24x7

Integrated services			Customer's low awareness of IT service's role	Lack good HR
In-house IT unit			Complex IT projects have to use foreign suppliers	Good IT lacks good business understan ding Hard to find high-level personnel
Recommen dation to the government	Support fund for jointly training. Legalize online-payment	Allow more competition in telecom. Increase the national bandwidth. Improve the telecom & DSL in provinces	Enhance the capacity of IT managers in government in IT project management, procurement. Tightly supervise IT bid in government. Raising awareness of IT services & consulting. Review IT procurement to include consulting	Focus on top-level HR. Improve quality of IT faculty towards industry base. Partnershi p in training between universitie s and IT companies

DPT: Department of Posts and Telematics

Appendix 10: Summary of challenges to and proposed solutions for the Vietnamese ICT industry (continued)

Services	IP/OSS	Regulation & management	Recommendations by the author
Net café	Buying copyright increases cost/price Using OSS if IP tightly controlled	Government control on Internet use not applicable	Review the management of Net-cafe Encourage opening net-cafe in countryside
Copyrighted software	No appropriate copyright law IP enforcement No real need on OSS OSS not accountable, documented, cost recoverable, technical support	Government incentive IT policy not effective Low capacity in departments of Posts and Telematics.	Review number of s/w enrolment in university to have balance. Review training curriculum. Protect domestic copyright Raising awareness of copyright
Open source software	Awareness of IP has been increased in Vn. No appropriate copyright law IP enforcement No real need on OSS OSS not accountable, documented, cost recoverable, technical support	GPL does not regulates accountability	Encourage the use of OSS in government. Support OSS. Raise awareness of copyright and OSS
IT consulting		No procedure to allow IT companies to mortgage for	Raise awareness of IT services & consulting. Review IT procurement to include consulting

		getting loan	
Hardware retailer		Unclear procedure to get loan from the bank or financial support from Govt.	Publicize policy and guideline in MPT's website
E-content	Difficult to protect e-content and business idea No use of OSS e-content only sold when more protected		Legalize online-payment. Raise awareness. Pilot e-commerce. Institutionalize the role of banks, CA Relevant treatment policy to attract and retain qualified human personnel
IT Training			Review IT curriculum Support fund for join-training between universities & industry
Game Production	IP not abode -> no market in Vietnam No use of OSS due to business leakage or quality assurance	Legal tie to contractor not close in IP protection	Protect domestic copyright Add soft-skills in IT training
Network & cable		Capacity of provincial telecom manager low Unclear understanding of law on tax	Open more competition in telecom. Enhance the capacity of provincial DPTs Increase the national bandwidth. Improve the telecom & DSL in provinces

CPCP			
Integrated services			no specific comment
In-house IT unit	Do not accept risks raised by OSS.		Enhance the capacity of IT managers in government in IT project management, procurement. Tightly supervise IT bid in government
Recommen dation to the government	Encourage the use of OSS in government. Support OSS companies linkage Protect domestic copyright Negotiate with MS to have proper price policy	Concretize incentive policies for ICT industry, esp. IP, tax, loan and procurement. Publicize policy and guideline in MPT' and govt.'s websites Enhance the capacity of provincial DPTs	

References

1vn.com. (2006). Thân gửi các thành viên VN99.NET [Dear members of vn99.net]. Retrieved August 4, 2006, from http://1vn.com/

Agar, J. (2003). Constant Touch: A Global History of the Mobile Phone. UK: Icon Books.

Agence France Presse. (1995, April 4). News release Agence France Presse.

Alstyne, M. V., & Buckley, N. (2004). Why information should influence productivity. In M. Castells (Ed.), The Network Society: A Cross-cultural perspective (pp. 145-173). Cheltenham UK: Edward Elgar.

Anh Thư. (2007). Bắt nguyên phó chủ nhiệm Văn phòng chính phủ [Arrest a former vice minister of Government Office]. Sep 14 Retrieved Sep 14, 2007, from http://vnexpress.net/Vietnam/Phap-luat/2007/09/3B9FA397/

Aplin, G. J. (2002). Australians and their environment : an introduction to environmental studies (2nd ed.). Melbourne: Oxford University Press.

Appenzeiller, T. (2004). The End of Cheap Oil. National Geographic, June, 84-113.

ARD/ZDF, & Multimedia, A. (1999). Nichtnutzer von online: Einstellungen und zugangsbarrieren. Ergebnisse der ARD/ZDF-offline-studie 1999 [Online nonusers: Attitudes and access barriers. Results of the ARD/ZDF offline study 1999]. Media Perspektiven, 8, 415-422.

Arrighi, G. (1994). The long twentieth century : money, power, and the origins of our times. London ; New York: Verso.

Arrighi, G., & Silver, B. J. (1999). Chaos and Governance in the Modern World System. London: University of Minnesota Press.

Asian Development Bank. (2005, December 1). ADB Loan to Boost Viet Nam's Poverty Reduction Program. Retrieved April 17, 2007, from http://www.adb.org/media/Articles/2005/8914_Viet_Nam_poverty_reduction/

AT&T. (2006). Networking In Vietnam...Industry Growth Through Liberalization and Incentives Retrieved September 16, 2006, from http://www.att.com/globalnetworking/docs/networking_in_vietnam_att.pdf#search=%22vietnam%20national%20telecom%20infrastructure%22

Baker, C., Tumbarello, P., & Ahmed, F. (2006). Vietnam: Statistical Appendix (No. IMF Country Report No. 06/423). Washington, D.C.: International Monetary Fund.

Barbajosa, A. (2004). Oil Reaches $54 a Barrel on Concern Winter Will Drain Supplies. Energy Bulletin(October 14).

Barlow, M., & Clarke, T. (2002). Blue gold : the battle against corporate theft of the world's water. London: Earthscan.

Barrat, B. M. (1988). Away with all the Great Arches: Anderson's History of British Capitalism. New Left Review, 167, 22-51.

Barrett, S. (2003). Environment and statecraft : the strategy of environmental treaty-making. Oxford ; New York: Oxford University Press.

Barrow, J. D. (1988). The world within the world. Oxford [England]: Clarendon Press.

Bartelmus, P., & van Tongeren, J. (1994). Environmental Accounting: An Operational Perspective (No. Working Paper Series, No. 1, ST/ESA/1994/WP.I,). New York: United Nations.

Bauer, M. (1995). 'Technophobia': a misleading conception of resistance to new technology. In M. Bauer (Ed.), Resistance to new technology: nuclear power, information technology and biotechnology (pp. 97-117). Cambridge: Cambridge University Press.

Baym, K. N. (2003). Online communities, communication in. In K. Christensen & D. Levinson (Eds.), Encyclopedia of community: from the village to the virtual world (Vol. 3, pp. 1015-1017). California: Thousand Oaks.

BBC. (2007, February 27). Gore accused of energy hypocrisy BBC News.

Beck, U. (1992). Risk Society. London: SAGE Publications.

Beckman, T. J. (1999). The Current State of Knowledge Management. In J. Liebowitz (Ed.), Knowledge Management Handbook (pp. 1-5). Boca Raton: CRC Press.

Bell, D. (1973). The coming of post-industrial society : venture in social forecasting. New York: Basic Books.

Bell, D., & Kristol, I. (1981). The Crisis in economic theory. New York: Basic Books.

Benkler, Y. (2002). Coase's Penguin, or, Linux and The Nature of the Firm. The Yale Law Journal, 112, 367-378.

Bergstein, B. (2007, March 5). So much data, relatively little space: The world created 161 billion gigabytes of digital information last year. Retrieved March 8, 2007, from http://www.msnbc.msn.com/id/17472946/

Birdsall, N. (2001). Government, Population, and Poverty: A Win-Win Tale. In J. M. Harris (Ed.), A Survey of sustainable development : social and economic dimensions (pp. 128-130). Washington: Island Press.

Boisot, M. H. (1995). Information Space. London: Rouledge.

Bologna, S. (2004). The social pattern of knowledge workers: myth and reality. Paper presented at the Humanities Conference, Prato.

Bostrom, N. (2002). Existential Risks: Analyzing Human Extinction Scenarios and Related Hazards. Journal of Evolution and Technology, 9, (March).

BP. (2007a). BP Statistical Review of World Energy June 2007. Retrieved July 27, 2007, from http://www.bp.com/liveassets/bp_internet/globalbp/globalbp_uk_english/reports_and_publications/statistical_energy_review_2007/STAGING/local_assets/downloads/pdf/statistical_review_of_world_energy_full_report_2007.pdf

BP. (2007b). Statistical Review 2006. Retrieved March 27, 2007, from http://www.bp.com/multipleimagesection.do?categoryId=9011001&contentId=7021619

Braudel, F. (1984). The perspective of the world. New York: Harper & Row.

Braun, P. (2004, April 13-14). Government-Industry ICT Partnering in Regional Australia: issues surrounding top down initiatives. Paper presented at the the Australian Electronic Governance Conference, Melbourne.

Brooks, D. R., & Wiley, E. O. (1988). Evolution as entropy : toward a unified theory of biology (2nd ed.). Chicago: University of Chicago Press.

Bryman, A., & Burgess, R. G. (Eds.). (1999). Qualitative research (Vol. 1). London: SAGE.

BSA, & IDC. (2006, May). Third Annual BSA and IDC Global Software: Piracy Study. Retrieved April 12, 2007, from http://www.bsa.org/global study/upload/2005%20Piracy%20Study%20-%20Official%20Version.pdf

Bui, D. (2005). Công nghiệp game Việt Nam – "Con gà đẻ trứng vàng" [Game industry in Vietnam – a hen with gold egg] [Electronic Version]. Vietnamnet, 24/10/2005. Retrieved October 24 from http://vnn.vn /cntt2005/10/503727.

Burke, A., Franklin, H. B., Urata, K., Searcy, C., & Cutting, S. M. (2002, July 26-28). Long-term consequences of the Vietnam war: Ethical, legal policy issues. Paper presented at the The Environmental Conference on Cambodia, Laos and Vietnam, Stockholm.

Caffentzis, C. G. (1989). Clipped Coins, Abused Words & Civil government: John Locke's philosophy of money. New York: Autonomedia.

Cairncross, F. (1997). The death of distance: how the communications revolution will change our lives (1st ed.). London: Texere.

Cairncross, F. (2001). The death of distance : how the communications revolution is changing our lives (New ed.). Boston: Harvard Business School Press.

Castells, M. (2000). The Rise of the Network Society. vol.1 of the Information Age: Economy, Society and Culture (2nd ed.). Oxford ; Malden, MA: Blackwell Publishers.

Castells, M., Fernandez-Ardevol, M., Qui, J. L., & Sey, A. (2007). Mobile communication and society : a global perspective : a project of the Annenberg Research Network on international communication. Cambridge, Mass.: MIT Press.

Central Intelligence Agency. (2007). World Fact Book. Washington D.C.: Potomac Books.

Charmaz, K. (2001). Grounded Theory: Methodology and Theory Construction. In N. J. Smelser & P. B. Baltes (Eds.), International Encyclopedia of the Social & Behavioral Sciences. Amsterdam: Elsevier.

Charmaz, K. (2005). Grounded theory in the 21st century: Applications for Advancing Social Justice Studies. In N. K. Denzin & Y. S. Lincoln (Eds.), The SAGE handbook of qualitative research (3rd ed., pp. 507-535). Thousand Oaks: Sage Publications.

Chau, N. (2007). FDI vượt 15 tỷ USD [FDI over USD15B]. Retrieved Dec 7, 2007, from http://vnexpress.net/Vietnam/Kinh-doanh/2007/11/3B9FCC32/

Choi, C. Q. (2007, 25 March 2007). Sweet New Battery Runs on Sugar. Retrieved March 30, 2007, from http://www.livescience.com/technology/070325_sugar_batteries.html

Commission for Vietnamese Overseas. (n.d.). Que Huong [Motherland]. Retrieved July 16, 2005 from http://www.quehuong.org.vn/vi/nr0503 07131435/

Communications and Information Magazine. (2006). Từ 1/6: VNPT giảm cước viễn thông [From June 1, VNPT reduces telecom prices]. Retrieved June 6, 2007, from http://www.tapchibcvt.gov.vn/vi-vn/vandesukien/2006/6/ 16567.bcvt

Corbitt, B., Peszynski, K., & Intranond, S. (2004). Culture, information and code systems. Journal of Global Information Management, 12(3), 65-85.

Costanza, R., & Daly, H. E. (1992). Natural Capital and Sustainable Development. Conservation Biology, 6(1), 37-46.

Costanza, R., & Daly, H. E. (2001). Summary of Natural Capital and Sustainable Development. In J. M. Harris, T. A. Wise, K. P. Gallagher & N. R. Goodwin (Eds.), A Survey of Sustainable Development: Social and Economic Dimensions (Vol. 6, pp. 14-17). Washington: Island Press.

Costanza, R., & Folke, C. (1997). Valuing Ecosystem Services with Efficience, Fairness, and Sustainability as Goals. In G. C. Daily (Ed.), Nature's services : societal dependence on natural ecosystems (pp. xx, 392 p.). Washington, DC: Island Press.

Dahlman, C. J., Andersson, T., Organisation for Economic Co-operation and Development, & World Bank Institute. (2000). Korea and the knowledge-based economy : making the transition. Washington, D.C.: The World Bank : Organisation for Economic Co-Operation and Development.

Dahlman, C. J., Aubert, J.-E., & World Bank Institute. (2001). China and the knowledge economy : seizing the 21st century. Washington, D.C.: World Bank.

Dahlman, C. J., Utz, A., & World Bank Institute. (2005). India and the knowledge economy : leveraging strengths and opportunities. Washington, D.C.: World Bank.

Dang, B. (2005). Circular 02 on Internet Cafe Control: Positive Reactions. July 28 Retrieved August 15, 2005, from http://www.vtv.vn/vi-VN/cntt/2005 /7/60414.vtv

Dang, H. (2000). Introduction Report: Knowledge economy with the development strategy of Vietnam. In M. T. Tran & N. L. Ho (Eds.), Proceedings of Conference on Knowledge Economy and its Issues to Vietnam (pp. 17-32). Hanoi: Communist Publishing.

Dang, N. D. (2000). Knowledge economy and the goal of industrialization and modernization of Vietnam in a 2020 vision. In M. T. Tran & N. L. Ho (Eds.), Proceedings of Conference on Knowledge Economy and its Issues to Vietnam (pp. 94-99). Hanoi: Communist Publishing.

Dang, N. P. (2004, December 16). Phát triển đô thị bền vững về môi trường ở Việt Nam [Urban development to sustain environment in Vietnam] Paper presented at the National Conference on Sustainable Development, Hanoi.

Dang, N. V., & Le, D. D. (2004). Dân số và phát triển ở vùng dân tộc thiểu số [Population and development with ethnic miorities people]. In P. Gubry, Nguyen Huu Dung. & P. T. Huong (Eds.), Dân số và phát triển ở Việt Nam [Population and Growth in Vietnam] (pp. 535-556). Hanoi: Nha xuat ban the gioi, ILSSA.

Dang, V. N., & Le, D. D. (2004). Dân số và phát triển ở vùng dân tộc thiểu số [Population and development with ethnic miorities people]. In P. Gubry, H. D. Nguyen & T. H. Pham (Eds.), Dân số và phát triển ở Việt Nam [Population and Growth in Vietnam] (pp. 535-556). Hanoi: Nha xuat ban the gioi, ILSSA.

Dapice, D. O., & Fellow, S. (2003). Vietnam's Economy: Success Story or Weird Dualism? A SWOT Analysis. Retrieved April 17, 2007, from http://ase.tufts.edu/econ/papers/200315.pdf

Deardorff, A. V. (1990). Should Patent Protection be Extended to All Developing Countries? The World Economy, 13(4), 497-507.

Debowski, S. (2006). Knowledge Management. Singapore: John Wiley & Sons Australia, Ltd.

Department of Foreign Affairs and Trade, A. (2007). Australia's Export Performance 2006. Retrieved Dec 10, 2007. from http://www.dfat.gov.au/publications/stats-pubs/downloads/Aus_export_performance06.pdf.

Diaz-Bordenave, J. (1976). Communication for Agricultural Innovations in Latin America. In E. M. Rogers (Ed.), Communication and Development: Critical Perspectives. Beverly Hills: Sage Publications.

DiBona, C., Cooper, D., & Stone, M. (2006). Open Sources 2.0 : the continuing evolution. Beijing ; London: O'Reilly.

Dicken, P. (1992). Global shift : the internationalization of economic activity (2nd ed.). London: Paul Chapman Pub.

Dinh, H. (2004). Mười năm, chín lần giảm cước di động...[Over ten years, nine times telecom tariff dropped]. Retrieved June 9, 2007, from http://vietnamnet.vn/cntt/2004/08/226932/

Do, T. T. (2006). Thư của Bộ trưởng Bộ Bưu chính, Viễn thông gửi CBCC CNV toàn Ngành nhân kỉ niệm 65 năm lời dạy của Bác Hồ đối với Thông tin liên lạc và 61 năm ngày Truyền thống Ngành (15/8/1945-15/8/2006) [Letter of the Minister of Posts and Telematics to staff and employees in the industry on occasion of the 65th years to realize Uncle Ho's instruction about information and communications and the 61st years of the industry anniversary day (15/8/1945-15/8/2006)]. Retrieved August 29, 2006, from http://mic.gov.vn/details.asp?Object=211054830&news_ID=2882045

Donath, J. (1999). Communities in cyberspace: Routledge.

Dowlah, C. A. F. (2004). Backwaters of global prosperity : how forces of globalization and GATT/WTO trade regimes contribute to the marginalization of the world's poorest nations. Westport, Conn.: Praeger.

Drachmann, A. G. (1963). Fragments from Archimedes in Heron's Mechanics. Centaurus, 8, 91-146.

Droege, P. (1997). Intelligent environments : spatial aspects of the information revolution. Amsterdam ; New York: Elsevier.

Drucker, P. F. (1969). The age of discontinuity : guidelines to our changing society. London: Heinemann.

Drucker, P. F. (1993). Post-Capitalist Society. Oxford: Butterworth-Heinemann Ltd.

Drucker, P. F. (1998). From Capitalism to Knowledge Society. In D. Neef (Ed.), The Knowledge Economy (pp. 15-34). Boston: Butterworth-Heinemann.

Dutta, S., & Jain, A. (2004). The Networked Readiness Index 2003-2004. Retrieved September 27, 2004, from http://www.weforum.org/pdf/Gcr/GITR_2003_2004/Framework_Chapter.pdf

Economist Intelligence Unit. (2007). Vietnam: Telecoms and technology forecast. Retrieved July 31, 2007, from http://www.ebusinessforum.com/index.asp?layout=rich_story&doc_id=10993&country_id=VN&title=Vietnam%3A+Telecoms+and+technology+forecast&channelid=4&categoryid=30&country=Vietnam

Eilperin, J. (2005, July 2, 2005). Climate Plan Splits U.S. and Europe.

Einstein, A. (1951). The meaning of relativity (5th ed.). London: Methuen.

EIU. (2006). The 2006 e-readiness rankings. Retrieved July 1, 2006, from www.eiu.com/2006eReadinessRankings

El Serafy, S. (1979). The Oil Price Revolution of 1973-1974. The Journal of Energy and Development,, 4(2), 273-290.

El Serafy, S. (1981). Absorptive Capacity, the Demand for Revenue and the Supply of Petroleum. The Journal of Energy and Development,, 7(1), 73-88.

Elliott, D., & Armson, R. (1986). Kondratieff cycles. Milton Keynes [Buckinghamshire]: Open University Press.

Elliott, L. M. (2004). The global politics of the environment (2nd ed.). Washington Square, N.Y.: New York University Press.

Elmer, L. (2002). ICTs for Sustainable Development in Vietnam: a Situational Analysis and Conceptual Framework. Working paper for UNDP Hanoi. Hanoi: UNDP.

Emmanuel, A. (1972). Unequal exchange; a study of the imperialism of trade. New York: [Monthly Review Press.

Energy and Environment Directorate. (2002). U.S. Energy Flow — 2002. Retrieved June 15, 2007, from https://eed.llnl.gov/flow/02flow.php

Enns, H. G., Ferratt, T. W., & Arasad, J. P. (2006). Beyond stereotypes of IT professionals: Implications for IT HR practices. Communications of the ACM, 49(4), 105-109.

Fallows, D. (2004). The Internet and Daily Life: Many Americans use the Internet in everyday activities, but traditional offline habits still dominate. Retrieved July 1, 2006, from http://www.pewinternet.org/pdfs/PIP_Internet_and_Daily_Life.pdf

Fine, M. (1992). Disruptive voices: The possibilities of feminist research. Ann Arbor: University of Michigan Press.

Flick, U. (1998). An Introduction to Qualitative Research. London: SAGE Publications.

Flick, U. (2006). An Introduction to Qualitative Research (3dr ed.). London: SAGE Publications.

Forbes, R. J. (1958). Power to 1850. In C. Singer (Ed.), A History of Technology (Vol. 4). Oxford: Oxford University Press.

Ford, D. P., & Chan, Y. E. (2003). Knowledge Sharing in a Multi-cultural Setting: a Case Study. Knowledge Management Research & Practice, 1, 11-27.

Foster, B. J., & Clark, B. (2004). Ecological Imperialism: The Curse of Capitalism. In L. Panitch & C. Leys (Eds.), Socialist Register 2004: The Imperialism Challenge (pp. 182-200). London: Merlin Press.

Fukuyama, F. (1995). Trust: The Social Virtues and the Creation of Prosperity. New York: Free Press.

GAIA Corporation. (2001, July 31). Vietnam-Canada Information Technology Project, Final Report - November 14, 1995 to March 31, 2001. Prepared for Canadian International Development Agency (CIDA Project Number: 975/19031). Retrieved April 1, 2007, from http://www.gaia.ca/completereport.pdf

General Statistics Office of Vietnam. (2006a). Average population by sex and by residence. November 15, 2006, from http://www.gso.gov.vn/default_en.aspx?tabid=467&idmid=3&ItemID=4704

General Statistics Office of Vietnam. (2006b). Number of administrative units as of 31 December 2005 by province. November 15, 2006, from http://www.gso.gov.vn/default_en.aspx?tabid=466&idmid=3&ItemID=4653

Georgescu-Roegen, N. (1971). The entropy law and the economic process. Cambridge: Harvard University Press.

Giddens, A. (1984). The Constitution Society: Outline of a Theory of Structuration. Cambridge, England: Polity Press.

Giddens, A. (1990). The Consequences of Modernity. California: Stanford University Press.

Giddens, A. (1999). Runaway world, lecture two - "Risk" - Hongkong. BBC Reith lectures Retrieved August 8, 2005, from http://news.bbc.co.uk/hi/english/static/events/reith_99/week2/week2.htm

Gilder, G. (1989). Microcosm: the quantum revolution in economics and technology. New York: Simon and Schuster.

Gillis, M. (2005). Some Neglected Aspects of Sustainable Development. In S. Asefa (Ed.), The economics of sustainable development (pp. 19-30). Kalamazoo, Michigan: W.E Upjohn Institute for Employment Research.

Glaser, B. G., & Strauss, A. L. (1967). The discovery of grounded theory; strategies for qualitative research. Chicago,: Aldine Pub. Co.

Goettmann, F., Thomas, A., & Antonietti, M. (2007). Metal-Free Activation of CO2 by Mesoporous Graphitic Carbon Nitride. Angewandte Chemie International Edition, Early View, Published online by Wiley Interscience(DOI:10.1002/anie.200603478).

Golding, P. (2000). Forthcoming Features: Information and Communication Technologies and the Sociology of the Future. Sociology, 34(1), 165-184.

Gourou, P. (1940). L'utilisation du sol en Indochine française. Paris: Paul Hartmann.

Government of Vietnam. (2000). Chiến lược dân số Việt Nam 2001 – 2010 [Population Strategy of Vietnam 2001-2010]. Retrieved Oct 11, 2006. from http://www.cpv.org.vn/details_cd.asp?topic=52&subtopic=150&id=BT1130058098.

Government of Vietnam. (2001). Nghị định 55/2001/NĐ-CP của Chính phủ về quản lý, cung cấp và sử dụng dịch vụ Internet [Government Decree No. 55/2001/NĐ-CP on Management, supply and utilization of Internet Service]. Retrieved Sep 16, 2006. from http://mic.gov.vn/details_law.asp?LawDoc_ID=100021.

Government of Vietnam. (2002). Quyết định của Thủ tướng chính phủ Phê duyệt Kế hoạch tổng thể về ứng dụng và phát triển công nghệ thông tin ở Việt Nam đến năm 2005 [Prime Minister's Decision No 95/2002/QĐ-Ttg on approval of the Master Plan on Use and Development of IT to 2005]. Retrieved Sep 16, 2006. from http://www.most.gov.vn/c_vbqp/cb_tracuu/tap6/TVTextrule.2004-10-19.4019/0410194322.doc/view.

Government of Vietnam. (2004a). Quyết định của Thủ tướng Chính phủ phê duyệt chương trình phát triển nguồn nhân lực về công nghệ thông tin đến năm 2010 [Prime Minister's decision to approve the program on IT human resource development to 2010], 331/QĐ-TTG C.F.R. (2004a).

Government of Vietnam. (2004b). Quyết định của Thủ tướng Chính phủ Phê duyệt Dự án tổng thể "Ứng dụng và Phát triển phần mềm nguồn mở ở Việt Nam giai đoạn 2004-2008 [Prime Minister's Decision No. 235/QĐ-CP on Use and development of Open source software in Vietnam in 2004-2008]. Retrieved Sepr 16, 2006. from http://mic.gov.vn/details_law.asp?LawDoc_ID=100323.

Government of Vietnam. (2004c). Quyết định số 235/QĐ-TTg về ứng dụng và phát triển phần mềm nguồn mở ở Việt Nam giai đoạn 2004-2008 [Decision # 235/QD-TTg on the Usage and Development of open source software in Vietnam in period 2004-2008]. Retrieved Aug 1, 2006. from http://www.mic.gov.vn/details.asp?Object=281033263&News_ID=26435000.

Government of Vietnam. (2004d). The Strategic Orientation for Sustainable Development (Vietnam Agenda 21). Hanoi: Northern Package Printing, Trading and Service Ltd. Co.

Government of Vietnam. (2005a). Decision of Prime Minister on the establishment of the National Council of Sustainable Development. Retrieved Sep 9, 2006. from http://www.va21.org/uploads/File/NCSD/Decision_NCSD_1032.pdf.

Government of Vietnam. (2005b). Nghị quyết của Chính phủ về xây dựng và phát triển công nghiệp phần mềm giai đoạn 2000 - 2005 [Decree of the Government on development of the software industry in 2000-2005]. Retrieved Apr 20, 2007. from http://mic.gov.vn/lawfiles/VB1180363615.doc.

Government of Vietnam. (2005c). Phê duyệt Đề án "Hỗ trợ doanh nghiệp ứng dụng công nghệ thông tin phục vụ hội nhập và phát triển giai đoạn 2005 -

2010 [Decision of Prime Minister on Proposal supporting IT application in businesses for global integration and development in 2005-2010]. Retrieved Sep 9, 2006. from http://mic.gov.vn/details_law.asp?LawDoc_ID=181178176.

Government of Vietnam. (2005d). Quyết định của Thủ tướng Chính phủ Phê duyệt Chiến lược phát triển công nghệ thông tin và truyền thông Việt Nam đến năm 2010 và định hướng đến năm 2020 [Decision of Prime Minister on approval of the strategy of information and telecommunication to 2010 and vision to 2020]. Retrieved Sep 9, 2006. from http://mic.gov.vn/details_law.asp?LawDoc_ID=181178176.

Government of Vietnam. (2006a). Quyết định của Thủ tướng Chính phủ Phê duyệt Quy hoạch phát triển viễn thông và Internet Việt Nam đến năm 2010 [Prime MInister's Decision on approval of telecommunication and Internet plan in Vietnam to 2010]. Retrieved Sep 9, 2006. from http://mic.gov.vn/lawfiles/100003.doc.

Government of Vietnam. (2006b). Quyết định của Thủ tướng chính phủ về thành lập, tổ chức và hoạt động của Quỹ Dịch vụ viễn thông công ích Việt Nam [Prime Minister's Decision of establishment, structure and operation of the Vietnam Public Telecom Service Fund]. Retrieved Sep 16, 2006. from http://mic.gov.vn/details_law.asp?LawDoc_ID=6574900.

Government of Vietnam. (2006c). Quyết định số 169/2006/QĐ-TTg về Quy định về việc đầu tư, mua sắm các sản phẩm công nghệ thông tin của các cơ quan, tổ chức sử dụng nguồn vốn ngân sách nhà nước [Decision # 169/2006/QĐ-TTg on the Regulation of IT procurement by government budget in agencies and organizations]. Retrieved Sep 9, 2006. from http://mic.gov.vn/details_law.asp?LawDoc_ID=181178176.

Government of Vietnam. (2007a). Chỉ thị của Thủ tướng chính phủ về việc tăng cường bảo hộ quyền tác giả đối với chương trình máy tính [Prime Minister's Directive on strengthening the protection of software copyright]. Retrieved Jul 2, 2007. from http://mic.gov.vn/lawfiles/22338877.DOC.

Government of Vietnam. (2007b). Nghị định 64/2007/NĐ-CP của Chính phủ về Ứng dụng công nghệ thông tin trong hoạt động của cơ quan nhà nước [Government Decree No. 64/2007/NĐ-CP on the use of Information technology in government agencies]. Retrieved Sep 16, 2006. from http://mic.gov.vn/details_law.asp?LawDoc_ID=19432044.

Government of Vietnam. (2007c). Nghị định Chính phủ số 26/2007/NĐ-CP về Quy định chi tiết về chữ ký số và chứng thư số; việc quản lý, cung cấp và sử dụng dịch vụ chứng thực chữ ký số [Government Decree No. 26/2007/NĐ-CP on detailed regulation of digital signature for enforcement of Law on electronic transaction]. Retrieved Sep 16, 2006. from http://mic.gov.vn/details_law.asp?LawDoc_ID=27246518.

Government of Vietnam. (2007d). Nghị định về Ứng dụng công nghệ thông tin trong hoạt động của cơ quan nhà nước [Decree on use of information technology in government agencies]. Retrieved Apr 20, 2007. from http://mic.gov.vn/lawfiles/19432044.doc.

Government of Vietnam. (2007e). Quyết định của Thủ tướng Chính phủ Phê duyệt Chương trình phát triển công nghiệp phần mềm Việt Nam đến năm

2010 [Decision of Prime Minister on approval of the Program of software industry in Vietnam to 2010]. Retrieved Sep 9, 2006. from http://mic.gov.vn/details_law.asp?LawDoc_ID=181178176.

Government of Vietnam, & Government of the United States of America. (1997). Hiệp định giữa Chính phủ CHXHCN Việt Nam và Chính phủ Hợp chủng quốc Hoa Kỳ về Quan hệ thương mại [Bilatal Agreement between the Government of Vietnam and the Government of the United States of America on Trade]. Retrieved Apr 20, 2007. from http://www.cov.org.vn/vietnam/viewNew.asp?newId=42.

Grainger, A. (1992). The role of biophysical indicators in evaluating national progress in sustainable development. Paper presented at the Global Environmental Indicators, Washington, DC.

Grainger, A. (1997). Assessing the environmental impacts of national development. In C. H. Kirkpatrick & N. Lee (Eds.), Sustainable development in a developing world: integrating socio-economic appraisal and environmental assessment (pp. 61-87). Northampton, MA: Edward Elgar Publ.

Groeschl, G., & Doherty, L. (2000). Conceptualizing culture. Cross-Cultural Management, 7(4), 12-17.

GTZ, & MPI. (2004). Phát triển bền vững các làng nghề truyền thống: Làng gốm Bát Tràng [Sustainable development of traditional trade villages: Battrang Ceramics Village]. In GTZ & MPI (Eds.), Nỗ lực hướng tới phát triển bền vững : Những kinh nghiệm thực tiễn từ các địa phương ở Việt Nam [Efforts towards sustainable development: Experiences from localities of Vietnam] (pp. 15-18). Hanoi: GTZ German & Ministry of Planning and Investment of Vietnam (MPI).

Ha, V., & Phan, A. (2005, August 18). Vụ cá da trơn: Ngành thuỷ sản lại lên dây cót' [Basa fish: The aquaculture is at risk again]. VnExpress.net Retrieved June 2, 2006, from http://vnexpress.net/Vietnam/Kinh-doanh /2005/08/3B9E13E1/

Hammond, A. L. (1998). Which world? : scenarios for the 21st Century : global destinies, regional choices. Washington, DC: Island Press.

Hardin, G. (1968). The tradegy of the commons. Science, 162(3859), 1243-1248.

Harris, J. M. (2001). Economics of Sustainability: The Environmental Dimension. Overview Essay. In J. M. Harris, T. A. Wise, K. P. Gallagher & N. R. Goodwin (Eds.), A Survey of sustainable development : social and economic dimensions (pp. 3-14). Washington: Island Press.

Harris, J. M., & Goodwin, N. R. (2001). Volume Introduction. In J. M. Harris, T. A. Wise, K. P. Gallagher & N. R. Goodwin (Eds.), A Survey of sustainable development : social and economic dimensions (pp. xxvii-xxxvii). Washington: Island Press.

Harris, J. M., Wise, T. A., Gallagher, K. P., & Goodwin, N. R. (2001). A Survey of Sustainable Development : Social and Economic Dimensions. Washington: Island Press.

Hawken, P. (1983). The next economy (1st ed.). New York (383 Madison Ave., New York 10017): Holt, Rinehart, and Winston.

Heads of G8 Nations. (2000). Okinawa Charter on Global Information Society. Retrieved February 16, 2007. from http://www.g7.utoronto.ca/summit/2000okinawa/gis.htm.

Heeks, R. (2005, September 7-9). ICTs and the MDGs: On the Wrong Track? Paper presented at the UK Development Studies Association Cenference, Milton Keynes.

Henwood, D. (2004). After the new economy. Carlton North, Vic.: Scribe Publications.

Heskett, J. (1986). Managing in the Service Economy. Cambridge: Harvard Business School Press.

Hien Tram. (2006, July 12, 2006). DN cung cấp game online sẽ không cần giấy phép OSP! [Business that provides online game needs not OSP license]. Vietnamnet.

Hirst, P. H. (2001). War and power in the 21st century : the State, military conflict, and the international system. Cambridge, UK: Polity Press; Blackwell Publishers.

Ho Tai, H.-T. (Ed.). (2001). The Country of Memory: Remarking the past in late Socialist Vietnam. Berkeley: University of California Press.

Hoang, H. (2005, August 15). 15/8: VNPT đồng loạt giảm cước các dịch vụ viễn thông! [From August 15, VNPT reduces telecom tariff]. Retrieved June 8, 2007, from http://vietnamnet.vn/cntt/2005/08/479311/

Hoang, T. (2000). Strengths and weaknesses in Vietnamese psychology in the entry into the intellectual civilization. In M. T. Tran & N. L. Ho (Eds.), Proceedings of Conference on Knowledge Economy and its Issues to Vietnam (pp. 151-154). Hanoi: Communist Publishing.

Hoang, T. Q., Constable, J., Noller, B., Dao, N. P., & Hay, A. W. M. (2002, July 26-28). Long-term consequences of the Vietnam war: Public health. Paper presented at the Environmental Conference on Cambodia, Laos and Vietnam, Stockholm.

Hoang, X. (2004). Quá độ dân số [Population transition]. In P. Gubry, D. H. Nguyễn & H. T. Phạm (Eds.), Dân số và Phát triển ở Việt Nam [Population and Development in Vietnam] (pp. 67-91). Hanoi: Thegioi [World Publishing].

Hobson, J. A. (1938). Imperialism, a study. New York,: Macmillan.

Hochiminh Computer Association. (2006). Toan Canh CNTT Vietnam 2006 [Vietnam ICT Outlook 2006]. Retrieved July 28, 2006, from www.hca.org.vn/tin_tuc/hd_hoi/nam2006/thang7/vio2006/baocaotoancanh_2006.pdf

Hof, R. D. (2005, April 27). Amazon: Revenue River, Profit Trickle. BusinessWeek Online Retrieved March 5, 2007, from http://www.businessweek.com/technology/content/apr2005/tc20050427_3567_tc024.htm

Hofstede, G. (1988). Confucius and economic growth: new trends in culture's consequences. Organizational Dynamics, 16(4), 4-21.

Hofstede, G. (1997). Cultures and organizations : software of the mind ([Rev. ed.). New York: McGraw-Hill.

Hồng Anh. (2006, Dec 26). Việt Nam có 22 triệu thuê bao di động [There are 22 million of mobile cellular in Vietnam] Retrieved May 24, 2007, from http://vnexpress.net/Vietnam/Kinh-doanh/2006/12/3B9F1AE4/

Horrigan, J. B., & Rainie, R. (2001). Online Communities: Networks that nurture long-distance relationships and local ties. Retrieved July 15, 2006, from http://www.pewinternet.org/pdfs/PIP_Communities_Report.pdf

Houghton, J. (2007). ICT adds 21bn to our trade deficit. Information Age, July, 3-8.

Houghton, J. T., Jenkins, G. J., & Ephraums, J. J. (1990). Climate change : the IPCC scientific assessment. Cambridge ; Sydney: Cambridge University Press.

Howard, P., Rainie, L., & Hones, S. (2001). Days and nights on the Internet: The impact of a diffusing technology. In B. Wellman & C. Haythorthwaite (Eds.), The Internet in everyday life (pp. 45-73). Oxford: Blackwell.

Huang, Y., Dickinson, R. E., & Chameides, W. L. (2006). Impact of aerosol indirect effect on surface temperature over East Asia Proceedings of the National Academy of Sciences of the United States of America, 103(12), 4371–4376.

Hugenholtz, P. B. (2006). Dutch Court upholds Creative Commons licence. Retrieved March 23, 2007

Hung, H. (2006, June 11). Tháng 3/2008 Việt Nam sẽ có cáp nối trực tiếp đến Mỹ [Vietnam will have a direct cable to the USA by March 2008]. VietnamNet Retrieved April 24, 2007, from http://vnexpress.net/Vietnam/Vi-tinh/2007/02/3B9F3361/

Hung, H. (2007). Chống vi phạm bản quyền bằng phần mềm giá rẻ [Anti copyright piracy by cheap software]. Retrieved April 22, 2007, from http://vnexpress.net/Vietnam/Vi-tinh/2007/03/3B9F46A9/

Hung, K. (2007). Mua bản quyền phần mềm của Microsoft [Buy the software copyright from Microsoft]. Retrieved June 3, 2007, from http://www .tuoitre.com.vn/Tianyon/Index.aspx?ArticleID=202124&ChannelID=16

Huws, U. (2003). The making of a cybertariat : virtual work in a real world (1st ed.). New York: Monthly Review Press.

ILO/JIL, & National Institute for Labour Studies. (2004). Determining the Impact of Information and Communication Technology on Decent work in the Asian and Pacific Region-VIET NAM REPORT. Hanoi: Institute of Labour Science and Social Affairs

Intel. (2006). Intel Expands Investment in Vietnam. Retrieved April 19, 2007, from http://www.intel.com/pressroom/archive/releases/20061110corp.htm

International Labour Office. (1984). World labour report (pp. v.). Geneva: International Labour Office.

International Telecommunication Union. (1998). Charging and accounting in international telecommunication services. ITU-T Recommendation D.140. Retrieved August 29, 2007, from http://www.itu.int/osg/spu/intset/focus/chrdft8.doc

International Telecommunication Union. (2006). ICT Statistics. Retrieved August 29, 2006, from http://www.itu.int/ITU-D/ict/statistics/

Internet-World-Stats. (2004). Korea: Internet Usage Stats and Marketing Report. Retrieved July 25, 2006, from http://www.internetworldstats.com /asia/kr.htm

Internet World Stats. (2007). Internet Growth Statistics: Today's road to eCommerce and global trade Retrieved February 21, 2007, from http://www.internetworldstats.com/emarketing.htm

Ito, M., Okabe, D., & Matsuda, M. (2005). Personal, Portable, Pedestrian: Mobile Phones in Japanese Life. Cambridge, MA: MIT Press.

ITU. (2006a). ICT Statistics. Retrieved August 29, 2006, from http://www.itu. int/ITU-D/ict/statistics/

ITU. (2006b). Reaching the Next Billion Users. Retrieved February 21, 2007, from http://www.itu.int/WORLD2006/forum/next_billion_users.html

James, M. (2000). Oil as a finite resource: when is global production likely to peak? Retrieved March 29, 2007, from http://www.wri.org/climate/pubs_ description.cfm?pid=2688

Jamienson, N. L., Le, C. T., & Rambo, A. T. (1998). The developmement crisis in Vietnam's mountain. East-West Center Special Report, 6(November).

Jawara, F., & Kwa, A. (2003). Behind the scenes at the WTO : the real world of international trade negotiations. New York: Zed Books.

Jenkins, R. (1992a). Industrialization and the global economy. In T. Hewitt, H. Johnson & D. Wield (Eds.), Industrialization and Development (pp. 13-40). Oxford: Oxford University Press.

Jenkins, R. (1992b). Theoretical perspective. In T. Hewitt, H. Johnson & D. Wield (Eds.), Industrialization and Development (pp. 128-166). Oxford: Oxford University Press.

Jenkins, R. (2004). Vietnam in the Global Economy: Trade, Employment and Poverty. Journal of International Development, 16, 13-28.

Johnson, J. C. (1990). Selecting ethnographic informants. Newbury Park, CA: Sage.

K.D. (2006, November 7). Công bố toàn văn cam kết WTO [Publication of full text of WTO commitment agreement]. Retrieved May 10, 2007, from http://vnexpress.net/Vietnam/Kinh-doanh/Duong-vao-WTO/2006/11/3B9F0224/

Kelly, K. (1999). New Rules for the New Economy: 10 Radical Strategies for a Connected World. New York: Penguin.

Kelly, T., & Minges, M. (2002). Vietnam Internet Case Study. from http://www.itu.int/itudoc/gs/promo/bdt/cast_int/79480.html

Kennedy, R. F. (2004). Crimes against nature : how George W. Bush and his corporate pals are plundering the country and hijacking our democracy (1st ed.). New York: HarperCollins.

Kessler, M. (2007, 8 March). Days of officially drowning in data almost upon us. USA Today.

Khanh, H. (2007, February 27). Nhiệt độ tháng 2 cao nhất trong lịch sử [Temperature in February hits a record in the history]. Retrieved March 20, 2007, from http://vnexpress.net/Vietnam/Xa-hoi/2007/02/3B9F384D/

Khong, D. (2002). Một số vấn đề dân số với phát triển các dân tộc thiểu số và miền núi Việt Nam [Some issues on population and developement of ethnic minorities in mountains of Vietnam]. In Phát triển bền vững miền núi Việt Nam, 10 năm nhìn lại và những vấn đề đặt ra [Sustainable development in mountainous areas of Vietnam: 10 years review and considerations]. Hanoi: Agricultural Publishing.

Kibirige, H. M. (1983). The Information Dilemma. A Critical Analysis of Information Pricing and the Fees Controversy, . Westport: Greenwood Press.

Kingsoft. (2006). Online Games. Retrieved July 18, 2006, from http://www.kingsoft.com/en/game.shtml

Kinh Hoang Dao. (2004, December 16). Nhà sinh thái - chiêm nghiệm và ý tưởng [Ecological houses - Experiment and thought] Paper presented at the National Conference on Sustainable Development, Hanoi.

Klare, M. T. (2004). Blood for Oil: The Bush-Cheney Energy Strategy. London: Merlin Press.

Kranzberg, M. (1985). The information Age: Evolution or Revolution. In B. R. Guile (Ed.), Information Technologies and Societal Transformation. Washington D.C.: National Academy of Engineering.

Labelle, R. (2005). ICT Policy Formulation and e-Strategy Development: A Comprehensive Guidebook. UNDP-APDIP ICT4D Series. Elsevier.

Landes, D. S. (1998). Homo Faber, Homo Sapiens: Knowledge, Technology, Growth, and Development. In D. Neef (Ed.), The Knowledge Economy (pp. 53-67). Boston: Butterworth-Heinemann.

Langlet, P. (2004). Lịch sử hình thành dân cư [History of population]. In P. Guby, H. D. Nguyen & T. H. Pham (Eds.), Dân số và Phát triển ở Việt Nam [Population and Development in Vietnam] (2nd ed., pp. 33-66). Hanoi: World Publishing, ILSSA, IMV, IRG.

Laodong. (2005, July 22, 2006). Sôi động nhập khẩu trò chơi trực tuyến [Eventful in importing online game]. Vietnamnet

Laozi, & Henricks, R. G. (1989). Lao-tzu : Te-tao ching : a new translation based on the recently discovered Ma-wang-tui texts (1st ed.). New York: Ballantine Books.

Laozi, & Merel, P. A. (1995). Tao Te Ching. Retrieved March 22, 2007, from http://www.religiousworlds.com/taoism/ttcmerel.html

Laurier, E. (2002). The region as a socio-technical accomplishment of mobile workers. In B. Brown, N. Green & R. Harper (Eds.), Wireless World: Social and Interactional Aspects of the Mobile Age (pp. 46-61). London: Springer.

Le, K. P. (2000). Tư tưởng Hồ Chí Minh soi sáLe Kha Phieu, Tư tưởng Hồ Chí Minh soi sáng con đường Đảng và nhân dân ta tiến vào thế kỷ 21 [Ho Chi Minh's ideology enlightens the way for our Party and people to enter the 21st century]. In Con đường Đảng và nhân dân ta tiến vào thế kỷ 21 [Ho Chi Minh's ideology enlightens the way for our Party and people to enter the 21st century]. Hanoi: National Politics Publishing.

Le, Q. A. (2002, March 6-8). Ngưỡng Phát triển và Quan điểm về Phát triển bền vững đối với Việt Nam [Threshold of development and viewpoint of sustainable development for Vietnam]. Retrieved May 1, 2007, from http://www.va21.org/vietnamese/index.php?param=NewsInfo&key=183

Le, Q. A., Le, T. C., Pham, N. D., Vo, Q., Nguyen, H. Y., & Nguyen, D. T. (Eds.). (2004). Vietnam: Environment and Life. Hanoi: National Political Publisher

Le, T. B. (2000). Contribution of Biotechnology to knowledge economy: status and solution. In M. T. Tran & N. L. Ho (Eds.), Proceedings of Conference on Knowledge Economy and its Issues to Vietnam (pp. 174-182). Hanoi: Communist Publishing.

Le, V. T. (2004). Chính sách dân số [Population policy]. In P. Gubry, H. D. Nguyen & T. H. Pham (Eds.), Dân số và phát triển ở Việt Nam [Population and Growth in Vietnam] (pp. 241-265). Hanoi: Nha xuat ban the gioi, ILSSA.

Le, X. T. (2000). Welcome Speech. Paper presented at the Proceedings of Conference on Knowledge Economy and its Issues to Vietnam, Hanoi.

Leadbeater, C. (1999). Living on Thin Air: the New Economy. London: Viking.

Lebo, H., Cole, J. I., Suman, M., Schramm, P., Bel, D. v., Lunn, B., et al. (2002). The UCLA Internet Report: Surveying the digital future, year one. Retrieved July 5, 2006, from http://www.digitalcenter.org/pdf/Internet ReportYearOne.pdf

Lebo, H., Cole, J. I., Suman, M., Schramm, P., Bel, D. v., Lunn, B., et al. (2004). The UCLA Internet Report: Surveying the digital future, year four: Ten years, ten trends. Retrieved July 4, 2006, from http://www.digital center.org/downloads/DigitalFutureReport-Year4-2004.pdf

Lebo, H., Cole, J. I., Suman, M., Schramm, P., Lunn, R., Aquino, J.-S., et al. (2003). The UCLA Internet Report: Surveying the digital future, year three. Retrieved July 1, 2006, from http://www.digitalcenter.org/pdf/Internet ReportYearThree.pdf

Lenhart, A., Horrigan, J., Rainie, L., Allen, K., Boyce, A., Madden, M., et al. (2003). The ever-shifting Internet population: A new look at Internet access and the digital divide. Retrieved June 20, 2006, from http://www.pew internet.org/pdfs/PIP_Shifting_Net_Pop_Report.pdf

Lenin, V. I. (1974). Imperialism, the highest stage of capitalism. In Selected Workds (Vol. 1). Moscow: Foreign Languages Publishing House.

Little, B. H., & Trepanie, C. W. (1997). Untangling the Intellectual Property Rights of Employers, Employees, Omventors amd Independant Contractors. Employee Relation Law Journal, 22(4), 49-77.

Lundin, J. (2005, June 1-3). Coordination in mobile police work. Paper presented at the Hongkong Mobility Roundtable Hongkong.

Luxemburg, R. (1951). The Accumulation of Capital. London: Routledge & Kegan Paul.

Machlup, F. (1973). The Production and Distribution of Knowledge in the United States. Princeton: Princeton University Press.

Mai, K. N., Vu, T. L., Pham, V. B., Pham, T. Y., & Nguyen, X. T. (2005). Báo cáo kết quả thực hiện đề tài nghiên cứu, đề xuất chương trình đào tạo, bồi dưỡng kỹ thuật nghiệp vụ cho nhân viên điểm bưu điện văn hoá xã [Report on the project achievements of research and recommendation on training on technical skills for employees at communal post and cultural points] (No.

31-HV-2005-RD-DT). Hanoi: Posts and Telecommunications Institute of Technology.

Malthus, T. R. (1798). An essay on the principle of population; and, A summary view of the principle of population. Harmondsworth,: republished in Penguin Books 1970.

Marx, K. (1915). Capital: a Critique of Political Economy. Volume 1. Chicago: Charles H. Kerr.

Marx, K. (1969). Theories of surplus value. London,: Lawrence & Wishart.

Marx, K., & Engels, F. (1954). Capital : a critique of political economy. Volume 3. Moscow: Progress Publishers.

Marx, K., & Engels, F. (1975). Karl Marx, Frederick Engels : Collected works, volume 7. London: Lawrence & Wishart.

Maskus, K. E. (2000). Intellectual property rights and economic development. Case Western Reserve. Journal of International Law, 32, 471.

Maslow, A. H. (1954). Motivation and personality ([1st ed.). New York,: Harper.

Mathew, D. (2004). The Fallacies of Shrimp Protectionism. June 27 Retrieved 27/05/2004, from http://www.mises.org/fullstory.aspx?control =1551

Matos, G., & Wagner, L. (1998). Consumption of Materials in the United States, 1900-1995 in Annual Reviews of Energy and the Environment, 23, 107-122.

May, C. (2000). A global Political Economy of Intellectual Property Rights: the New Enclosures? . New York: Routledge.

May, C. (2002). The Information society: a sceptical view. Cambridge: Polity.

Meadows, D. H., Meadows, D. L., Randers, J., & Behrens, W. W. (1972). The limits to growth : a report for the Club of Rome's project on the predicament of mankind. London: Earth Island, Ltd. : Pan Books.

Mensch, G. (1979). Stalemate in technology : innovations overcome the depression. Cambridge, Mass.: Ballinger Pub. Co.

Mikesell, R. F. (1993). Economic development and the environment : a comparison of sustainable development with conventional development economics. London ; New York: Mansell.

Millenniumassessment.org. (2005). Millennium Ecosystem Assessment Sysnthesis Report. Washington D.C.: Island Press.

Minh Sơn. (2004). Tất cả nước của sông Mekong đã đi đâu? [Where is the water of Mekong River?] [Electronic Version]. Retrieved March 24 from http://www.vietnamnet.vn/khoahoc/vande/2004/03/56380/.

Ministry of Agriculture and Rural Development. (2004). Báo cáo tổng kết: Chuyển dịch cơ cấu nông nghiệp và kinh tế nông thôn 20 năm đổi mới [Final report of the transition of the agricultural structure and rural economy in 20 years of Doimoi]. Ministry of Agriculture and Rural Development, Vietnam.

Ministry of Information and Communications. (2007). Tổng quan về Internet Việt Nam [Overview of the Internet in Vietnam]. Retrieved August 29, 2006, from http://mic.gov.vn/details.asp?Object=211055477&News_ID=28432725

Ministry of Information and Communications. (2009). Vietnam information and communication technology – White book 2009. Hanoi: Information and Communications Publishing House.

Ministry of Information and Communications. (2010). Vietnam information and communication technology – White book 2010. Hanoi: Information and Communications Publishing House.

Ministry of Natural Resources and Environment. (2005). Quy định chức năng, nhiệm vụ, quyền hạn và cơ cấu tổ chức của Cục Bảo vệ môi trường [Decision of the mandate, responsibility, right and structure of the Vietnam Environment Protection Agency], 14/2005/QĐ-BTNMT C.F.R. (2005).

Ministry of Public Security. (2007, March 16). Ban hành Chỉ thị phòng chống tội phạm môi trường trong thời kỳ hội nhập [Issuance of a directive on prevention and control of environmental crimes in the integration period]. Retrieved May 1, 2007, from http://www.va21.org/vietnamese/index.php ?param=NewsInfo&key=544

Ministry of Planning and Investment of Vietnam, UNDP, & Danish International Development Agency. (2003). Định hướng chiến lược tiến tới phát triển bền vững ở Việt Nam: Kỷ yếu các hội thảo tại địa phương [Strategic Orientation toward sustainable development in Vietnam: Proceedings of local conferences]. Hanoi: VIE/01/021 Project.

Mokyr, J. (1990). The lever of riches : technological creativity and economic progress. New York: Oxford University Press.

Morse, J. M. (1995). The significance of saturation. Qualitative Health Research, 5, 147-149.

Munda, G. (2001). Environmental Economics, Ecological Economics and the Concept of Sstainable Development. In J. M. Harris (Ed.), A Survey of sustainable development : social and economic dimensions (pp. 17-21). Washington: Island Press.

National Assembly of Vietnam. (1992). The Constitution 1992. Retrieved Jan 2, 2007. from http://www.luatvietnam.vn/noidung2k.asp?id=2394.

National Assembly of Vietnam. (1996). Luật Ban hành Văn bản quy phạm pháp luật [Law on promulgation of legal document]. Retrieved Jul 2, 2007. from http://www.chinhphu.vn/portal/page?_pageid=33,638900&_dad =portal&_schema=PORTAL&docid=3261.

National Assembly of Vietnam. (2001). The Constitution 1992. Amendment in 2001. Retrieved Nov 2, 2007. from http://www.luatvietnam.vn/noidung 2k.asp?id=13276.

National Assembly of Vietnam. (2005a). Bộ luật Dân sự 2005 [Civil Code 2005]. Retrieved Dec 1, 2007. from http://www.wincolaw.com.vn/vn/ resources/legaldocs/Bo%20Luat%20dan%20su%20(sua%20doi).doc.

National Assembly of Vietnam. (2005b). Luật giao dịch điện tử [Law on Electronic Transaction]. Retrieved Dec 2, 2007. from http://mic.gov.vn details_law.asp?LawDoc_ID=100321.

National Assembly of Vietnam. (2005c). Luật Sở hữu trí tuệ [Law of Intellectual Property]. Retrieved Dec 1, 2007. from http://www.most.gov.vn/most/c_vbqp/ca_luatkhcn/LUAT_SHTT-2005.doc.

National Assembly of Vietnam. (2005d). Pháp lệnh Bưu chính, Viễn thông [Ordinance of Posts and Telecommunication]. Retrieved Dec 2, 2007. from http://mic.gov.vn/details_law.asp?LawDoc_ID=100036.

National Assembly of Vietnam. (2006). Luật Công nghệ thông tin [Law of Information Technology]. Retrieved July 25, 2006. from http://mic.gov.vn/lawfiles/20734118.DOC.

National Environmental Agency. (2002). State of Environment in Vietnam 2001. Retrieved April 18, 2007. from http://www.nea.gov.vn/html/Baocao_hientrangVN_SoE/issues/biodiversity /index.htm.

Neef, D. (1998). The knowledge economy. Boston: Butterworth-Heinemann.

Nelkin, D. (1984). Science as Intellectual Property: Who Controls Research? . New York: Macmillan.

Ngo Duc Thinh. (2002). Status and some Issues of Social Life in Ethnic Minorities of Vietnam in the last 10 Years. In Sustainable Development in the Mountainous Areas of Vietnam, 10 Years Review. Hanoi: Agricultural Publishing.

Ngo, T. D. (2000). Biotechnology in Vietnam: status, perspective and solution. In M. T. Tran & N. L. Ho (Eds.), Proceedings of Conference on Knowledge Economy and its Issues to Vietnam (pp. 174-182). Hanoi: Communist Publishing.

Nguyen, C. H. (2002). Tiếp cận phát triển bền vững vùng ven biển và nghề cá nước ta [Sustainable approach to coastal regions and fishing industry]. Tạp chí Thuỷ sản [Aquaculture Journal], 6, 23-25.

Nguyen, D. (1983). The Tale of Kieu: a bilingual edition of Truyen Kieu. New Haven, [Conn.]: Yale University Press.

Nguyen, D. H. (2000). Trends of knowledge economy and impacts to international relations. In M. T. Tran & N. L. Ho (Eds.), Proceedings of Conference on Knowledge Economy and its Issues to Vietnam (pp. 39-47). Hanoi: Communist Publishing.

Nguyen, H. (2004). Sẽ có PC rẻ hơn máy tính Thánh Gióng [There will be PC cheaper than Thanhgiong computer] [Electronic Version]. VnExpress.net. Retrieved August, 10, 2007 from http://vnexpress.net/Vietnam/Vi-tinh/2004/08/3B9D53BE/.

Nguyen, H. (2006a, November 29). Những dự án máy tính giá rẻ VN 'đi về nơi xa lắm' [Cheap computer projects have gone away]. Retrieved June 5, 2007, from http://vnexpress.net/Vietnam/Vi-tinh/2006/11/3B9F0A70/

Nguyen, H. (2006b, June 29). 'Thời loạn' - game Việt Nam hay nhất năm 2006 [Wartime, the best Vietnamese game in 2006]. Retrieved June 15, 2007

Nguyen, H. (2007). Sẽ kế thừa những kết quả của Đề án 112 [Inherit the 112 Project]. Retrieved May 23, 2007, from http://vnexpress.net/Vietnam/Vi-tinh/2007/05/3B9F5E20/

Nguyên Hải., & Hằng Ngọc. (2006). Blog VN - lập ngẫu hứng, viết cũng ngẫu hứng [Vietnamese blog - Create in sudden inspiration, write also in sudden inspiration] [Electronic Version]. VnExpress.net. Retrieved June 19 from http://vnexpress.net/Vietnam/Vi-tinh/2006/06/3B9EAF53/.

Nguyen, L. T. P. (2003). Status of on-send tiger shrimp culture in the central provinces and sustainable management of the environment. In International Instutute for Sustainable Development., Internatonal Centre for Trade and Sustainable Development., Thuysan & The World Conservation Union. (Eds.), Expanding shrimp aquaculture on sandy land in Vietnam: Challenges and Opportunities (pp. 31-37). Hanoi: IUCN & IISD.

Nguyen, M. X. (n.d.). Giới thiệu về Mạng Trái Tim Việt Nam [Introduction of the Vietnam Heart Network]. Retrieved October 22, 2006, from http://www5.ttvnol.com/

Nguyen, N. C., & Sachs, D. (2003). Two cakes fit for a king: folktales from Vietnam. Honolulu: University of Hawaii Press.

Nguyen, N. T. (2001). Kinh tế tri thức, sở hữu trí tuệ và chiếm hữu tri thức: Báo cáo Quốc hội [Knowledge Economy, Intellectual Property and Knowledge Enclosure. Report to the National Assembly of Vietnam]. Hanoi: the National Assembly of Vietnam.

Nguyen, Q. A. (2000). Knowledge economy: the role of business. In M. T. Tran & N. L. Ho (Eds.), Proceedings of Conference on Knowledge Economy and its Issues to Vietnam (pp. 197-204). Hanoi: Communist Publishing.

Nguyen, Q. D. (2003). Issues shrimp production and trade, on-sand shrimp culture and arising problems. In International Instutute for Sustainable Development., Internatonal Centre for Trade and Sustainable Development., Thuysan & The World Conservation Union. (Eds.), Expanding shrimp aquaculture on sandy land in Vietnam: Challenges and Opportunities (pp. 38-47). Hanoi: Camau Publisher.

Nguyen, Q. T. (2000). Industrialization, Modernization and Knowledge based economy. In M. T. Tran & N. L. Ho (Eds.), Proceedings of Conference on Knowledge Economy and its Issues to Vietnam (pp. 89-93). Hanoi: Communist Publishing.

Nguyen, T. (1993). Binh Ngo Dai Cao [Proclamation of Victory over the Wu]. In S. L. Ngo (Ed.), Dai Viet Su Ky Toan Thu [The Complete Work of the Historical Records of Great Viet]. Hanoi: reprinted by Social Science Publishing.

Nguyen, T. T. (2002). Tentative Plan for the IT/PIAP from March to December 2002. Vietnamese Executing Agency.

Nguyen, T. T. (2003a). The Current Situation of ICT in Vietnam. Discussion paper prepared upon UNDP's request, . Hanoi: UNDP Hanoi.

Nguyen, T. T. (2003b). A Digital Opportunity Initiative: How Could ICT be Harnessed for Development and Knowledge Based Economy in Vietnam. Discussion paper prepared upon UNDP's request. Hanoi: UNDP Hanoi.

Nguyen, T. T., & Johanson, G. (2006). Culture, Knowledge and Development in Vietnam: Local Oasis or Global Empire. In F. Burstein & H. Linger (Eds.), The local and Global in Knowledge Management: Why Culture Matters (pp. 95-119). Melbourne: Australian Scholarly.

Nguyen, V. D. (2000). Vai tro cua khoa hoc co ban trong nen kinh te tri thuc [Role of basic sciences in a knowledge economy]. In M. T. Tran & N. L.

Ho (Eds.), Proceedings of Conference on Knowledge Economy and its Issues to Vietnam (pp. 142-150). Hanoi: Communist Publishing.

Nguyen, X. T. (2000). Knowledge economy: Experience of some developed countries. In M. T. Tran & N. L. Ho (Eds.), Proceedings of Conference on Knowledge Economy and its Issues to Vietnam (pp. 65-78). Hanoi: Communist Publishing.

Nhan Dan. (2007, May 25). Two Chinese software products introduced in Vietnam. Retrieved June 10, 2007, from http://english.vietnamnet.vn/tech/2007/05/698990/

Nora, S., & Minc, A. (1978). L'informatisation de la Societe. Paris: La Documentation francaise.

Norris, P. (2001). Digital Divide: Civic Engagement, Information Poverty, and the Internet Worldwide. Cambridge: Cambridge University Press.

Okun, L. B. (1989). The concept of mass. Physics Today, 42(6), 34-36.

Organization for Economic Cooperation and Development. (1997). Towards a Global Information Society. Paris: OECD.

Oxfam International. (2005). Extortion at the gate: Will Viet Nam join the WTO on pro-development terms? Retrieved July 2, 2005, from http://www.oxfam.org.uk/what_we_do/issues/trade/bp67_viet_nam_wto.htm

Pearce, D. W., Barbier, E., Markandy, A., Barrett, S., Turner, R. K., & Swanson, T. (1991). Blueprint 2 : greening the world economy. London: Earthscan.

Pearce, D. W., Markandya, A., & Barbier, E. (1989). Blueprint for a green economy. London: Earthscan.

Perry, M., O'Hara, K., Sellen, A., Brown, B., & Harper, R. (2001). Dealing with mobile technology: understanding access anytime, anywhere. ACM Transactions on Computer-Human Interaction, 8(4), 323-347.

Pham, D. C. (2000). Intellectual Property Right - A Condition for knowledge economy in Vietnam. In M. T. Tran & N. L. Ho (Eds.), Proceedings of Conference on Knowledge Economy and its Issues to Vietnam (pp. 213-226). Hanoi: Communist Publishing.

Pham, H., & Anh, V. (2006, July 12). Nguyễn Lâm Thái hối lộ hơn 1 tỷ đồng cho lãnh đạo bưu điện [Nguyen Lam Thai corrupted more than 1 billion VND to posts managers]. VnExpress.

Pham, X. N., Galla, A., Logan, W., & Slatis, T. (2003). Culture and Development in Vietnam. Hanoi: UNDP Hanoi.

Phan, D. D. (2000). On the way to build a knowledge economy in our country. In M. T. Tran & N. L. Ho (Eds.), Proceedings of Conference on Knowledge Economy and its Issues to Vietnam (pp. 113-121). Hanoi: Communist Publishing.

Plato. (1955). The republic. Harmondsworth: Penguin.

Poirier, R. (1990). The Information Economy Approach: Characteristics, Limitations and Future Prospects. The Information Society, 7(4), 245-285.

Polanyi, M. (1958). Personal knowledge: towards a post-critical philosophy. Chicago: University of Chicago Press.

Porter, M. (1985). Competitive advantage: creating and sustaining superior performance. from http://images.lib.monash.edu.au/mgx5560/04078642.pdf Online Resource (restricted access): Value chain and competitive advantage [chapter 2] p33-61

Porter, M., & Millar, V. E. (1985). How information gives you competitive advantage. Harvard Business Review(July-August).

Prebisch, R. (1962). The Economic Development of Latin America and its Principal Problems. Economic Bulletin for Latin America, 7(1), 1-22.

PTT. (2006). Hội thảo Chương trình phát triển Công nghiệp nội dung số đến 2010 [Workshop on Digital Content Industry of Vietnam by 2010]. Retrieved June 2, 2006, from http://mpt.gov.vn/tintuc/detail.asp?thucdon=tt&id=BT860640362&chude=B1&ngaycuoi=6/8/2006%2011:09:27%20AM

Redclift, M. R. (2006). Sustainable Development (1987-20050) - An oxymoron comes of age. Horizontes Antropológicos, 12(25), 65-84.

Rees, W. E. (2002). An ecological economics perspective on sustianability and prospects for ending poverty. Population and Environment, 24(1), 15-46.

Resnick, D. (1988). Politics on the Internet: The Normalization of Cyberspace. In C. Toulouse & T. W. Luke (Eds.), The Politics of Cyberspace. New York: Routledge.

Reuters. (2007, Friday, February 02). Romanian President Thanks Bill Gates for Easily Pirated Software. FoxNews.com.

Ricardo, D. (1821). On the principles of political economy and taxation microform (3rd ed.). London: John Murray.

Ritter, W. (2004). Knowledge management - Made in Hong Kong. Paper presented at the Intangible Assets - The Wealth of Knowledge-Based Economies, Hongkong.

Robinson, S. S. (2004). Networking Communities in the South - Challenges for Diverse Actors.

Rogers, E. M. (2003). Diffusion of Innovations (5 ed.). New York: Free Press.

Rogers, P. (2004). The US Military Posture: 'A Uniquely Benign Imperialism'? In L. Panitch & C. Leys (Eds.), Socialist Register 2004: The New Imperial Challenge (pp. 146-165). London: Merlin Press.

Romer, P. (1990). Endogenous Technological Change. Journal of Political Economy, 98(5), 71-102.

Rossen, E. (2003). Cross-cultural Inequality in IS: a Preliminary Exploration of Scandinavia and France. Paper presented at the Cross Cultural Workshop at the International Conference of Information Systems, Seattle.

Sachs, W. (2002). Fairness in a fragile world: the Johannesburg agenda. Development, 45(3), 12-17.

Sartre, J.-P. (1992). Being and nothingness : a phenomenological essay on ontology. New York: Washington Square Press.

Saunders, R. (1999). Business the amazon.com way : secrets of the world's most astonishing Web business. Oxford: Capstone.

Sayer, D. (1991). Capitalism and modernity : an excursus on Marx and Weber. London ; New York: Routledge.

Schauder, D. E. (2006). Good vs Evil: the Interet and Society. Paper presented at the IEEE Social Implications of Technology inaugural seminar

Schumacher, E. F. (1973). Small is beautiful: a study of economics as if people mattered. London,: Blond and Briggs.

Serafy, S. E. (1997). Green Accounting and Economic Policy. Ecological Economics, 21(1997), 217-229.

Shapiro, C., & Varian, H. R. (1999). Information Rules. Boston: Harvard Business School Press.

Sharma, A., & Adkins, R. (2006). OSS in India. In C. DiBona, D. Cooper & M. Stone (Eds.), Open Sources 2.0 : the continuing evolution (pp. 189-196). Beijing ; London: O'Reilly.

Shy, O. (2001). The Economics of Network Industries. Cambridge: Cambridge University Press.

Simonite, T. (2007, March 15). Catalyst could help turn CO2 into fuel. NewScientist.com, from http://www.newscientisttech.com/article/dn11390-catalyst-could-help-turn-cosub2sub-into-fuel.html

Simpson, G. G. (1967). The meaning of evolution : a study of the history of life and of its significance for man (Rev. (i.e. 2nd) ed.). New Haven: Yale University Press.

Singer, H. W. (1950). U.S. Foreign Investment in Underdeveloped Areas: The Distribution of Gains Between Investing and Borrowing Countries. American Economic Review, 40(Papers and Proceedings), 473-485.

Sitarz, D. (1993). AGENDA 21: the Earth Summit strategy to save our planet. Boulder, Colo.: EarthPress.

Smith, A., & Seligman, E. R. A. (1970). The wealth of nations. London: Dent.

Smith, J. W. (2000). Economic democracy : the political struggle of the twenty-first century. Armonk, NY: M.E.Sharpe.

Smith, P., Toulmin, L., & Qiang, C. Z. W. (2003). In-Focus: ICT in Vietnam - Accelerating ICT Development in Vietnam. I-Ways, Digest of Electronic Commerce Policy and Regulation, 26 31-40.

Spector, R. (2000). Amazon.com : get big fast. New York: HarperBusiness.

Stallman, R. (2006). The GNU Manifesto. Retrieved March 23, 2007, from http://www.gnu.org/gnu/manifesto.html

Steiner, G. A., & Steiner, J. S. (2001). Corporate Social Responsibility. In J. M. Harris (Ed.), A Survey of sustainable development : social and economic dimensions (pp. 283-285). Washington: Island Press.

Stevens, C. (1996). The Knowledge-Driven Economy. In D. Neef (Ed.), The Knowledge Economy (pp. 87-94). Oxford: Butterworth-Heinemann Ltd.

Stiglitz, J. E. (2002). Globalization and its discontents. London: Penguin.

Strategic Policy and Research Division. (2004). Indicators of Community Strength in Victoria. Retrieved June 15, 2007. from http://indicators.moreland.vic.gov.au/documents/399_File.pdf.

T.V. (2005). FPT trở thành nhà cung cấp độc quyền game MU Online [FPT becomes the monopolistic distributor of MU online game] [Electronic Version]. Vnexpress.net. Retrieved July 1, 2007 from http://www.vnexpress.net/Vietnam/Vi-tinh/2005/05/3B9DE77E/.

Tapscott, D. (1996). The digital economy : promise and peril in the age of networked intelligence. New York: McGraw-Hill.

Tapscott, D., & Caston, A. (1993). Paradigm shift : the new promise of information technology. New York: McGraw-Hill.

Tapscott, D., Lowy, A., Ticoll, D., & Klym, N. (1998). Blueprint to the digital economy : creating wealth in the era of e-business. New York: McGraw-Hill.

Taylor, E. F., & Wheeler, J. A. (1992). Spacetime physics : introduction to special relativity (2nd ed.). New York: W.H. Freeman.

Taylor, P. J. (1996). The Way the Modern World Works: World Hegemony to World Impasse. New York: Wiley.

Thai, T. N. D. (2004). Phụ nữ Việt Nam [Vietnamese Women]. In P. Gubry, H. D. Nguyen & T. H. Pham (Eds.), Dân số và phát triển ở Việt Nam [Population and Growth in Vietnam] (pp. 457-492). Hanoi: Nha xuat ban the gioi, ILSSA.

Thanh, H. (2005, September 7). Dự luật Phòng chống tham nhũng chưa khả thi [Anti-corruption Bill is not feasible]. VnExpress.net., Retrieved July 1, 2006 from http://vnexpress.net/Vietnam/Phap-luat/2005/09/3B9E1DAD/

The Group of Lisbon. (1995). Limits to Competition. Cambridge: The MIT Press.

The Phong. (2006). Liệu VinaGame sẽ kiện FPT [If VinaGame sues FPT] [Electronic Version]. Vietnamnet, 23/01/2006. Retrieved July 1, 2006 from http://www.vietnamnet.vn/cntt/2006/01/528750/.

Thurow, L. C. (1996). The future of capitalism : how today's economic forces shape tomorrow's world. St Leonards, N.S.W.: Allen & Unwin.

Thurow, L. C. (1998). From Capitalism to Knowledge Society. In D. Neef (Ed.), The Knowledge Economy (pp. 199-218). Boston: Butterworth-Heinemann.

Tiến, D. (2006). Dân chơi Hà Thành phơi bày trên blog [Hanoi players on blog]. Retrieved October 20, 2006, from http://vnexpress.net/Vietnam/Xa-hoi/2006/09/3B9EE6A2/

Tien Phong. (2007). Vietnam poor for water resources. Retrieved April 18, 2007, from http://english.vietnamnet.vn/tech/2007/03/677726/

Toffler, A. (1980). The third wave (1st ed.). New York: Morrow.

Tran, A. (2006, October 6). Doanh nghiệp Việt Nam trước vấn đề đổi mới công nghệ [Vietnamese enterprises and technology innovation]. Radio the Voice of Vietnam.

Tran, D. (2006). Nidec Ups Investment in Vietnam to US$1 Billion. Retrieved April 19, 2007, from http://techon.nikkeibp.co.jp/article/HONSHI/20061108/123276/

Tran, D. T. (2000). Knowledge economy and the issue of selecting a development model for Vietnam: Treatises. In M. T. Tran & N. L. Ho (Eds.), Proceedings of Conference on Knowledge Economy and its Issues to Vietnam (pp. 100-112). Hanoi: Communist Publishing.

Tran, M. T., & Ho, N. L. (2000a). Information technology: Drive of a Knowledge economy. In M. T. Tran & N. L. Ho (Eds.), Proceedings of

Conference on Knowledge Economy and its Issues to Vietnam (pp. 155-173). Hanoi: Communist Publishing.

Tran, M. T., & Ho, N. L. (Eds.). (2000b). Kỷ yếu hội nghị về kinh tế tri thức và những vấn đề đặt ra đối với Việt Nam [Proceedings of the Conference on Knowledge Economy and its Issues to Vietnam]. Hanoi: Communist Publishing.

Tran, N. C. (2005). Impact of Policy on Development of E-Commerce in Vietnam In R. Lafond & C. Sinha (Eds.), E-commerce in the Asian context: selected case studies (pp. 70-94). Singapore: Institute of Southeast Asian Studies & International Development Research Centre.

Tran, Q. V. (2001). Văn hoá Việt Nam: Tìm Tòi và Suy nghĩ [Vietnamese culture: Researches and Reflections]. Hanoi: Nhà xuất bản Văn hoá [Literature Publishing].

Tran, S. C. (2003). International export implications of expanding shrimp aquaculture in Vietnam. In International Institute for Sustainable Development., Internatonal Centre for Trade and Sustainable Development., Thuysan & The World Conservation Union. (Eds.), Expanding shrimp aquaculture on sandy land in Vietnam: Challenges and Opportunities (pp. 48-63). Hanoi: IUCN & IISD.

Tran, T. X. (2004). Tài nguyên nước mặt Việt Nam và những thách thức trong tương lai [Water ground resource in Vietnam and future challenges]. Retrieved April 19, 2007, from http://www.monre.gov.vn/monreNet/Default.aspx?tabid=220&ItemID=390

Tran, V. P. (2000). Kinh tế tri thức và những vấn đề đặt ra đối với Việt Nam [Knowledge economy and its problems to Vietnam]. In M. T. Tran & N. L. Ho (Eds.), Proceedings of Conference on Knowledge Economy and its Issues to Vietnam (pp. 79-84). Hanoi: Communist Publishing.

Triandis, H. C. (1972). The analysis of subjective culture. New York: Wiley-Interscience.

Trong, K. V. (2004). Phát triển nông thôn Việt Nam từ làng xã truyền thống đến văn minh hiện đại [Development of Vietnamese Rurality from traditional commune to modern civilization]. Hanoi: Agriculture Publish.

Tuoi Tre. (2007, April 19). Dọn nhà đón 20 tỷ USD [Prepare to receive $20 billion]. Vnexpress.net.

UNDP. (1996). Human Development Report 1996. New York: Oxford University Press.

UNDP, & Ministry of Agriculture and Rural Development. (2003). Achieving MDSs through ICT: Experiences and Challenges in Vietnam. Retrieved August 20, 2004. from http://www.undp.org.vn/themes/ict4d/index.htm.

UNEP. (2002). Global Environment Outlook 3. London: Earthscan.

UNESCO. (2001). UNESCO Universal Declaration on Cultural Diversity. Paris: UNESCO.

UNESCO. (2003). Convention for the Safeguarding of the Intangible Cultural Heritage. Retrieved June 11, 2006, from http://unesdoc.unesco.org/images/0013/001325/132540e.pdf

UNESCO. (n.d.). Best Practices on Indigenous Knowledge. Retrieved June 16, 2006, from http://www.unesco.org/most/bpindi.htm#_ftn3

UniCorn. (2000). VCIT End of Project Performance Assessment (No. CIDA Project Number 975/19031). Montreal: UniCorn.

United Nations. (1998). Kyoto Protocol to the United Nations Framework: Convention on Climate Change. Retrieved August 5, 2007, from http://unfccc.int/resource/docs/convkp/kpeng.pdf

United Nations, European Commission, International Monetary Fund, Organization for Economic Co-operation and Development, & World Bank. (2003). Handbook of National Accounting: Integrated Environmental and Economic Accounting 2003. New York: United Nations.

United Nations Development Program, Accenture, & Markle Foundation. (2001). Digital Opportunity Initiatives Report - Creating a Development Dynamic. Retrieved March 03, 2005, from http://www.opt-init.org/framework/DOI-Final-Report.doc

United Nations Development Programme. (2003). Achieving MDSs through ICT: Experiences and Challenges in Vietnam. Hanoi: UNDP.

United Nations Population Division. (2006). Urban and Rural Areas 2005. Retrieved November 22, 2006, from http://www.un.org/esa/population/publications/WUP2005/2005urban_rural.htm

United Nations. National Assembly. (2001). Resolution adopted by the General Assembly: 56/183. World Summit on the Information Society. A/RES/56/183, from http://www.itu.int/wsis/docs/background/resolutions/56_183_unga_2002.pdf

United Nations. Statistical Division, & United Nations Environment Programme. Economics and Trade Unit. Division of Technology Industry and Economics. (2000). Integrated environmental and economic accounting : an operational manual. New York: United Nations.

United States. National Energy Policy Development Group, & Bush, G. W. (2001). Reliable, affordable, and environmentally sound energy for America's future : report of the National Energy Policy Development Group. Washington, DC: The Group.

van-Dijk, J. A. G. M. (2005). The Deepening Divide: Inequality in the Information Society. Thousand Oaks: SAGE.

van Dijk, J. A. G. M. (2005). The Deepening Divide: Inequality in the Information Society. Thousand Oaks: SAGE.

Viet Nam National Administration of Tourism, UNDP, & World Tourism Organization. (2001). Revised Tourism Development Plan 2001-2010. Summary Plan and Action Programme, Volume I (No. UNDP VIE/00/012). Ha Noi.

Vietnam Communist Party Poliburo. (2000). Directive No.58-CT/TW: Accelerating the Use and Development of Information Technology for the Course of Industrialization and Modernization. Retrieved November 22, 2006 from http://www.most.gov.vn/c_vbqp/cb_tracuu/tap4/TVTextrule.2004-07-28.4926/view.

Vietnam Environment Protection Agency. (2007). Báo cáo môi trường quốc gia 2006: Hiện trạng môi trường nước 3 lưu vực sông Cầu, Nhuệ - Đáy, hệ thống sông Đồng Nai [National Environmental Report 2006: Status of water environment in the valleys of Cau, Nhue-Day and Dongnai rivers]. Hanoi: Ministry of Natural Resources and Environment.

Vietnam Farmer Association. (2005, July 03). Farmers lost land and became jobless due to urbanization. Vnexpress.net, Retrieved November 22, 2006 from http://vnexpress.net/Vietnam/Xa-hoi/2005/07/3B9DFB49/

Vietnam Government Portal. (2007, January 30). Không thể cứ xuất khẩu dầu thô mãi [Can't export crude oil longer]. Vnexpress.net Retrieved April 19, 2007, from http://vnexpress.net/Vietnam/Kinh-doanh/2007/01/3B9F2D30/

Vietnam Internet Center. (2006). Thong bao so lieu phat trien Internet Vietnam 4/2006 [Data of the Internet Development in Vietnam by 4/2006]. Retrieved September 2, 2006, from http://www.thongkeinternet.vn/jsp/trangchu/index.jsp

Vietnam Internet Center. (2007a). Dung lượng kênh và hướng kết nối quốc tế quí 4/2006 [Capacity of International Internet Connection by Quarter 4, 2006]. Retrieved April 24, 2007, from http://thongkeinternet.vn/jsp/dungluong/dungluong.jsp

Vietnam Internet Center. (2007b). Thong bao so lieu phat trien Internet Vietnam 5/2007 [Statistic data of Internet growth in Vietnam by May 2007]. Retrieved May 31, 2007, from http://www.thongkeinternet.vn/jsp/trangchu/index.jsp

Vietnam Internet Center. (2007c). Tình hình phát triển số lượng tên miền .vn tháng 4/2007 [Number of Vietnamese doman names by April 2007]. Retrieved May 29, 2007, from http://www.thongkeinternet.vn/jsp/tainguyen/domain_tab_dt.jsp

Vietnam News. (2005). Mật độ dân số VN gấp 6 lần so với tiêu chuẩn [Population density of Vietnam is 6 times higher than the standard]. Retrieved September 21, 2006, from http://vnexpress.net/Vietnam/Xa-hoi/2005/09/3B9E2477

Vietnam Posts and Communication Group. (n.d.). Vietnam Posts and Communication Group's organization structure. Retrieved Dec 12, 2007, from http://www.vnpt.com.vn/detail.asp?id=749&dataID=3763

Vietnamnet. (2005a, August 06). Đã đến lúc cần một Hiệp hội Game Việt Nam? [Is it time to have a Vietnamese Game Association]. Vietnamnet Retrieved May 1, 2007, from http://www.vnn.vn/cntt/2006/04/561869/

Vietnamnet. (2005b, October 25). Game Việt Nam quy mô lớn - tương lai không gần [Vietnam game – large scale but not near future]. Retrieved May 1, 2007, from http://www.vietnamnet.vn/cntt/2005/10/504086/

VietnamNet. (2006a). Bảo tồn di sản văn hoá phi vật thể: đường còn dài [Preservation of Intangible cultural heritage: the long way] [Electronic Version]. VietnamNet. Retrieved June 11, 2006 from http://www.vietnamnet.vn/vanhoa/tintuc/2006/02/541800/.

Vietnamnet. (2006b, July 19). Đại lý Internet công cộng:không vi phạm mới lạ [Public Interne dealer: how can not violate]. Vietnamnet Retrieved May 1, 2007, from http://www.vietnamnet.vn/cntt/206/07/592721

VietnamnetBridge. (2005, May 20). Gaming servers lock horns over market share. Retrieved July 11, 2006, from http://english.vietnamnet.vn/spf/2005/05/443151/

Vinagame. (n.d.). VoLamtruyenky. Retrieved August 3, 2006, from http://www.volam.com.vn/shxt/gioithieutrochoi/boicanh.html

VNE. (2006, July 31). MPI introduces new FDI attraction policy. VietnamNet Bridge, from http://english.vietnamnet.vn/reports/2006/07/596830/

Vu, D. C. (2000). Contribution to Understand Knowledge Economy. In M. T. Tran & N. L. Ho (Eds.), Proceedings of Conference on Knowledge Economy and its Issues to Vietnam (pp. 59-60). Hanoi: Communist Publishing.

Vu, K. M., & Jones, R. (2006). VNCI Policy paper #7: E-Government and Business Competitiveness: a Policy Review. Retrieved November 1, 2006, from http://www.vnci.org/default.asp?act=Download&downloadID=57

Vu, N. Q., & Nguyen, L. P. (2004). Chính sách dân số [Population policy]. In P. Gubry, Nguyen Huu Dung. & P. T. Huong (Eds.), Dân số và phát triển ở Việt Nam [Population and Growth in Vietnam] (pp. 631-662). Hanoi: Nha xuat ban the gioi, ILSSA.

Vu, X. N. H. (Ed.). (2004). Vấn đề hạch toán môi trường trong tài khoản quốc gia [Issues of environmental accounting in the system of national account]. Hanoi: Science and Technology Publishing.

Wallerstein, I. (1984). The Politics of the World Economy: the States, the Movements and the Civilizations. Cambridge: Cambridge University Press.

Wallerstein, I. (1997, April 3-5). Ecology and Capitalist Costs of Production: No Exit. Paper presented at the PEWS XXI: The Global Environment and the World-System, University of California, Santa Cruz.

Wallerstein, I. (1998a). A Long-Term View of the Trajectory of the World-System. Retrieved June 20, 2007, from http://fbc.binghamton.edu/iwtrajws.htm

Wallerstein, I. (1998b). The So-called Asian Crisis: Geopolitics in the Longue Durée. Retrieved June 20, 2007, from http://fbc.binghamton.edu/iwasncrs.htm

Walsham, G. (2002). Cross-cultural Software Production and Use: a Structurational Analysis. MIS Quarterly, 26(4), 359-380.

Watt, K. E. F. (1982). Understanding the Environment. Boston: Allyn and Bacon.

Waud, R. N. (1996). Microeconomics (3rd Australian ed.). South Melbourne: Longman.

Welsh, A. (2007, March 29). Brazil to offer free internet to Amazon tribes. Reuters.

Westing, A. H., Phuong, T. B., Vo, Q., Bui, T. L., & Dwernychuk, L. W. (2002, July 26-28). Long-term consequences of the Vietnam war: Ecosystems. Paper presented at the The Environmental Conference on Cambodia, Laos and Vietnam, Stockholm.

Williams, J. L. (2004). Oil Price History and Analysis. Retrieved September 16, 2004, from www.wtrg.com/prices.htm

Williamson, K. (2002). Research techniques: sampling. In K. Williamson & A. Bow (Eds.), Research methods for students, academics and professionals: information management and systems (pp. 225-234). Wagga Wagga, NSW: Centre for Information Studies, Charles Sturt University.

Wills, I. R. (2006). Economics and the environment : a signalling and incentives approach (2nd ed.). Crows Nest, NSW: Allen & Unwin.

Wilson, T. D. (2002). The nonsense of 'knowledge management. Information Research, 8(1).

Wise, T. A. (2001). Global Perspectives: The North/South Imbalance. Overview Essay. In J. M. Harris, T. A. Wise, K. P. Gallegher & N. R. Goodwin (Eds.), A Survey of sustainable development : social and economic dimensions (pp. 79-90). Washington: Island Press.

World-Bank. (2002). Accelerating Information & Communication Technologies Development in Vietnam.

World-Bank. (2003). Vietnam Development Report 2004: Poverty. from http://www.worldbank.org.vn/news/VDR04%20Poverty.pdf

World Bank. (1998). Peasant Household Welfare and Transition in Vietnam: The World Bank.

World Bank. (2001). Vietnam: Country Assistance Evaluation (No. 23288). Washington, D.C.: World Bank.

World Bank. (2002). Accelerating Information & Communication Technologies Development in Vietnam. Hanoi.

World Bank. (2003a). Vietnam Development Report 2004: Poverty (No. 27130-VN). Hanoi: Poverty Reduction and Economic Management Unit, East Asia and Pacific Region, the World Bank.

World Bank. (2003b). World Development Report. New York: Oxford University Press.

World Bank. (2004). Vietnam Development Report 2005: Governance. Retrieved Oct 25, 2006, from http://siteresources.worldbank.org/INTVIETNAM/Data%20and%20Reference/20292720/VDR05-governance.pdf

World Commission on Environment and Development. (1987). Our Common Future. Oxford: Oxford University Press.

World Economic Forum. (2004). Global Information Technology Report 2003-2004. Retrieved March 03, 2005, from http://www.weforum.org/site/homepublic.nsf/Content/Global+Competitiveness+Programme%5CGlobal+Information+Technology+Report

World Summit on Information Society. (2003). Declaration of Principles - Building the Information Society: a global challenge in the new Millennium. Retrieved August 08, 2004, from http://www.itu.int/wsis/basic/about.html

wowwiki.com. (n.d.). Gold Farmer. Retrieved Oct 12, 2007, from http://www.wowwiki.com/Gold_Farmer

Wu, J. C. (2006). The Mineral Industry of Vietnam. In U.S. Department of the Interior & U.S. Geological Survey (Eds.), U.S. Geological Survey Minerals Yearbook 2005. Washington D.C.: USGS.

WWF. (2006a). Rich countries, poor water. Retrieved Agust 16, 2006, from http://www.wwf.org.uk/filelibrary/pdf/richcountriespoorwater.pdf

WWF. (2006b, August 16). Water crisis hits rich countries. Retrieved August 17, 2006, from http://www.wwf.org.uk/news/n_0000002898.asp

Index of Names

Subject Index

P

Q

R

S

Printed and bound by CPI Group (UK) Ltd, Croydon, CR0 4YY
13/09/2026
14961091-0004